Flood Pulsing in Wetlands

Flood Pulsing in Wetlands: Restoring the Natural Hydrological Balance

edited by
Beth A. Middleton
National Wetlands Research Center,
USGS, Lafayette, Louisiana

John Wiley & Sons, Inc.

This book is printed on acid-free paper. ♾

Published simultaneously in Canada.

Library of Congress Cataloging-in-Publication Data:

Middleton, Beth.
Flood pulsing in wetlands: restoring the natural hydrological balance / by Beth A. Middleton.
p. cm.
Includes bibliographical references.
ISBN 0-471-41807-2
1. Floodplain ecology—North America. 2. Wetland restoration—North America.

QH541.5.V3 M54 2002
333.91'8153'097—dc21

2001045615

Printed in the United States of America.

10 9 8 7 6 5 4 3 2 1

Contents

Contributors

GREGOR T. AUBLE, U.S. Geological Survey, Midcontinent Ecological Science Center, Fort Collins, Colorado

JOANNE CHAMBERLAIN, Kissimmee Division, Watershed Management Department, South Florida Water Management District, West Palm Beach, Florida

M. K. CHEW, Arizona State University, Tempe, Arizona

LISA M. ELLIS, University of New Mexico, Albuquerque, New Mexico

CLIFFORD S. CRAWFORD, University of New Mexico, Albuquerque, New Mexico

JOSEPH W. KOEBEL JR., Kissimmee Division, Watershed Management Department, South Florida Water Management District, West Palm Beach, Florida

P. METTLER, Southern Illinois University, Carbondale, Illinois

BETH A. MIDDLETON, National Wetlands Research Center, USGS, Lafayette, Louisiana

MANUEL C. MOLLES JR., University of New Mexico, Albuquerque, New Mexico

MICHAEL L. SCOTT, U.S. Geological Survey, Midcontinent Ecological Science Center, Fort Collins, Colorado

M. SMITH, Southern Illinois University, Edwardsville, Illinois

JULIE C. STROMBERG, Arizona State University, Tempe, Arizona

LOUIS A. TOTH, Kissimmee Division, Watershed Management Department, South Florida Water Management District, West Palm Beach, Florida

ANDREW G. WARNE, U.S. Geological Survey, Water Resources Division, Caribbean District, GSA Center, Guaynabo, Puerto Rico

Preface

Beth A. Middleton, *National Wetlands Research Center, USGS, Lafayette, Louisiana*

This book is a first-of-its-kind compilation of the research of leaders in the field of restoration ecology whose work involves the use of flood pulsing in the restoration of wetlands. The contributed chapters give regional examples of wetland restoration projects in which flood pulsing was a critical part of restoring the hydrodynamic setting for the plants and animals of floodplains. They thus provide an argument for the widespread incorporation of this approach in restoration projects. Restoration practitioners, academics, and students will find this book invaluable for the information it draws together from cutting-edge ideas in the technology of restoration.

Each contributed chapter makes its own case for the importance of flood pulsing in restoration within its own regional setting, as based on research and monitoring of the projects reported here.

- Chapter 1, "The Flood Pulse Concept in Wetland Restoration," provides the basic argument for the use of flood pulsing in restoration projects.
- Chapter 2, "Flood Pulses and Restoration of Riparian Vegetation in the American Southwest," describes the impacts of projects that reduce flood pulsing in the southwestern United States. It also describes research that demonstrates the importance of flood pulsing for the regeneration of trees and other vegetation in Sonoran Desert communities.
- Chapter 3, "The Role of the Flood Pulse in Ecosystem-Level Processes in Southwestern Riparian Forests: A Case Study from the Middle Rio Grande," outlines the decline of cottonwoods and fish species along the Rio Grande. These changes are related to ecosystem processes that have been affected by the absence of the flood pulse

and changes in insect communities in riparian settings that have led to the buildup of organic debris.

- Chapter 4, "The Role of the Flood Pulse in Maintaining *Boltonia decurrens,* a Fugitive Plant Species of the Illinois River Floodplain: A Case History of a Threatened Species," documents the changes in hydrology of the Illinois River and their relationship to the decline of *Boltonia decurrens* following the construction of navigation dams, and agricultural levees that have disrupted the annual flood pulse.
- Chapter 5, "Conservation and Restoration of Semiarid Riparian Forests: A Case Study from the Upper Missouri River, Montana," documents regeneration events in *Populus* forests along the Missouri River as related to the timing of flood pulsing events.
- Chapter 6, "Implications of Reestablishing Prolonged Flood Pulse Characteristics of the Kissimmee River and Floodplain Ecosystem," describes the most famous case concerning the use of flood pulsing in the restoration of an entire landscape. The chapter discusses the history of its changes, attempts to restore the original flood pulse, and the projected reestablishment of communities (fish, insects, birds, and vegetation) along the Kissimmee River.
- Chapter 7, "Flood Pulsing in the Regeneration and Maintenance of Species in Riverine Forested Wetlands of the Southeastern United States," describes river regulation projects across the southeastern United States. Based on the early life history dynamics of plants, it makes the case for the incorporation of flood pulsing in riverine forested wetlands and describes the Brushy Lake, Arkansas, project, where a levee was breached to reconnect the channel to the floodplain by flood pulsing.
- An extensive reference section is included in each chapter as an aid to wetland restorationists and researchers.

The authors extend their special thanks to the many librarians who helped to locate the volumes of information that contribute to the success of such an endeavor.

1

The Flood Pulse Concept in Wetland Restoration

Beth A. Middleton
National Wetlands Research Center, USGS, Lafayette, Louisiana

The reestablishment of flood pulsing in riverine and tidal systems is becoming recognized as an essential step in the restoration of wetlands worldwide. Especially in North America, monitoring of projects that have incorporated more natural water regimes is now under way. In most instances, researchers are still collecting the essential life history data that will aid in building a case for the need to recreate flood-pulsed hydrology in wetland restoration projects. In this book, each chapter examines a case history of one these projects, written by a field researcher close to the heart of this rapidly developing field.

The flood pulse concept was first developed to describe seasonal changes in water levels on Amazonian floodplains and their relationships to functional dynamics and the maintenance of species diversity (Junk, 1982, 1997; Junk and Howard-Williams, 1984; Junk et al., 1989; National Research Council, 1992; Bayley, 1995) (Figure 1-1). The interconnection of the river channel and floodplain is critical because functions such as production, decomposition, and consumption are driven by the flood pulse (Grubaugh and Anderson, 1988; Sparks et al., 1990) and water fluctuation drives succession (van der Valk, 1981; Finlayson et al., 1989; Niering,

Flood Pulsing in Wetlands: Restoring the Natural Hydrological Balance, edited by Beth A. Middleton.
ISBN 0 471-41807-2

1994; Middleton, 1999a). Although this idea emerged from the study of large river ecosystems, there is growing recognition that tidal pulsing is also important in salt marshes (Niering, 1994; Turner and Lewis, 1997; Zedler and Callaway, 1999) and mangrove swamps (McKee and Faulkner, 1999). In addition, isolated restored sedge meadows in the Prairie Pothole Region have lower species richness than natural wetlands that developed while large floods still occasionally interconnected them (van der Valk, 1999).

Although the importance of the flood pulse is recognized for a variety of wetland types worldwide, the idea that it is necessary to reestablish a functional flood or tidal pulse in damaged systems has been adapted rather slowly by wetland restorationists. In fact, it is not yet known whether the restoration of flood pulsing restores function (Brookes et al., 1996), and at least some evidence shows that in and of itself, flood pulsing is not enough. For example, merely reopening a tidal channel may not restore salt marsh function if the soil structure is altered and/or too saline (Haltiner et al., 1997). Regardless of what else may have to be adjusted, reestablishment of the original water dynamics (and sometimes soil conditions) is a critical aspect of wetland restoration, even more than reestablishing the vegetation.

To restore a wetland, most often what is required is a reversal of the engineering that dried the wetland in the first place—that is, dam removal, dechannelization, remeandering, addition of debris, rediversion of water, cessation of water extraction, levee or polder removal. The reengineering at a landscape level that is often required for such change is not easy, either physically or politically. However, simpler and widely used approaches such as damming create static water levels and so are not adequate restoration approaches (Middleton, 1999b, 2000).

The alteration of riverine and coastal ecosystems worldwide is so widespread as to leave us few examples of systems that still have a natural hydrologic regime (Sparks et al., 1990; Petts et al., 1992; Junk, 1999). This is especially true in temperate areas of the world; in the 139 largest river systems in Europe, the republics of the former Soviet Union, and regions north of Mexico, 77 percent of their total discharge is affected by dam and reservoir operation, interbasin diversion, and irrigation (Dynesius and Nilsson, 1994). Water extraction along rivers is also causing salt water intrusion in fresh and brackish water coastal systems (Muñoz and Prat, 1989; Prat and Ibañez, 1995). Along rivers in industrialized countries,

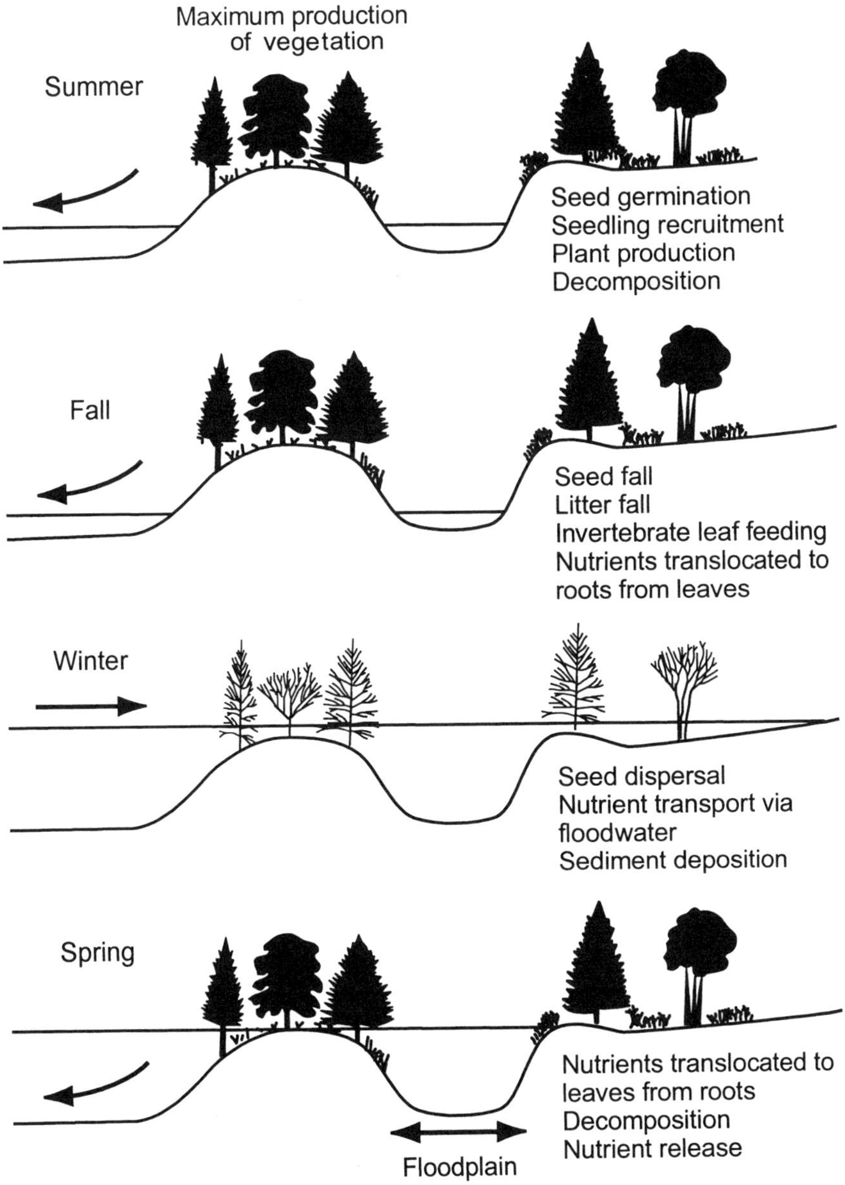

Figure 1-1. *Flood pulsing across a forested floodplain in various seasons in North America, related functional dynamics and biotic adaptations. (Adapted from Bayley, 1991, as derived from Junk et al., 1989, in Middleton, 1999b.)*

natural flood regimes are almost absent as a result of the reengineering of waterways (Bayley, 1995). Nevertheless, a few northern rivers that have been reengineered have portions that still flood pulse—for example, the Illinois (Sparks et al., 1998) and the Danube Rivers (Heiler et al., 1995).

After levees were constructed along major rivers such as the Mississippi, floodplains were converted to other uses, such as agriculture (Allen, 1997). If people move onto a floodplain after the completion of a water control project, it is often politically impossible to initiate the types of reengineering measures necessary for flood pulsing on the flood or tidal plain. Yet sudden, destructive floods sometimes occur on reengineered floodplains, so that a certain amount of rethinking is occurring recently. Is it really wise for us to restrict a river to its immediate channel and thus allow the encroachment of the floodplain, which exposes people to the threat of dangerous floods (Interagency Floodplain Management Review Committee, 1994; Junk, 1999)? In cases where the threat of future flooding is likely, portions of flood or tidal plains may be designated as nature areas to provide for flood storage (Zinke and Gutzweiler, 1990; Lathbury, 1996). Chronically flooded sites present some opportunities for the use of flood pulsing in restoration, albeit on a small scale. Nevertheless, there are some recent examples where flood-pulsed conditions have been (or are being) restored on a regional or landscape scale because of public demand, such as on the Kissimmee River (see Chapter 6). Unfortunately, because of the danger of flooding private property, restoration projects have usually been limited to ineffective measures, such as impounding waterways, that do not provide the biota with the pulsing environment to which they are adapted (Middleton 1999b).

The importance of reestablishing water regimes in sync with seasonal climate fluctuation and water flow in riverine and tidal systems has not been fully appreciated in wetland restoration. Organisms have specific adaptations that allow them to tolerate the wet/dry conditions that are a part of a flood-pulsed environment (Junk, 1997; Middleton, 1999a). Not only does each species have different water requirements and tolerances, these differ for each life stage—seed, seedling, and adult (see Chapters 2, 4, 5 and 7).

Damming, one of the most common river regulation procedures, is illustrative of the problems created by altered environments for biota (Middleton 1999b). Upstream, the reservoir above the dam becomes permanently impounded, resulting in a replacement of riparian vegetation with algal or submerged communities. Downstream from the dam, flows in the stream channel are altered, which changes the nature of the pulse transmitted to the floodplain (Middleton 1999b). Sediments become trapped behind the dam, so downcutting and erosion occur in the downstream channel, further cutting off the channel from the floodplain (Petts and Lewin, 1979; Hickin, 1983; Petts, 1984).

Permanent flooding lowers the overall species richness along regulated rivers because the sites never draw down (Nilsson et al., 1997). The dry phase of the flood pulse is critical, because even the most flood-tolerant species will eventually die in anaerobic conditions (Crawford, 1983; Armstrong et al., 1994) even though such species possess many mechanisms to survive periods of inundation (Crawford and Braendle, 1996; McKevlin et al., 1998). The long-term effects of impoundment in reservoirs indicate that when a river margin is permanently flooded, many species are lost, as was demonstrated in a study of eight Swedish rivers (Jansson et al., 2000b).

Impoundment is often used in restoration as a means of increasing water levels in a dried wetland, but because of the lack of a flood pulse, regeneration—from seed dispersal to the seedling recruitment stage—is problematic (Middleton, 1999b, 2000). Dams inhibit the movement of hydrochorous seeds because of fragmentation and low current velocity, and this affects seed availability along the corridor; as a result, each impoundment develops a distinctive flora (Jansson et al., 2000a) (Figure 1-2). In addition, because impoundment reduces dispersal distance, impoundments are likely inhabited by individuals that are more closely related to each other (Jansson et al., 2000a). The impacts of dams on flora were still apparent 65 km downstream of dams along six rivers in Virginia (Schneider et al., 1989). River regulation also has severe impacts on fauna; it desynchronizes environmental cycles and thus disrupts reproductive cycles of fish (Welcomme, 1989; Gehrke et al., 1995; Junk, 1997) and migrations of invertebrates (Adis et al., 1996).

Successful restoration depends on a better understanding of the life history requirements of plants and animals (Chapters 2 to 6). Seed germination can be critically dependent on flood pulsing, with the high phase of the pulse necessary for dispersal and drawdown necessary for germination (Junk and Piedade, 1997; Middleton, 1999b, 2000). Without a flood pulse, the dispersal of some species, such as *Taxodium distichum* and *Populus* spp., to suitable elevations for germination during the growing season is

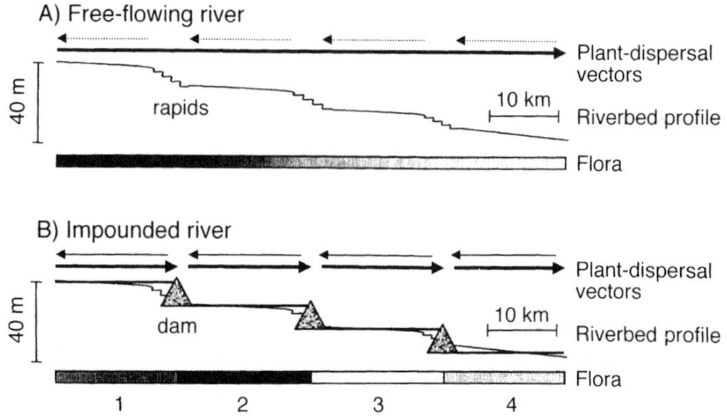

Figure 1-2. *Hypothesized relationships between plant-dispersal vectors, riverbed profile, and composition of the riparian flora in free-flowing versus impounded rivers.* (A) *The flora in the free-flowing river is hypothesized to describe a gradual change downstream, whereas* (B) *in the regulated river, each impoundment is projected to develop an individual flora (denoted 1–4). (From Jansson et al., 2000a, copyright © Ecological Society of America; reprinted by permission.)*

hampered (Chapters 7 and 5, respectively). Certain endangered species such as *Boltonia decurrens* on the Illinois River cannot germinate and set seed without a flood pulse (Chapter 4). Seed germination can also be sensitive to other environmental factors, such as salinity (Galinato and van der Valk, 1986; Baldwin et al., 1996), temperature, substrate, pH, and light quality (Baskin and Baskin, 1998). At the same time, floods remove the debris that sometimes decreases the germinability of seeds (Chapter 3). Species become increasingly tolerant of flooding as plants mature (Chapter 7). By the adult stage, water tolerance is widely variable between species and forms the basis for the compositional differences of wetlands (Harris et al., 1975; Whitlow and Harris, 1979; Hook, 1984; Theriot, 1993; Middleton, 1999b). Unfortunately, water for restoration purposes in the arid West may be so limited by competing demands by humans that restoration may be nearly impossible (Chapter 2).

River regulation impacts the flood pulsing environment experienced by flora and fauna on flood and coastal plains, which is critical in the life history dynamics of these species. Without proper attention to the hydrologic setting created, attempts at wetland restoration will fail. This book reviews the case histories of restoration situations where either flood pulsing has been reestablished as part of the project, or extensive studies of the life history requirements of species that are likely to need flood pulsing are being conducted.

REFERENCES

Adis, J., S. I. Golovatch, and S. Hamann. 1996. Survival strategy of the terricolous millipede *Cutervodesmus adisi* Golovatch (Fuhrmannodesmideae, Polydesmida) in a blackwater inundation forest of central Amazonia (Brazil) in response to the flood pulse. *Memoires du Museum National d'Histoire Naturelle* **169:** 523–532.

Allen, J. A., and V. Burkett. 1997. Bottomland hardwood forest restoration: overview of techniques, successes and failures, In *Wetlands and Watershed Management: Science Applications and Public Policy,* ed. by J. A. Kusler, D. E. Willard, and H. C. Hull, Jr., pp. 328–332, April 23–26, 1995. Tampa, FL., Association of State Wetland Managers, Berne, NY.

Armstrong, W., R. Brändle, and M. B. Jackson. 1994. Mechanisms of flood tolerance in plants. *Acta Botanica Neerlandica* **43:** 307–358.

Baldwin, A. H., K. L. McKee, and I. A. Mendelssohn. 1996. The influence of vegetation, salinity, and inundation on seed banks of oligohaline coastal marshes. *American Journal of Botany* **83:** 470–479.

Baskin, C. C., and J. M. Baskin. 1998. Seeds: Ecology, biogeography, and evolution of dormancy and germination. Academic Press, San Diego, CA.

Bayley, P. B. 1991. The flood pulse advantage and the restoration of river-floodplain systems. *Regulated Rivers: Research and Management* **6:** 75–86.

Bayley, P. B. 1995. Understanding large river-floodplain ecosystems. *BioScience* **45:** 153–158.

Brookes, A., J. Baker, and C. Redmond. 1996. Floodplain restoration and riparian zone management. In *River Channel Restoration: Guiding Principles for Sustainable Projects,* ed. by A. Brookes and F. D. Shields Jr., pp. 201–229. John Wiley & Sons, Chichester, U.K.

Crawford, R. M. M. 1983. Root survival in flooded soils. In *Ecosystems of the World 4A; Mires: Swamp, Bog, Fen and Moor; General Studies,* ed. by A. J. P. Gore, pp. 257–283. Elsevier, Amsterdam, The Netherlands.

Crawford, R. M. M., and R. Braendle. 1996. Oxygen deprivation stress in a changing environment. *Journal of Experimental Botany* **47:** 145–159.

Dynesius, M., and C. Nilsson. 1994. Fragmentation and flow regulation of river systems in the northern third of the world. *Science* **266:** 753–762.

Finlayson, C. M., B. J. Bailey, and I. D. Cowie. 1989. *Macrophyte Vegetation of the Magela Creek Floodplain, Alligator Rivers Region, Northern Territory.* Supervising Scientist for the Alligator Rivers Region, Australian Government Publishing Service, Canberra.

Galinato, M. I., and A. G. van der Valk. 1986. Seed germination traits of annuals and emergents recruited during drawdowns in the Delta Marsh, Manitoba, Canada. *Aquatic Botany* **26:** 89–102.

Gehrke, P. C., P. Brown, C. B. Schiller, D. B. Moffatt, and A. M. Bruce. 1995. River regulation and fish communities in the Murray-Darling River System, Australia. *Regulated Rivers: Research and Management* **11:** 363–375.

Grubaugh, J. W., and R. V. Anderson. 1988. Spatial and temporal availability of floodplain habitat: Long-term changes at Pool 19, Mississippi River. *American Midland Naturalist* **119:** 402–411.

Haltiner, J., J. B. Zedler, K. E. Boyer, G. D. Williams, and J. C. Callaway. 1997. Influence of physical processes on the design, functioning and evolution of restored tidal wetlands in California (USA). *Wetlands Ecology and Management* **4:** 73–91.

Harris, R. W., A. T. Leiser, and R. E. Fissell. 1975. *Plant Tolerance to Flooding,* RWH-200-7/1/75. Department of Environmental Horticulture, Davis, CA.

Heiler, G., T. Hein, and F. Schiemer. 1995. Hydrological connectivity and flood pulses as the central aspects for the integrity of a river-floodplain system. *Regulated Rivers: Research and Management* **11:** 351–361.

Hickin, E. J. 1983. River channel changes: Retrospect and prospect. In *Modern and Ancient Fluvial Systems,* ed. by J. D. Collinson and J. Lewin, pp. 61–83. Blackwell Scientific Publications, Oxford, U.K.

Hook, D. D. 1984. Waterlogging tolerance of lowland tree species of the South. *Southern Journal of Applied Forestry* **8:** 136–149.

Interagency Floodplain Management Review Committee. 1994. *Sharing the Challenge: Floodplain Management into the 21st Century.* Administration Floodplain Management Task Force, Washington, DC.

Jansson, R., C. Nilsson, and B. Renöfält. 2000a, Fragmentation of riparian floras in rivers with multiple dams. *Ecology* **81:** 899–903.

Jansson, R., C. Nilsson, M. Dynesius, and E. Andersson. 2000b, Effects of river regulation on river-margin vegetation: A comparison of eight boreal rivers. *Ecological Applications* **10:** 203–224.

Junk, W. J. 1982. Amazonian floodplains: Their ecology, present and potential use. In *Proceedings of the International Scientific Workshop on Ecosystem Dynamics in Freshwater Wetlands and Shallow Water Bodies,* pp. 98–126. Scientific Committee on Problems of the Environment (SCOPE), United Nations Environment Program (UNEP), New York.

Junk, W. J. 1997. Structure and function of the large central Amazonian River floodplains: Synthesis and discussion. In *The Central Amazonian Floodplain,* ed. by W. J. Junk, pp. 455–472. Springer-Verlag, Berlin.

Junk, W. J. 1999. The flood pulse concept of large rivers: Learning from the tropics. *Archiv für Hydrobiologie* **115:** 261–280.

Junk, W. J., and C. Howard-Williams. 1984. Ecology of aquatic macrophytes in Amazonia. In *The Amazon: Limnology and Landscape Ecology of a Mighty Tropical River and Its Basin,* ed. by H. Sioli, pp. 269–309. Dr. W. Junk Publishers, Dordrecht, The Netherlands.

Junk, W. J., and M. T. F. Piedade. 1997. Plant life in the floodplain with special reference to herbaceous plants. *In The Central Amazon Floodplain: Ecology of a Pulsing System,* ed. by W. J. Junk, pp. 147–185. Springer, Berlin.

Junk, W. J., P. B. Bayley, and R. E. Sparks. 1989. The flood pulse concept in river-floodplain systems. In *Proceedings of the International Large River Symposium (LARS),* ed. by D. P. Dodge, pp. 110–127. Canadian Special Publication of Fisheries and Aquatic Sciences, Ottawa, Canada.

Lathbury, M. E. 1996. Toward natural flood control: Floodplain wetlands. *Wetland Journal* **8:** 10–13.

McKee, K. L., and P. L. Faulkner. 1999. Biogeochemical Functioning of Restored and Natural Mangrove Forests in Southwest Florida. *Final Report to NOAA/NERRS/SRD.* National Wetlands Research Center, Lafayette, LA.

McKevlin, M. R., D. D. Hook, and A. A. Rozelle. 1998. Adaptations of plants to flooding and soil waterlogging. In *Southern Forested Wetlands Ecology and Management,* ed. by M. G. Messina and W. H. Conner, pp. 173–203. Lewis Publishers, Boca Raton, FL.

Middleton, B. A. 1999a, Succession and herbivory in monsoonal wetlands. *Wetlands Ecology and Management* **6:** 189–202.

Middleton, B. A. 1999b. *Wetland Restoration, Flood Pulsing and Disturbance Dynamics.* John Wiley & Sons, New York.

Middleton, B. A. 2000. Hydrochory, seed banks, and regeneration dynamics along the landscape boundaries of a forested wetland. *Plant Ecology* **146:** 169–184.

Muñoz, I., and N. Prat. 1989. Effects of river regulation on the lower Ebro River (NE Spain). *Regulated Rivers: Research and Management* **3:** 345–354.

National Research Council. 1992. *Restoration of Aquatic Ecosystems.* National Academy Press, Washington, DC.

Niering, W. 1994. Wetland vegetation change: A dynamic process, *Wetland Journal* **6:** 6–15.

Nilsson, C., A. Ekblad, M. Gardfjell, and B. Carlberg. 1997. Long-term effects of river regulation on river margin vegetation. *Journal of Applied Ecology* **28:** 963–987.

Petts, G. E. 1984. *Impounded Rivers: Perspectives for Ecological Management.* John Wiley & Sons, Chichester, U. K.

Petts, G. E., and J. Lewin. 1979. Physical effects of reservoirs on river systems. In *Man's Impact on the Hydrological Cycle in the United Kingdom,* ed. by G. E. Hollis, pp. 79–91. Geo Abstracts Ltd., Norwich, U. K.

Petts, G. E., A. R. G. Large, M. T. Greenwood, and M. A. Bickerton. 1992. Floodplain assessment for restoration and conservation: Linking hydrogeomorphology and ecology. In *Lowland Floodplain Rivers: Geomorphological Perspectives,* ed. by P. A. Carling, and G. E. Petts, pp. 217–234. John Wiley & Sons, Chichester, U. K.

Prat, N., and C. Ibañez. 1995. Effects of water transfers projected in the Spanish National Hydrological Plan on the ecology of the Lower River Ebro (NE Spain) and its delta. *Water Science and Technology* **31:** 79–86.

Schneider, R. L., N. E. Martin, and R. R. Sharitz. 1989. Impact of dam operations on hydrology and associated floodplain forests of southeastern rivers. In *Freshwater Wetlands and Wildlife,* ed. by R. R. Sharitz and J. W. Gibbons, pp. 1113–1122. Office of Scientific and Technical Information, Oak Ridge, TN.

Sparks, R. E., J. C. Nelson, and Y. Yin. 1998. Naturalization of the flood regime in regulated rivers. *BioScience* **48:** 706–720.

Sparks, R. E., P. B. Bayley, S. L. Kohler, and L. L. Osborne. 1990. Disturbance and recovery of large floodplain rivers. *Environmental Management* **14:** 699–709.

Theriot, R. F. 1993. *Flood Tolerance of Plant Species in Bottomland Forests of the Southeastern United States.* WRP-DE-6. U.S. Army Corps of Engineers, Vicksburg, MS.

Turner, R. E., and R. R. Lewis III. 1997. Hydrologic restoration of coastal wetlands. *Wetlands Ecology and Management* **4:** 65–72.

van der Valk. A. G. 1981. Succession in wetlands: A Gleasonian approach. *Ecology* **62:** 688–696.

van der Valk, A. G. 1999. Succession theory and wetland restoration. In *Proceedings of INTECOL's V International Wetlands Conference,* ed. by A. J. McComb and J. A. Davis, pp. 657–667. Gleneagles Press, Adelaide, Australia.

Welcomme, R. L. 1989. Floodplain fisheries management. In *Alternatives in Regulated River Management,* ed. by J. A. Gore and G. E. Petts, pp. 209–233. CRC, Boca Raton, FL.

Whitlow, T. H., and R. W. Harris. 1979. *Flood Tolerance in Plants: A State-of-the-Art Review.* Technical Report E-79-2. U.S. Army Corps of Engineers, Vicksburg, MS.

Zedler, J. B., and J. C. Callaway. 1999. Tracking wetland restoration: Do mitigation sites follow desired trajectories? *Restoration Ecology* **7:** 69–73.

Zinke, A., and K.-A. Gutzweiler. 1990. Possibilities for regeneration of floodplain forests within the framework on the flood-protection measures on the Upper Rhine, West Germany. *Forest Ecology and Management* **33/34:** 13–20.

2

Flood Pulses and Restoration of Riparian Vegetation in the American Southwest

Julie C. Stromberg and M. K. Chew
Arizona State University, Tempe, Arizona

In mesic regions of North America, to the casual observer, riparian trees are relatively indistinguishable from their nearby upland counterparts. In arid parts of the southwestern United States, the transition from riparian to upland zones is striking and often abrupt. Forests with multistory tree and shrub canopies along water courses give way to sparse grassland, subshrub, and succulent communities, often within a few meters of the floodplain. Riparian forests along desert rivers provide the water, shelter, and food that sustain obligate riparian animal species and provide a critical buffer for upland species during dry seasons and droughts. Beginning in the late 1800s, however, regional water and land management practices substantially reduced the acreage suitable or available for riverine marsh-

Flood Pulsing in Wetlands: Restoring the Natural Hydrological Balance, edited by Beth A. Middleton
ISBN 0 471-41807-2

lands, Populus-Salix *forests, and other Sonoran riparian plant communities (Hendrickson and Minckley, 1984; Brown, 1994; Patten, 1998). In addition to their direct replacement by irrigated fields and urban areas, riparian ecosystems have been degraded by dams that disrupt water and sediment flow, by diversion structures and groundwater wells that dewater streams and aquifers, and by the multiple impacts of livestock grazing.*

Many riparian restorations have been undertaken to overcome such degradations. Flood pulsing has expedited, or has been the major feature of, many ecologically successful projects. Some flood pulsing events fall within the category of managed or designer flood releases for the express purpose of ecosystem restoration. Others are "accidents," whereby floods released from a reservoir for one purpose have collaterally promoted riparian restoration. Natural floods have facilitated the restoration of riparian ecosystems along free-flowing rivers following the removal of stressors such as groundwater pumping and livestock grazing.

This chapter discusses examples of flood-assisted restoration efforts, organized by specific restoration goals. It addresses the restoration of regenerative processes for dominant trees (Populus, Salix), *the management complications arising from the presence of exotic woody species* (Tamarix), *and the broader goal of restoring a large complement of riparian plant species. The role of flood pulses in restoring plant productivity, minimizing fire disturbance, and restoring resilience is discussed. The chapter focuses on rivers of the Sonoran and Mojave Deserts, occasionally ranging beyond these geographic limits. It begins by summarizing regional climatic flood patterns, managerial changes to flood regimes, and associated ecological responses of riparian vegetation.*

FLOOD PATTERNS AND RIPARIAN VEGETATION IN THE DESERT SOUTHWEST

Natural Flood Regimes

The rivers of the southwestern United States are prone to rapid but seasonally consistent changes in flow. Most floods in the rivers of the Sonoran and Mojave Deserts result from rainfall, with snowmelt contributing to a lesser degree. Localized but intense convective rainstorms usually develop during the July-August monsoon season. Regionwide precipitation and runoff are again high in December and January, following Pacific frontal storms, declining through early spring. If tropical storms or hurricanes penetrate inland, floods occur in late fall, but less frequently. Following a southeastward gradient from the Mojave Desert, across the Sonoran Desert, toward the Chihuahuan Desert, summer rains become more frequent and summer floods more intense (Figure 2-1).

Floods in southwestern desert rivers are extreme in comparison with their low flows (Graf, 1988). Channels may carry no surface flows for short periods in May or June, yet experience instantaneous peak flows of several thousand cubic feet per second during the subsequent monsoon season. Flood patterns also vary over longer temporal scales (Molles et al., 1992). Both mean and peak annual river flows documented in the region over the past century have varied by several orders of magnitude between

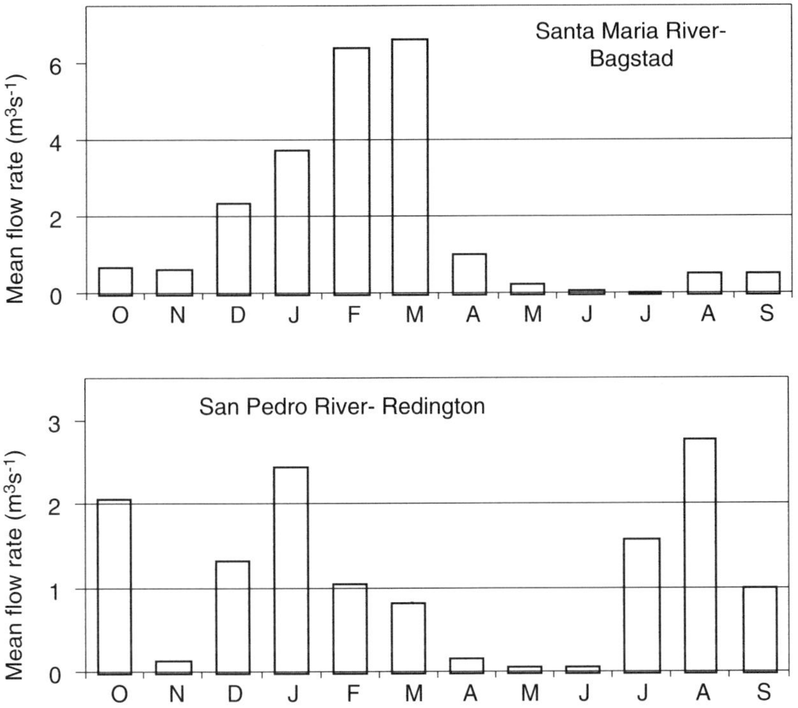

Figure 2-1. *Mean monthly flow rates (m^3s^{-1}) for streams in the Sonoran-Mojave Desert transition zone (Santa Maria River-Bagstad) and the Sonoran-Chihuahuan Desert transition zone (San Pedro River-Redington). Values shown are based on U. S. Geological Survey (USGS) historical daily streams flow data and are averages for a 25-year time period.*

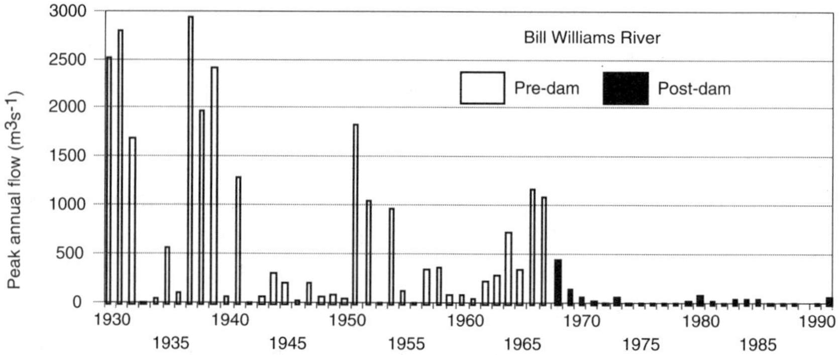

Figure 2-2. *Annual flood magnitude size, pre-dam and post-dam, for the Bill Williams River, in western Arizona. Values shown are based on USGS peak flow data (instantaneous discharge, station number 09426000).*

"dry years" and "wet years" (Figure 2-2). The patterns are nonstationary and reflect drought cycles and the El Niño-Southern Oscillation. In some rivers of the Southwest, large winter floods were relatively more frequent near the end of the nineteenth century and again toward the end of the twentieth century. During the 1980s and 1990s, some rivers experienced back-to-back 100-year return floods in winter. Such floods were less common in the interim. The Santa Cruz River in southern Arizona, for example, had no large winter floods from the 1930s through the 1960s (Webb and Betancourt, 1992). This type of flood pattern results in "boom and bust" cycles, with episodes of intensive scour and channel widening followed by rapid and dense regrowth of the vegetation.

The high-energy floods influence patch dynamics and reinitiate succession. Along Arizona's free-flowing (undammed) Hassayampa River, for example, riparian vegetation changed dramatically after large El Niño winter floods in 1993, which widened the channel from 5 to 50 m and eroded floodplains to an elevation very near the water table (Stromberg et al., 1997) (Figure 2-3). Emergent marshlands characterized by *Juncus articulatus, Typha domingensis,* and *Scirpus americanus* increased fivefold after the flood. Many mature *Populus fremontii–Salix gooddingii* forests and *Prosopis velutina* woodlands were replaced by young stands of *Populus-Salix* (Figure 2-4). The long period of flood recession created establishment opportunities for spring-germinating *Populus* and *Salix* as well as summer-germinating *Baccharis salicifolia, Tessaria sericea,* and exotic *Tamarix ramosissima* (Stromberg, 1997) (Figure 2-5). At the same time,

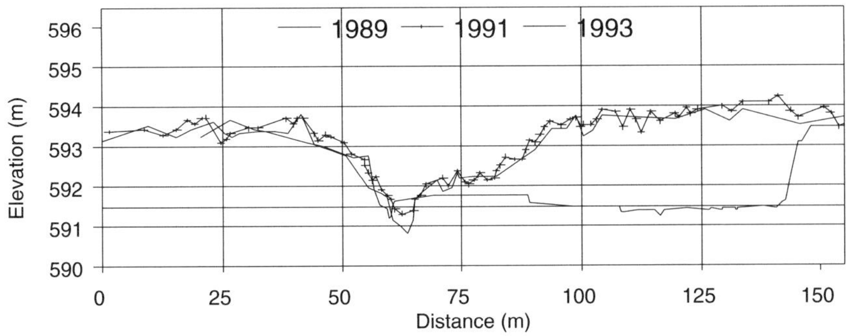

Figure 2-3. *Changes in topography of the Hassayampa River floodplain after two major flood events: a flood in 1991 with instantaneous discharge of 368 m^3s^{-1} and a flood in 1993 with instantaneous discharge of 745 m^3s^{-1}. (Figure is modified from Stromberg et al., 1997.)*

parts of the pre-1993 channel were buried by more than 1 m of sediment. Riparian shrublands dominated by the stress-tolerant *Hymenoclea monogyra* established on the coarse flood deposits, too high and dry for marsh plants or *Populus-Salix* seedlings.

Low-energy floods, in contrast, influence successional processes by effecting other changes, such as depositing layers of fine sediments around the boles of trees and redistributing leaf litter. Rivers in the Sonoran Desert, as in other arid regions, carry high sediment loads. In the Hassayampa River floodplain, annual rates of sedimentation have ranged from less than 1 to 10 cm, with the sedimentation rate increasing with flood magnitude (Stromberg et al., 1993b). Deposition declines with distance from the active channel for smaller floods (Figure 2-6). Root crowns of pioneer trees can eventually be buried under several meters of fine sediment. Such sediments may maintain nutrient availability near the soil surface. Large-seeded, shade-tolerant, competitive trees (Grime, 1979), including *Prosopis velutina, P. glandulosa, P. pubsecens,* and *Juglans major,* establish in the aggraded floodplain soils, aided by summer monsoon rains and the resultant flood spikes that disperse seeds and moisten safe-sites (Stromberg and Patten, 1990; Stromberg et al., 1991).

As the predominant natural disturbances, floods are clearly a major force structuring southwestern riparian vegetation (Junk et al., 1989; Poff et al., 1997). Floods create a riparian ecosystem that is dynamic in space and time by mobilizing, eroding, and depositing sediment, causing river channels to relocate and meander, creating backwater depressions, scour-

Figure 2-4. *Riparian vegetation along the Hassayampa River during a nonflood year (1988, top photo) and one year after the large El Niño floods (1993, bottom photo). (Photographs by J. Stromberg.)*

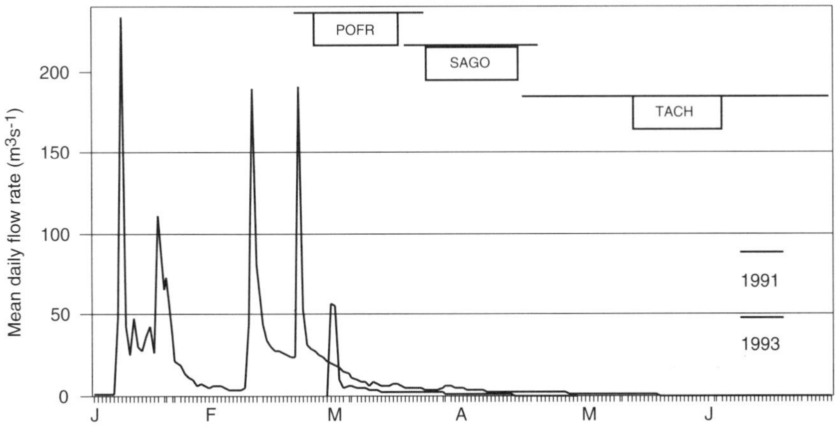

Figure 2-5. *Mean daily flows in the Hassayampa River (Morristown gauge) during January through June of 1991 and 1993. Values shown are based on USGS historical daily stream flow data. Also shown are approximate seed dispersal periods for* Populus fremontii, Salix gooddingii, *and* Tamarix chinensis.

ing vegetation, and dispersing propagules. Diverse microhabitats are created that vary in depth to the water table, soil moisture, texture, nutrient content, and light availability. These habitat characteristics help to produce and sustain a diverse mosaic of native plant communities (Naiman et al., 1993). Floods also maintain high rates of plant productivity by recharging

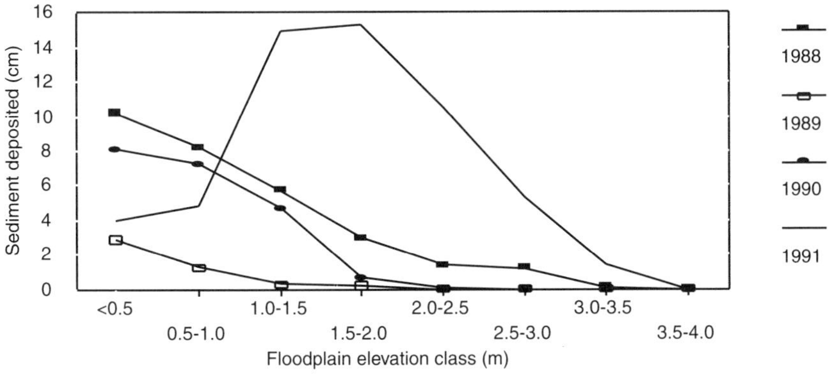

Figure 2-6. *Spatial distribution of sediment deposited on the Hassayampa River floodplain during four flood events of varying sizes (from 14 m^3s^{-1} in 1989 to 368 m^3s^{-1} in 1991). Values shown are means within classes based on floodplain elevation relative to the channel bed. (Data are modified from Stromberg et al., 1993b. With permission of* Great Basin Naturalist © *1993, Brigham Young University, Provo, UT 84602.)*

aquifers, moistening surface soils, and depositing limiting nutrients on the floodplain.

Human-Altered Flood Patterns

A system of dams and high-capacity reservoirs has been developed on the rivers of the southwestern United States. The amount of water captured and stored per capita (within the service area) is among the highest in the nation (Graf, 1999). The Colorado River, largest in the region, has been transformed from a powerfully dynamic waterway to a string of storage pools. The completion of Hoover Dam in 1935 effectively disconnected the upper and lower Colorado watersheds. Multipurpose dams, such as Hoover, store water, control floods, and generate hydropower. Major and minor tributaries of the lower Colorado, including the Gila, Salt, Verde, and Agua Fria, are similarly impounded. Reservoirs such as Lake Pleasant, formed by New Waddell Dam on the Agua Fria, as part of the Central Arizona Project, receive and store water that is pumped upgrade and cross-country from the Colorado (Springer et al., 1999). Diversion structures, including New Waddell Dam and Granite Reef Dam on the Salt River, completely reroute flows into canals, leaving a dry riverbed below (Figure 2-7). Small rivers, such as the Sonoita and Aravaipa Creeks, have been dammed solely for recreational purposes. Only a few small perennial Sonoran Desert rivers, including the Hassayampa, Santa Cruz, and San Pedro, remain undammed.

Flood and low-flow patterns are now fundamentally different as a result of reservoir development and water diversion for off-channel uses. In many cases, water that would have arrived in rivers as spring runoff now appears (if at all) as the sustained high flow of summer water delivery to downstream diversion structures, tailing off and exposing near-channel sediments only as the irrigation season wanes in the autumn. Dammed rivers usually flood less frequently, and the rare floods that do occur tend to be drastic ones that overwhelm water storage or flood control capabilities. Dynamic fluvial processes, such as channel migration, erosion, and sedimentation, can become static (Shields et al., 2000). Salt levels in stream waters and floodplain soils increase as water evaporates from the large reservoirs and "flushing" floods are eliminated.

Nonimpoundment flood control measures, like channelization, expedite water movement through "improved" parts of river systems, minimizing

Figure 2-7. A dewatered reach of the Agua Fria River below New Waddell Dam, in central Arizona. (Photograph by J. Stromberg.)

overbank flows, riparian sediment deposition, and bank storage recharge. Regionwide watershed degradation has also modified flood patterns. Widespread soil compaction, reduced plant cover, and loss of "A" horizon and organic matter in soils have followed overgrazing, poor timber management, and rapid urbanization. As a secondary effect, rapid runoff has increased "flashiness" in the remaining undammed fragments of these already flashy systems, resulting in less infiltration and more runoff. The net effect is larger flood peaks, higher sedimentation rates, and reduced base flows (Trimble and Mendel, 1995).

The range of ecological responses to flood flow alterations has varied with the nature and extent of hydrologic alteration, the geomorphic setting, and the interactive effects of other stressors, including livestock grazing (Scott et al., 1996). In the narrow Glen Canyon reach of the Colorado River, flood suppression has allowed narrow strips of marshland and riparian forest to persist along the channel, which during pre-dam times was scoured by annual floods (Stevens et al., 1995). Many ecological responses have been negative, however. Farther downstream along the Colorado River, for example, the cumulative effects of river damming and diversion have nearly obliterated the most extensive riparian wetlands in the desert Southwest, leaving only a few scattered ecological remnants amid a vast expanse of unvegetated and degraded land (Glenn et al., 1992; Briggs and Cornelius, 1998). The Lower Colorado River riparian vegetation zone has been narrowed, and the vegetation mosaic has been simplified (Ohmart et al., 1988). As a consequence of the reduced extent and quality of riparian habitat in the Southwest, many animal species that are riparian obligates have undergone population declines. The southwestern willow flycatcher (*Empidonax traillii extimus*), a small neotropical migrant songbird that nests in dense thickets of *Salix* spp. and other structural analogs along perennial streams, is now on the federal endangered species list (Unitt, 1987). The western yellow-billed cuckoo (*Coccyzus americanus occidentalis*), another neotropical migrant that nests in dense riparian vegetation, including *Populus-Salix* forests, also has declined in numbers. The Yuma clapper rail (*Rallus longirostris yumanensis*) and the least Bell's vireo (*Vireo bellii pusillus*) are also on the federal endangered species list (Kus, 1998). Without widespread changes in river management, the prognosis for these and many other species is grim (Graf et al., in press).

FLOOD PULSES AND RIPARIAN RESTORATION

Restoration of riparian ecosystems has become a major enterprise in the desert Southwest. Federal and state agencies, Indian tribes, cities, and private landowners are spending money and time on efforts to reverse past degradations. Goals and approaches vary widely. A decade ago in the Southwest, goals were somewhat limited, and riparian restoration was synonymous with cottonwood pole planting (Briggs, 1996). Although plantings are still widely used today (Alpert et al., 1999), the practice of riparian restoration is maturing beyond single-species plantings to encompass an ecosystem approach and a goal of self-sustainability (Goodwin et al., 1997). There is increasing recognition by restoration managers that hydrologic regimes and fluvial processes are prime determinants of riparian community structure and that restoration of native biodiversity and ecosystem complexity depends on the restoration of fluvial dynamism. There is increasing acceptance of the need to restore flows of water, sediment, and nutrients in sufficient quantities and with appropriate temporal patterns (Poff et al., 1997; Hill and Platts, 1998; Taylor et al., 1999).

Full restoration of riparian ecosystems requires removing all impediments to natural flow regimes (Schmidt et al., 1998). However, there are few cases in the desert Southwest in which dams have been removed for the purpose of habitat restoration. An exception involves Fossil Creek in central Arizona; the impending decommissioning of this small hydropower dam is a significant undertaking. Where dams remain in place, creative ways are being found to rehabilitate, if not fully restore, below-dam ecosystems while still allowing for some degree of municipal or agricultural water supply, hydropower production, or flood control (Whittaker and Shelby, 2000). Along several free-flowing rivers, ecological stressors have been eliminated or reduced. At these sites, natural flood pulses are facilitating passive recovery of the riparian vegetation following removal of the stressors.

Restoring Dominant Species

Populus fremontii and *Salix gooddingii* historically were the dominant pioneer forest species along Sonoran and Mojave Desert rivers. Precise estimates are impossible because rangewide baseline data do not exist, but losses as high as 90 percent have been postulated for *Populus-Salix* forests on various rivers such as the lower Colorado River (e.g., Ohmart et al., 1997). These forests are considered a globally imperiled ecosystem type by The Nature Conservancy (Anderson et al., 1998; Stein et al., 2000).

The population dynamics of *Populus fremontii* and *Salix gooddingii* are influenced by many aspects of a flood regime, including the timing, magnitude, duration, and rate of change of any given flood. Both are short-lived pioneer trees that exploit and depend on the flooding cycle through temporally specialized reproduction strategies. Plants establish from seed during occasional ideal years when appropriate flood conditions are present (Turner, 1974; Everitt, 1995; Scott et al., 1997; Mahoney and Rood, 1998; Stromberg, 1998a). The timing of flood flows is critical, as the species have evolved to produce seeds that are viable during the brief period when high spring flows are declining and exposing bare, damp sediments (Horton et al., 1960; Fenner et al., 1984). Large winter floods scour and redeposit floodplain sediments, creating the patchwork of bare mineral soils on which plants can establish without competition from overstory trees. *S. gooddingii* disperse seeds somewhat later in the season than *P. fremontii* (although the dispersal periods overlap) and, as the flood waters recede, establish on sites that are lower and closer to the stream. Small differences in the timing of spring flood pulses can influence the relative success of recruitment of these and other pioneer trees and shrubs (e.g., *Salix exigua*) because of phenological differences (Stromberg et al., 1991). Fenner et al. (1985) were among the first to call attention to the fact that *Populus fremontii* were not regenerating below dams where flood patterns had been altered. They studied the Salt River upstream of Phoenix, Arizona, and attributed the lack of *Populus* regeneration to the lack of spring floods, as well to water depletion caused by pumping groundwater. The 1980s and 1990s saw a spate of research on the patterns and causes of *Populus* declines below dams throughout the western United States and Canada (Rood and Mahoney, 1990). Mahoney and Rood (1998) observed that one cause of *Populus* seedling mortality was rapid recession of floodwaters. Friedman et al. (1998) and Johnson (1998) noted that in some cases, forests of *Populus* and other woody pioneers expanded immediately following dam closure, only to be replaced by later successional species without the return of appropriate regeneration floods.

Research on the autecology of *Populus* led to the development of recruitment models, sometimes called "recruitment boxes," that indicate when waters should be released from dams and at what drawdown rate, to allow for seedling establishment (Mahoney and Rood, 1998). Several projects have been implemented that have used these models as a basis for restoring flows to regenerate *Populus.* However, suitable flows of water and sediment have yet to be restored in the dammed river reaches that constituted the study area of Fenner et al. (1985). Landowners and river advocates along the Salt River continue to be concerned about the lack of adequate *Populus* regeneration, some because of concern for loss of nesting habitat for bald eagles.

The Truckee River in Nevada provides an example of river rehabilitation via restoration of a more natural stream flow pattern. Dams, channelization, and water diversions had contributed to a loss of age class and structural diversity within the *Populus fremontii* forests and a collapse of native fish populations, including the endangered cui-ui fish (*Chasmistes cujus*). To stimulate spawning of the fish populations, the U.S. Fish and Wildlife Service began managing the Stampede Reservoir in the late

1980s for spring flood release. An ancillary benefit was the establishment of *Populus* seedlings. However, many seedlings were at possible risk of mortality, having established on sites that could be flooded in the future. Subsequently, in 1995 a collaboration between The Nature Conservancy, federal agencies, and university researchers resulted in the release of a flood tailored from recruitment models, expressly for the purpose of *Populus* regeneration (Mahoney and Rood, 1998). Although the prescribed flood resulted in added costs to the Bureau of Reclamation, the Nevada Nature Conservancy speculates that there will be long-term economic savings due to improved ecosystem functioning: "With greater shading of the river, water temperature will be lower, and less water will be needed to manage the endangered cui-ui fish, reducing the long-term cost of the species' recovery program" (Chisholm, 1996). Another take-home message here is that "a whole array of ecosystem components may begin to recover when restoring a basic ecosystem process," such as the natural flow regime of a river (Gourley, 1998). Although it is impossible to manage directly for every species, we increase our odds of providing sustainable ecosystem improvement if we take an approach that allows for natural cycles of flood disturbance (Bayley, 1991; Stanford et al., 1996).

Flood pulses have accidentally helped to restore small patches of *Populus-Salix* forests to some Southwestern rivers. The Gila River and its major tributaries, the Salt and Verde Rivers, are impounded by seven large water supply dams as they traverse Arizona before meeting with the Colorado River near Yuma. The Colorado River itself is one of the most extensively dammed and diverted rivers in the United States, with flows at the Mexican border now only a small fraction of what they were a century ago. Most of the water in the Gila and Colorado reservoirs is stored for agricultural use; vast portions of the lower floodplains of these two rivers are farmed for lettuce, cotton, alfalfa, and other irrigated crops. During El Niño years in the late 1970s, 1980s, and 1990s, heavy upland precipitation, particularly rain on deep snowpack, forced large flows to be released through some of the dams during winter and spring (Figure 2-8). Releases may have become more frequent in recent decades, as the dams have aged and filled with water and sediment (Zamora-Arroyo et al., 2001). Such releases were not conceived to provide downstream ecological benefits. However, some of the water releases fortuitously corresponded to the regeneration needs of the trees while (presumably) flushing accumulated agricultural tailwater salts and recharging aquifers. Several young popula-

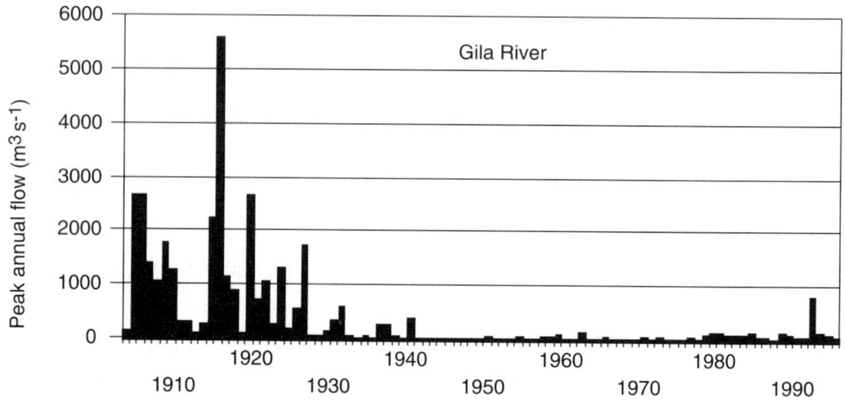

Figure 2-8. *Annual flood magnitude in the lower Gila River near Yuma, Arizona. Coolidge Dam, the main water storage dam on the Gila, was completed in 1928. Completion dates for dams on major tributaries range from 1911 (Roosevelt Dam) to 1946 (Horseshoe Dam). Values shown are instantaneous discharges (m^3s^{-1}) based on USGS peak flow data (station number 09520500).*

tions of *Populus* and *Salix* now occur along portions of the lower Gila and Colorado Rivers, as well as the lower Salt and Verde Rivers (Rea, 1983; Briggs and Cornelius, 1998). Passive management of floods has also allowed for regeneration of *Populus fremontii* along portions of Utah's Fremont River, that are utilized for irrigation storage (Everitt, 1995). Yet trees on some of these rivers may die unless water sources needed for their survival are forthcoming. In addition to working toward securing minimum maintenance flows, there is a need to negotiate dam operating agreements that provide for the intentional release of flows during periodic wet years in order to mimic the natural flow regime.

Floods have also served as restorative agents for *Populus-Salix* forests on degraded undammed rivers. The Santa Cruz River, which originates in Mexico and flows northward through southern Arizona, has remained free of dams. Historically, the Santa Cruz was described as a small but lush river that flowed perennially near the U.S./Mexican border (Tellman et al., 1997). During the 1950s and 1960s, however, extensive portions of the Santa Cruz River in the vicinity of Nogales, Arizona, went completely dry. The river lost perennial flow, groundwater tables declined, and most of the *Populus-Salix* forests and marsh vegetation died. The 1950s were a period of extreme drought in the Southwest (Swetnam and Betancourt, 1998). Compounding the drought was extensive pumping of groundwater from the small alluvial aquifer lining the Santa Cruz River (Stromberg et al., 1993c). During recent decades, however, the Santa Cruz riparian ecosystem was revitalized. Abundant rains and runoff, together with the release of large amounts of treated municipal wastewater into the channel, contributed to aquifer recharge and restoration of the perennial stream flow. Some of this water originated in another watershed, resulting in a net increase to the Santa Cruz. Although the restoration of a high water table set the stage for a high rate of survivorship, it was the occurrence of winter and spring flood flows in the 1970s, 1980s and 1990s that triggered the establishment of the extensive young groves of *Populus* and *Salix* that line the riverbed. The sources of seeds for the natural recovery presumably were the trees that survived in upstream reaches not subjected to extreme groundwater pumping, or the trees along the edges of irrigated farm fields.

Exotic Species Management

Tamarix ramosissima and related species, large shrubs or trees that are native to Eurasia, have become dominant woody species along rivers of the desert Southwest (Harris, 1966). Vast portions of the Colorado River, Gila River, and Salt River flood plains, for example, are dominated by *Tamarix* species as well as by the native shrub *Pluchea sericea.* The replacement of species-rich communities by homogenous thickets of single species, be they native or exotic, can be symptomatic of dam-related reductions in fluvial disturbances and/or imposition of stressors that select for a small number of stress-tolerant species. Many management practices in riparian corridors, including reduced flood pulsing, have caused floodplain soils to become saltier, drier, and nutrient-poor (Stevens, 1989; Busch and Smith, 1995). These factors have favored stress-tolerant species such as *Tamarix* (Glenn et al., 1998; Busch, 1995). *Tamarix* are capable of great osmotic adjustment in saline soils (Decker, 1961; Shafroth et al., 1995; Glenn et al., 1998). They are deeper rooted and more able than native *Populus* and *Salix* trees to persist on the available soil moisture at sites where groundwater is below rooting depths (Horton, 1972; Busch and Smith, 1995; Stromberg, 1998b).

When rivers flood less frequently and at different seasons than their climatic legacies dictate, exotic species may gain a recruitment and survival advantage. Like the *Populus* and *Salix* they have widely replaced, *Tamarix* annually produce large crops of tiny, wind-dispersed seeds that germinate on bare, moist soil. Temporally, however, *Tamarix* are reproductive generalists and disperse seed over a much longer time period during the growing season (Shafroth et al., 1998; Roelle and Gladwin, 1999). Their reproductive strategy allows them to thrive on dammed rivers where high water flow is delayed by the timing of irrigation water storage and release schedules, as well as in the techno-littoral zone of reservoir edges, where seedbeds are exposed in midsummer during irrigation-driven drawdowns. There are some native riparian pioneer shrubs (e.g., *Baccharis salicifolia, Salix exigua*) that have a lengthy seed dispersal period. Unlike *Tamarix,* these seem not to have benefited by such alterations, perhaps partly because of lower fecundity.

Restoration of exotic-dominated sites can be approached by managing for ecological processes that favor native species and, if necessary, managing against the exotics. Restoration efforts at two National Wildlife Refuges provide examples of both approaches. The first involves the Bill Williams Wildlife Refuge, located on an alluvial tributary of the lower Colorado in west-central Arizona. Alamo Dam was constructed on the Bill Williams River in 1968 to minimize flood pulses into the Colorado River. Over the past 25 years, the size and frequency of winter floods in the Bill Williams River have decreased and summer floods have all but disappeared (Figure 2-2). Pre-dam, the ten-year recurrence interval flood was 1400 m^3s^{-1}. Plants would rapidly recolonize after the floods, but periodic scour maintained a more open system than occurs today. Today, the maximum possible controlled releases are 198 m^3s^{-1}, the ten-year flood is 148 m^3s^{-1}, and the mean flow rate is 4 m^3s^{-1}. The river is not diverted, and base flows have increased somewhat as the temporal pattern of flow release into the below-dam system has become less flashy. With fewer scouring flows and less water stress, vegetation has become more abundant than in the above-dam free-flowing Santa Maria River and covers more area than would be the case were the dam not present (Shafroth et al., in press) (Figure 2-9). However, most of the "infill" is *T. ramosissima.*

One goal of the Bill Williams Refuge managers is to restore *Populus* and *Salix* trees to dominance and thereby improve the quality of the riparian habitat. Other goals include maintaining flood control in the Colorado River and the recreational and wildlife benefits of Alamo Lake. To accomplish the first goal, the U.S. Fish and Wildlife Service, U.S. Army Corps of Engineers, and university scientists worked together to develop a flow-release plan that calls for high base flows and periodic flood flows. During

Figure 2-9. *Riparian vegetation along the free-flowing Santa Maria River* (above) *and flow-regulated Bill Williams* (below). *(Photographs by J. Stromberg.)*

1993 and 1995, high spring flows were released into the Bill Williams River. Recruitment models were retroactively tested, confirming that the recruitment box concepts can be used to enable seedling establishment of these target species (Shafroth et al., 1998). However, without the ability to release large scouring floods from the dam, extensive seedbeds for new generations of riparian trees will not be created. Small floods predictably produce smaller recruitment bands (Stromberg et al., 1993b). Rates of establishment of *Populus* and *Salix* are predicted to decline on the Bill Williams in the future, despite the release of spring flows (Shafroth et al., in press). Thus, *Tamarix* will remain a dominant species. Intervention in the form of mechanical clearing of seedbeds in *Tamarix*-dominated habitat, followed by removal of aggraded sediments, may be necessary to restore *Populus* and *Salix* to dominance. Active restoration measures that mimic the effects of large floods are needed if natural processes cannot be fully restored (Friedman et al., 1995).

Tamarix has also become a dominant species in the Bosque del Apache Wildlife Refuge, as on much of New Mexico's highly regulated Rio Grande (Everitt, 1998). Refuge managers here have mimicked the scouring effects of large floods by using bulldozers, herbicides, and fire to clear extensive stands of *Tamarix,* which has cost $750 to $1300 per hectare (Taylor and McDaniel, 1998). They released river water onto the bare floodplains in spring, with a seasonal timing and quantity that mimic the flood hydrograph of the Rio Grande and thereby favor a diverse assemblage of pioneer plant species. Long-term monitoring will determine whether the multilevel canopy, diversity of vegetation structure, and diversity of insect life that develop provide superior wildlife habitat to the *Tamarix* thickets that existed previously. This type of "wet soil management" is increasingly being used on other regulated rivers, notably in wildlife refuges along the lower Colorado River, where flood pulses can be released through water control structures to small, cleared areas of the floodplain or abandoned farm fields. These efforts do not completely eliminate *Tamarix*. However, there is growing consensus that total eradication is neither realistic nor necessary. For example, removal of *Tamarix* from *Populus* and *Salix* stands does not cause detectable change in bird abundance and diversity, suggesting that it is the presence of the structurally diverse and insect-rich natives, rather than the complete absence of the exotic, that is of key importance (Weintraub, 1993).

In addition to managing river flows for native species, there are ways to manage flows so as to decrease the birth rates or increase the death rates of *Tamarix*. Few such projects have been implemented, and more experimentation is warranted. Prolonged late-season inundation has been used as a technique to increase the mortality of *Tamarix* seedlings at a *Populus deltoides* ssp. *monolifera* restoration site (Roelle and Gladwin, 1999); the inundation time and duration were selected on the basis of experimental studies that compared mortality rates of native and exotic tree seedlings (Gladwin and Roelle, 1998). Restoration projects on dammed rivers could experiment with the release of flashy summer flood pulses to kill summer germinants of *Tamarix*. Although certainly able to survive flood scour and sedimentation as adults, there is evidence that *Tamarix* are less tolerant of flooding as juveniles than are *Populus* species. As compared with *P. fremontii, Tamarix ramosissima* juveniles exhibited higher mortality resulting from experimental sedimentation (Levine, 2001) and natural river flooding (Stromberg et al., 1993b). Less ability to tolerate flood scour may explain why *Tamarix* population levels are low relative to those of native species on some free-flowing, frequently flooded rivers and may contribute to their tendency to proliferate on flood-regulated rivers (D'Antonio et al., 1999; Shafroth et al., 2002). However, there are conflicting opinions on this matter, and certainly there is room for additional study.

A study on the free-flowing San Pedro River provides an example of a natural shift from exotic to native dominance over time and evidences a capacity for self-repair in degraded Sonoran riparian ecosystems, once stressors are eliminated or adequately reduced. The San Pedro flows northward from Sonora, Mexico, to the Gila River in southern Arizona. Stream flows vary from perennial to ephemeral, depending on local geology and tributary inputs and the extent of groundwater pumping. Historically, floodplain agriculture and cattle grazing have been the main land uses, but some reaches have recently been set aside by The Nature Conservancy and the Bureau of Land Management as conservation areas. Livestock grazing, gravel mining, crop irrigation, and pumping groundwater from the alluvial aquifer have been eliminated from the conservation areas. The composition of the San Pedro riparian plant community has also changed, coincident with changes in management and weather patterns. *Tamarix ramosissima* became established on the river in the 1950s (a drought period) and by the 1960s was the dominant woody pioneer in the central reaches of the river (Stromberg, 1998a). Since then, *Populus fremontii* and *Salix gooddingii* have been increasing in relative abundance. In the 1990s, young *Populus* and *Salix* outnumbered young *Tamarix* for the first time in more than 50 years in the central reaches.

Flood events were critical to the recovery of *Populus* and *Salix* on the San Pedro River. El Niño winter floods in the 1980s and 1990s were sufficiently large to scour vegetation and create opportunities for species replacement and were appropriately timed to favor the native pioneers (Figure 2-10). Under the conditions of livestock exclusion, reduced rates of upstream groundwater pumping, and frequent winter/spring floods, the native species were able to outcompete the exotic one. Studies have shown *Populus* spp. to be more competitive than *Tamarix* under conditions of ad-

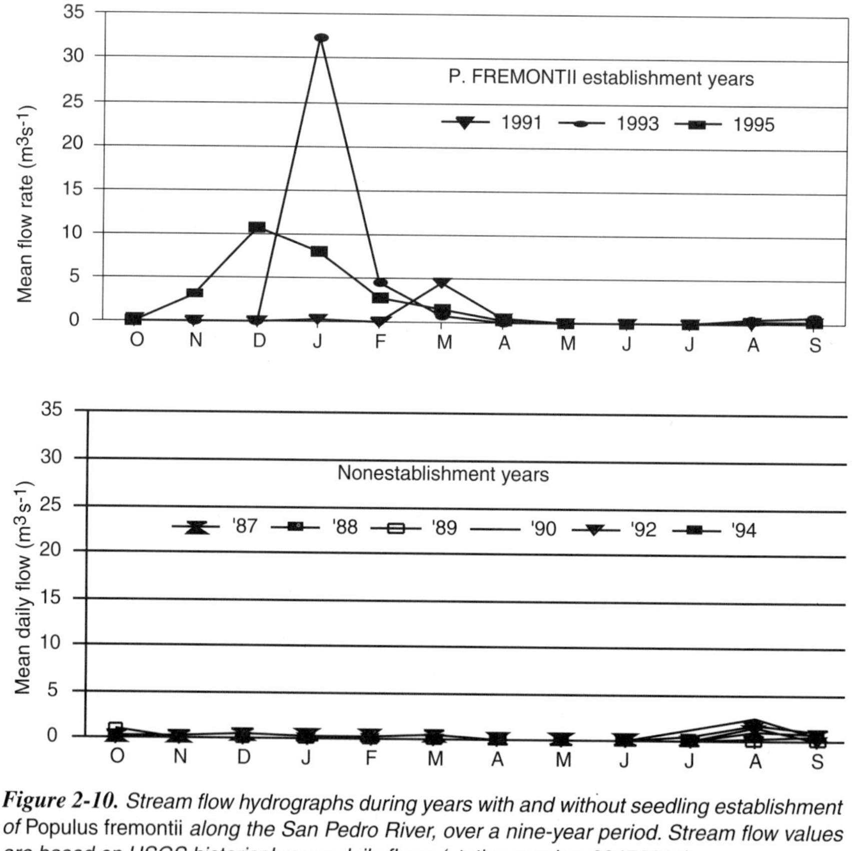

Figure 2-10. *Stream flow hydrographs during years with and without seedling establishment of* Populus fremontii *along the San Pedro River, over a nine-year period. Stream flow values are based on USGS historical mean daily flows (station number 09472000).*

equate water (Stevens, 1989; Sher et al., 2000) and livestock exclusion (with livestock grazing giving a competitive edge to the less palatable *Tamarix*) (Stromberg, 1997). Notably, these patterns of recovery were apparent only in the wetter, ungrazed reaches of the San Pedro River; *Tamarix* continued to dominate in reaches with ephemeral flow, deep water tables, and livestock grazing (Figure 2-11). There is a need for additional studies that assess the potential for natural recovery of native species on exotic-dominated sites upon removal of stressors and/or removal of the exotic species. Theoretically, by restoring natural flow regimes and herbivory patterns, we can tip the ecological balance in favor of native species (Poff et al., 1997).

Restoring Plant Productivity

Riparian forests, supplied with ample water and nutrients, are among the most productive vegetation types in arid regions. Shallow groundwater is the primary water source for many floodplain trees, including *Populus* spp., *Salix* spp., and *Platanus wrightii* (Busch et al., 1992; Smith et al., 1998; Scott et al., 1999; Stromberg, 2001a). Flood flows are an important source of recharge of the alluvial aquifer in many rivers (Workman and Serrano, 1999) and thus can contribute much of the water that sustains phreatophytic vegetation. The annual radial growth rate of *Platanus wrightii* along Sycamore Creek in Arizona increased as the number of winter floods increased, ranging from 0.1 cm yr^{-1} in dry years to 0.9 cm yr^{-1} in wet, flood years (Stromberg, 2001a). Winter flood flows in this small river are critical for elevating groundwater to within tree rooting zones (Marti et al., 1999; Stromberg, 2001b).

Flood pulses that occur during the growing season are used as a water source by some riparian trees in desert regions (Akeroyd et al., 1998). Summer flood pulses are critical in some areas to provide the shallow soil moisture that sustains seedlings or saplings of *Populus* species until their roots grow deep enough to extract water from the aquifer (Cooper et al., 1999). Summer floods provide an important water source for the facultative phreatophytic tree *Prosopis velutina* by wetting surface soils and recharging groundwater. This deep-rooted species (with one report of a 50 m taproot) attains its largest size when able to access shallow groundwater. Where groundwater is inaccessible, *Prosopis* trees undergo strong

Figure 2-11. Flood pulses have contributed to increased abundance of Populus fremontii *and* Salix gooddingii *in some parts of Arizona's San Pedro River (*above*).* Tamarix *spp. continue to dominate in drier reaches (*below*). (Photographs by J. Stromberg.)*

seasonal increases in productivity in response to the summer monsoon flood pulse (Stromberg et al., 1992; Stromberg et al., 1993a).

Flood pulses are often rich in organic matter and nutrients, (Grimm et al., 1981). Storm runoff mobilizes nutrients that have accumulated in and on upland desert soils, and flood waters then deposit them on floodplain soils (Grimm and Fisher, 1986). Floods also stimulate microbial activity and increase the rates of decomposition of organic matter, which can be low on the dry surface soils of floodplains in semiarid regions (Molles et al., 1998). A high frequency of summer floods has been linked to increased growth of the tree *Platanus wrightii,* presumably because summer floods replenish the limited supply of nutrients (Stromberg, 2001b) (Figure 2-12).

Stevens (1989) noted that waters in the Colorado River have become depleted in phosphorous because of sediment trapping in reservoirs. This nutrient depletion caused a reduced growth rate of woody riparian pioneer species in experimental studies. High salinity levels, such as can develop in infrequently flushed soils of regulated rivers, reduce the growth of some riparian tree species (Glenn et al., 1998). However, there have been no systematic studies of the effects of altered or suppressed flood pulses on

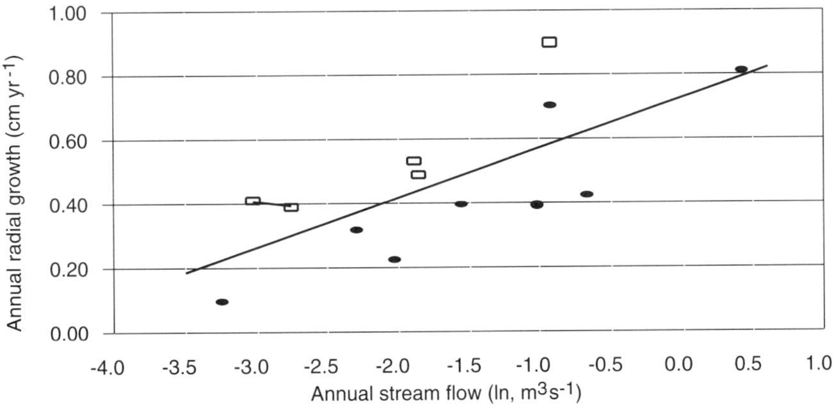

Figure 2-12. *Mean annual radial growth rate of young* Platanus wrightii *trees in relation to natural log-transformed annual stream flow rate in an intermittently flowing reach of Sycamore Creek, Arizona. Rectangular symbols indicate years with at least one summer flood. (Figure is modified from Stromberg, 2001b, with permission of Blackwell Science Ltd., Osney Mead, Oxford, U.K.)*

woody plant growth and productivity in below-dam reaches of Sonoran Desert rivers. Such studies would be useful as a basis for determining whether there is a need to restore productivity pulses, in addition to regeneration pulses.

Minimizing Fire Disturbance

The dynamic fluvial processes of flow-regulated rivers are altered in ways that can variously increase or decrease standing live and dead biomass. On the flood-suppressed Bill Williams River and portions of the lower Colorado River, post-dam increases in dense stands of fire-prone *T. ramosissima* have set the stage for frequent, intense, and large fires (Busch and Smith, 1995; Shafroth et al., in press). Along these rivers, more than a third of the studied riparian forests burned over a recent 12-year period (Busch, 1995). *Tamarix ramosissima* and the clonal shrub *Tesseria sericea* resprout prolifically after fires, but *Populus fremontii* does not, resulting in postfire shifts in vegetation structure.

Fires historically occurred only rarely in regularly flooded Sonoran riparian corridors, but they have replaced floods as a primary disturbance factor on many regulated rivers (Figure 2-13). When floods are suppressed, woody debris can accumulate; and vegetation can senescence and increase in extent, density, and homogeneity. Floods, in contrast, reduce fire potential by scouring vegetation and creating natural firebreaks, removing dead branches and other debris, increasing litter decomposition rates, and increasing the moisture content of the vegetation (Ellis et al. 1998; Ellis 2001). In addition, *Salix* and other pioneer species quickly revegetate flood-scoured areas, replacing older, senescent stands with young, "green" wood.

Restoration measures to reduce atypically high fire frequencies have yet to be implemented. On the Bill Williams River, for example, although small spring floods will be restored, the dense post-dam vegetation remains susceptible to fire damage because of constraints on flood size. Experiments are needed on this and other rivers to determine whether frequent, small summer flood pulses can promote fire resistance by increasing litter decomposition rates and dislodging dead vegetation. Ultimately, only dam removal may reliably shift the disturbance cycle from fire back to floods.

Figure 2-13. Fire in Tamarix-dominated habitat along the Rio Grande, New Mexico. This particular fire was a controlled burn. (Photograph by John Taylor.)

Restoring Herbaceous Biodiversity

Riparian ecosystems are linear oases of biodiversity in arid regions. The few dominant woody species provide the vertical and horizontal structure that composes a necessary habitat feature for many animal species. Much of the floristic diversity, however, lies in the herbaceous component of the plant communities. Hundreds of different grasses and forbs, many ruderal, can occur along free-flowing intermittent or perennial rivers (Wolden et al., 1994). The vegetation cycles yearly from cool-season species, such as the winter annual *Amsinckia intermedia* and the perennial vine *Marah gilensis,* through warm-season species, including *Amaranthus palmeri* and *Aristolochia watsonii.* The composition of a herbaceous community varies temporally in response to climate dynamics, and spatially across the floodplain. Obligate and facultative wetland plants on channel bars and off-channel depressions give way to upland species on aggraded floodplains and terraces (Figure 2-14). Patchy deposition of coarse and fine sediments in the floodplain further contributes to floristic diversity. Along the San Pedro River, herbaceous species that grew on coarse-textured, nutrient-poor soils included *Heterotheca subaxillaris, Helianthus petiolaris, Sporobolus contractus, S. cryptandrus, Chamaesyce hyssopifolia,* and *Bouteloua rothrockii.* Another group, which tended to grow on fine-textured, nutrient-rich

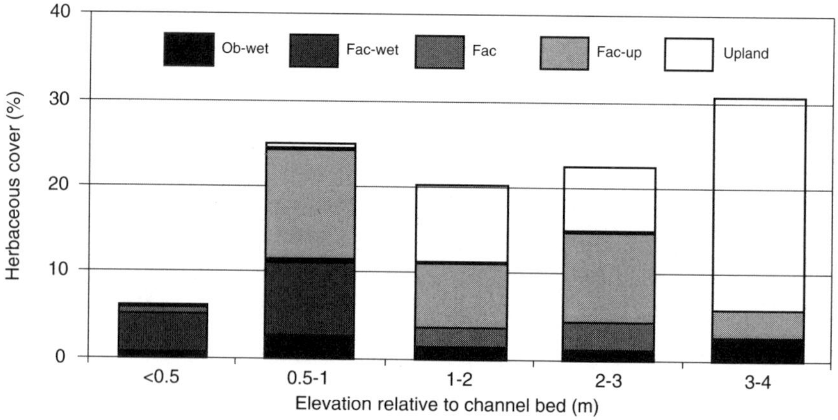

Figure 2-14. *Mean cover of herbaceous plants classified within wetland indicator categories (obligate wetland, facultative wetland, facultative, facultative upland, and upland), by floodplain elevation zone, during September, 1993, along the upper San Pedro River.*

soils, included *Viguiera dentata, Panicum obtusum, Plueraphis mutica, Sporobolus wrightii,* and *Sporobolus airoides* (Stromberg et al., 1996).

There has been little quantitative study of the effects of dams and flood-flow alteration on plant diversity patterns in Sonoran riparian ecosystems. Some studies have found plant biodiversity levels to be low in the dense thickets of *Tamarix* that can develop on flood-suppressed rivers (Brock, 1994). Some authors have attributed this to an autogenic effect of the *Tamarix,* but allogenic factors associated with river regulation may be at play as well. For example, litter accumulation and development of dense canopies (which to some degree are artifacts of flood suppression) can contribute to diminished understory biodiversity (Nilsson et al., 1999).

Plant species diversity can decline when flood disturbance becomes too infrequent or when the spatial and temporal diversity of flood disturbance and flood-influenced environmental conditions is reduced (Pollock et al., 1998). Cover and biodiversity of riparian herbs tend to increase on fine-textured soils (Stromberg 1998b; Jansson et al. 2000b) and, conversely, may decline below dams as "fines" are trapped in reservoirs and downstream sediments become uniformly coarse (Figure 2-15). Restoring a range of soil conditions on dammed rivers will require restoring sediment flows during flood pulses, possibly through the use of sediment-bypass structures. Restoring a range of temporal regeneration niches, to enable the establishment of warm-season and cool-season plants, will require restoring the full seasonal range of flood flows.

Dam and reservoir development can also influence biodiversity levels by impeding the dispersal of propagules (Jansson et al., 2000a). Studies are needed to determine whether dams are disrupting such plant dispersal processes along Sonoran Desert rivers. A study of dispersal patterns at the free-flowing Hassaympa River showed that most riparian plant species were adapted for dispersal by wind or animals (Drezner et al., 2001). Floods may, however, provide a secondary method of dispersing propagules laterally within the riparian zone and into downstream reaches. Dams and reservoirs diminish connectivity and may thus impede the travel of many animals, forming potential dispersal barriers without particular reference to flood effects.

Along many free-flowing rivers, the composition and diversity of the herbaceous vegetation has been altered by land uses, including livestock grazing and irrigated agriculture. Where such management disturbances have been removed, natural floods may be a practical (albeit slow) means

Figure 2-15. *A cobble bar along the Verde River, Arizona. Sediments have been trapped in upstream reservoirs, producing coarse soils in this below-dam reach. (Photograph by J. Stromberg.)*

of restoring biodiversity, as demonstrated in a study of the Hassayampa River. The Hassayampa is an alluvial river in central Arizona's Sonoran Desert. It has been free-flowing for more than a century, following a massive dam failure in 1890. The Nature Conservancy's Hassayampa River Preserve occupies an 8 km perennial segment of the river. Prior to 1986, the Preserve was a working cattle ranch and parts of the riparian corridor were used as a trailer park and for commercial orchards. These activities probably catalyzed development of the dense stands of exotic herbs that occur in the riparian habitats of the Preserve. Twenty-six percent of the 344 plant taxa collected here were exotic, 74 percent of which were annual or biennial grasses and forbs (Wolden et al., 1994). Most of the 88 exotic species covered only small areas and were encountered infrequently, but some became dominant. In 1989, areas adjacent to the channel were vegetated primarily by the native shrub *Baccharis salicifolia* and the exotic Bermuda grass (*Cynodon dactylon*). Exotic brome grasses (e.g., *Bromus rubens*), Bermuda grass, and wild barley (*Hordeum murinum* ssp. *glau-*

cum) dominated the *Populus-Salix* forest understory, and *Bromus* spp. and *H. murinum* ssp. *glaucum* dominated the understory of the *Prosopis* forests.

Populations of exotic species can persist for a long time after removal of the disturbance factor(s) that facilitated their invasion (Bartolome and Gemmill, 1981; Milchunas and Lauenroth, 1995). They may produce self-favoring conditions, may have a long life span, or may dominate the seed supply. Or abiotic conditions favoring these species may be persistent. Long periods may be required to reverse the physical effects of land use, such as soil compaction. We attempted to accelerate the recovery of native species to the Hassayampa forest understories by imposing several restoration treatments, including seeding natives and removing exotics by hoeing. We monitored treatment and control plots along stream banks, in mature *Populus-Salix* forests, and in *Prosopis* forests, for three years (Wolden and Stromberg, 1997). In separate studies, we monitored herbaceous cover in plots after a ten-year return flood (Stromberg et al., 1993b). The restoration treatments overall were not very effective. However, following the ten-year return flood, we did detect reductions in cover of many exotics (e.g., *Bromus rubens, Melilotus officinelis, Cynodon dactylon*), increases in cover of some native species (e.g., *Xanthium strumarium, Dicoria canescens*), and a general increase in diversity.

Other studies also show that floods can naturally restore native species to sites floristically altered by grazing (Chaneton et al., 1988). Floods accomplish many of the same goals of restoration treatments. Flood waters deposit a diverse seed mixture on fresh substrate while scouring or burying established exotics. Seed banks, distributed like a veneer in the upper layer of soil throughout the floodplain forests and shrublands, are a key source for postflood reestablishment of many native riparian herbs, including *Bowlesia incana, Calibrachoa parviflora, Centaurium calycosum, Mimulus guttatus,* and *Nicotiana obtusifolia* (Boudell and Stromberg 1999). Floods can create opportunities for exotic invasion in some circumstances, but can also facilitate recovery of natives when anthropogenic disturbances are removed and pressures, such as selection for grazing-tolerant species, are reduced.

Restoring Resilience and Self-Sustainability

Resilience, or resistance to flood disturbance, is considered to be a trait of "healthy" riparian ecosystems. Many riparian plant species have adaptations that allow them to resist flood damage or have sufficient regenerative ability to allow for rapid postflood redevelopment (Stromberg et al., 1991). However, if subject to stressors such as dewatering or overgrazing, plant communities may lose their capacity for self-repair after flood disturbance. Along the Carmel River in California, groundwater pumping caused the mortality of streamside vegetation. Without dense vegetation to stabilize banks during floods or to recolonize them after the fact, the result was extreme channel widening (Groeneveld and Griepentrog, 1985). Flood-related channel widening has also occurred along portions of the San Pedro River, where the combined influences of water diversion, beaver trapping, and river incision have constrained the resistance and resilience of the streamside vegetation (Fonseca, 1998). Livestock grazing, historically ubiquitous in the riparian West, can cause loss or reduction in streamside plant cover, loss of fine sediments and nutrients from the soil, channel downcutting, and lowering of the water table (Armour et al., 1991; Trimble and Mendel, 1995; Belsky et al., 1999). All of these factors reduce the potential of the riparian ecosystem to revegetate following flood disturbance. Floods, under such circumstances, can have adverse ecological effects.

Riparian restoration projects that do not effectively promote resilience will inevitably fail in the long-term. Some restoration projects are being designed to restore resilience and a capacity for self-recovery, by restoring perennial stream flows, regeneration-flood pulses, and connectivity to upstream riparian ecosystems to ensure a supply of propagules. In other cases, ecosystem processes are not being restored and notions of self-sustainability remain unaddressed. For example, the U.S. Army Corps of Engineers and municipal authorities are (as of October 2000) engaged in an ambitious and costly project to "restore" habitat quality to a watershed-decoupled reach of the Salt River through the heart of Phoenix, Arizona. Their congressionally funded technique entails planting some 75,000 nursery-grown *Populus fremontii, Salix gooddingii, Prosopis velutina,* and riparian shrubs along an engineered low-flow channel. Water will be supplied to plants over the long term through pumping from a shallow aquifer (derived from urban runoff) into surface pools and a high-pressure spray irrigation system. Because of the presence of an estimated 50 poorly documented and economically untreatable landfills along the adjacent historic floodplain, there is neither intention nor capability to restore the hydraulic dynamics that normally drive riparian plant community development. The project planners and managers are unable to encorporate major floods as a positive, renewing factor. Rather, they fear expensive damage to "parklike" plantings and the unlooked-for exposure of some toxic legacy of unregulated waste disposal (U. S. Army Corps of Engineers, 1998). The result must therefore be a sort of semistatic riparian diorama. There likely will be short-term benefits for some animal species. However, "restoration" in this instance resembles nothing so much as traditional urban structural flood control sheathed in a biotic veneer.

CONCLUSION

Free-flowing rivers in the arid Southwest can change within minutes from a string of quiet pools or a lazy trickle to wild, dangerous torrents. Despite engineering efforts to manage and contain stream waters in reservoirs, floods periodically overwhelm these systems and spill onto urban lands, reminding us that control is an illusion. Efforts to address the situation are being discussed and experiments initiated to "re-wild" some of the rivers, albeit in a semicontrolled fashion, to restore some of the ecological, social, economic, and psychological benefits of natural waters.

Further studies of ecological processes on free-flowing reaches, and of the effects of experimental floods on regulated reaches, are critically needed. The many demands and legal constraints on western waters necessitate substantial ecological documentation and justification for every drop released downstream. Sunbelt cities, including Las Vegas and Phoenix, are among the most rapidly growing in the United States, putting increasing pressure on diminishing desert water supplies. The problems of riparian management will worsen and the functions of riparian ecosystems will further decline unless we revise our expectations of water resources in the desert Southwest.

Templates such as the "natural flow regime" (Poff et al., 1997) provide only a starting point for managers. We need to experiment with the various components of the flow regime and quantify in-stream and over-bank flow budgets to construct and test integrated management models. Will an occasional winter/spring regeneration flood suffice for restoring *Populus-Salix* communities, or is it necessary to also release frequent summer floods to maintain bioproductivity and biodiversity? If so, how often? How large? Research at the landscape, ecosystem, community, and population levels must continue. The many National Wildlife Refuges on dammed rivers of the desert Southwest are desirable places to test our knowledge of restoration ecology by releasing experimental floods of varying magnitude, seasonal timing, and frequency. The support of river managers such as the U.S. Bureau of Reclamation, the U.S. Army Corps of Engineers, and the Salt River Project is needed if we are to include ecosystem restoration as one of the many multiple-purpose river management goals.

The widespread changes that southwestern riverine and riparian ecosystems have undergone in the past century are not attributable solely to the alteration of flood patterns. There is a long list of agents of change to be accounted for as we undertake future studies and assemble management proposals. But beneath this chaotic overlay of factors we can discern that maintaining or restoring a high degree of biotic integrity hinges on retaining appropriate flood disturbance as a fundamental organizing factor in these ecosystems. To our advantage, riparian ecosystems are inherently resilient. Given something like "normal" hydrologic conditions, we believe they often can rebound without much further intervention.

REFERENCES

Akeroyd, M. D., S. D. Tyerman, G. R. Walker, and I. D. Jolly. 1998. Impact of flooding on the water use of semi-arid riparian eucalypts. *Journal of Hydrology* **206:**104–117.

Alpert, P., F. T. Griggs, and D. R. Peterson. 1999. Riparian forest restoration along large rivers: Initial results from the Sacramento River Project. *Restoration Ecology* **7:**360–368.

Anderson, M., P. Bougeron, M. T. Bryer, R. Crawford, L. Engelking, D. Faber-Langendoen, M. Gallyoun, K. Goodin, D. H. Grossman, S. Landaal, K. Metzler, K. D. Paterson, M. Pyne, M. Reid, L. Sneddon, and A. S. Weakley. 1998. *Terrestrial Vegetation of the United States.* Vol. 2. *The National Vegetation Classification System: List of Types.* The Nature Conservancy, Arlington, VA.

Armour, C. L., D. A. Duff, and W. Elmore. 1991. The effects of livestock grazing on riparian and stream ecosystems. *Fisheries* **16:**7–11.

Bartolome, J. W., and B. Gemmill. 1981. The ecological status of *Stipa pulchra* (Poaceae) in California. *Madroño* **28:**172–184.

Bayley, P. B. 1991. The flood-pulse advantage and the restoration of river-floodplain systems. *Regulated Rivers: Research and Management* **6:**75–86.

Belsky, A. J., A. Matzke, and S. Uselman. 1999. Survey of livestock influences on stream and riparian ecosystems in the western United States. *Journal of Soil and Water Conservation* **54:**419–431.

Boudell, J. A., and J. C. Stromberg. 1999. Role of the seed bank in maintaining biodiversity in Southwestern riparian ecosystems. Society of Wetland Scientists Bulletin Supplement 16(2): A-113.

Briggs, M. K. 1996. *Riparian Ecosystem Recovery in Arid Lands.* University of Arizona Press, Tucson.

Briggs, M. K., and S. Cornelius. 1998. Opportunities for ecological improvement along the lower Colorado River and delta. *Wetlands* **18:**513–529.

Brock, J. H. 1994. *Tamarix* spp. (salt cedar), an invasive exotic woody plant in arid and semi-arid riparian habitats of western USA. In *Ecology and Management of Invasive Riverside Plants,* ed. by L. C. de Waal, L. E. Child, P. M. Wade, and J. H. Brock, pp. 27–44. John Wiley & Sons, Chichester, U.K.

Brown, D. E., ed. 1994. *Biotic Communities: Southwestern United States and Northwestern Mexico,* University of Utah Press, Salt Lake City.

Busch, D. E. 1995. Effects of fire on southwestern riparian plant community structure. *Southwestern Naturalist* **40:**259.

Busch, D. E., and S. D. Smith. 1995. Mechanisms associated with decline of woody species in riparian ecosystems of the southwestern U.S. *Ecological Monographs* **65:**347–370.

Busch, D. E., N. L. Ingraham, and S. D. Smith. 1992. Water uptake in woody riparian phreatophytes of the southwestern United States: A stable isotope study. *Ecological Applications* **2:** 450–459.

Chaneton, E. J., J. M. Facelli, and R. J. C. Leon. 1988. Floristic change induced by flooding on grazed and ungrazed lowland grasslands in Argentina. *Journal of Range Management* **41:**495–499.

Chisholm, G. 1996. *Managing Reservoirs for Ecosystem Needs in the Truckee River Basin.* http://consci.tnc.org/library/pubs/Stew96/truckee.html. The Nature Conservancy, Reno, NV.

Cooper, D. J., D. M. Merritt, D. C. Anderson, and R. A. Chimner. 1999. Factors controlling the establishment of Fremont cottonwood seedlings on the Upper Green River, USA. *Regulated Rivers: Research and Management* **15:**419–440.

D'Antonio, C. M., T .L. Dudley, and M. Mack. 1999. Disturbance and biological invasions: Direct effects and feedbacks. In *Ecosystems of Disturbed Ground,* ed. by L. R. Walker, pp. 413–452. Elsevier, Amsterdam, The Netherlands.

Decker, J. P. 1961. Salt secretion by *Tamarix pentandra* Pall. *Forest Science* **7:**214–217.

Drezner, T., P. Fall, and J. C. Stromberg. 2001. Plant distribution and dispersal mechanisms at the Hassayampa River Preserve, Arizona, USA. *Global Ecology and Biogeography.* 10: 149–162.

Ellis, L. M. 2001. Short-term response of woody plants to fire in a Rio Grande riparian forest, Central New Mexico, USA. *Biological Conservation* **97:**159–170.

Ellis, L. M., C. S. Crawford, and M. C. Molles. 1998. Comparison of litter dynamics in native and exotic riparian vegetation along the Middle Rio Grande of central New Mexico, USA. *Journal of Arid Environments* **38:**283–296.

Everitt, B. L. 1995. Hydrologic factors in regeneration of Fremont cottonwood along the Fremont River, Utah. In Natural and Anthropogenic Influences in Fluvial Geomorphology, ed. by E. Costa et al., pp. 197–208 American Geophysical Union, Washington, DC.

Everitt, B. L. 1998. Chronology of the spread of tamarisk in the central Rio Grande. *Wetlands* **18:**658–668.

Fenner, P., W. W. Brady, and D. R. Patton. 1984. Observations on seeds and seedlings of Fremont cottonwood. *Desert Plants* **6:**55–58.

Fenner, P., W. W. Brady, and D. R. Patton. 1985. Effects of regulated water flows on regeneration of Fremont cottonwood. *Journal of Range Management* **38:**135–138.

Fonseca, J. 1998. Vegetation changes at Bingham Cienega, the San Pedro River Valley, Pima County, Arizona, since 1879. *Journal of the Arizona-Nevada Academy of Science* **31:**103–116.

Friedman, J. M., M. L. Scott, and W. M. Lewis. 1995. Restoration of riparian forest using irrigation, artificial disturbance, and natural seedfall. *Environmental Management* **19:**547–557.

Friedman, J. M, W. R. Osterkamp, M. L. Scott, and G. T. Auble. 1998. Downstream effects of dams on channel geometry and bottomland vegetation: Regional patterns in the Great Plains. *Wetlands* **18:**619–633.

Gladwin, D. N., and J. E. Roelle. 1998. Survival of plains cottonwood (*Populus deltoides* subsp. *monilifer*a) and saltcedar (*Tamarix ramosissima*) seedlings in response to flooding. *Wetlands* **18:**669–674.

Glenn, E., R. Tanner, S. Mendez, T. Kehret, D. Moore, J. Garcia, and C. Valdes. 1998. Growth rates, salt tolerance and water use characteristics of native and invasive riparian plants from the delta of the Colorado River, Mexico. *Journal of Arid Environments* **40:**281–294.

Glenn, E. P., R. S. Felger, A. Burquez, and D. S. Turner. 1992. Cienega de Santa Clara: Endangered wetland in the Colorado River Delta, Sonora, Mexico. *Natural Resources Journal* **32:**817–824.

Goodwin, C. N., C. P. Hawkins, and J. L. Kershner. 1997. Riparian restoration in the western United States: Overview and perspective. *Restoration Ecology* **5**(4S):4–14.

Gourley, C. R. 1998. Restoration of the lower Truckee River ecosystem: Challenges and opportunities. *Journal of Land, Resources, & Environmental Law* **18:**113–121.

Graf, W. L. 1988. *Fluvial Processes in Dryland Rivers.* Springer-Verlag, New York.

Graf, W. L. 1999. Dam nation: A geographic census of American dams and their large-scale hydrologic impacts. *Water Resources Research* **35:**1305–1311.

Graf, W. L., J. C. Stromberg, and B. Valentine. In press. Rivers, dams, and willow flycatchers: Summary of their science and policy connections. *Geomorphology.*

Grime, J. P. 1979. *Plant Strategies and Vegetation Processes.* John Wiley & Sons, New York.

Grimm, N. B., and S. G. Fisher. 1986. Nitrogen limitation potential of Arizona streams and rivers. *Journal of the Arizona Nevada Academy of Science* **21:**31–43.

Grimm, N. B., S. G. Fisher, and W. L. Minckley. 1981. Nitrogen and phosphorus dynamics in hot desert streams of Southwestern USA. *Hydrobiologia* **83:**303–312.

Groeneveld, D. P., and T. E. Griepentrog. 1985. Interdependence of groundwater, riparian vegetation, and streambank stability: A case study. *United States Forest Service General Technical Report* **RM-120:**44–48.

Harris, D. R. 1966. Recent plant invasions in the arid and semi-arid Southwest of the United States. *Annals of the Association of American Geographers* **56:**408–422.

Hendrickson, D. A., and W. L. Minckley. 1984. Ciénegas—Vanishing climax communities of the American Southwest. *Desert Plants* **6:**131–175.

Hill, M. T., and W. S. Platts. 1998. Ecosystem restoration: A case study in the Owens River Gorge, California. *Fisheries* **23:**18–27.

Horton, J. S. 1972. Management problems in phreatophyte and riparian zones. *Journal of Soil and Water Conservation* **27:**57–61.

Horton, J. S., F. C. Mounts, and J. M. Kraft. 1960. Seed germination and seedling establishment of phreatophyte species. *Rocky Mountain Forest and Range Experiment Station Paper* **48:**1–16.

Jansson, R., C. Nilsson, and B. Renofalt. 2000a. Fragmentation of riparian floras in rivers with multiple dams. *Ecology* **81:**899–903.

Jansson R., C. Nilsson , M. Dynesius, and E. Andersson. 2000b. Effects of river regulation on river-margin vegetation: A comparison of eight boreal rivers. *Ecological Applications* **10:**203–224.

Johnson, W. C. 1998. Adjustment of riparian vegetation to river regulation in the Great Plains, USA. *Wetlands* **18:**608–618.

Junk, W. J., P. B. Bayley, and R. E. Sparks. 1989. The flood-pulse concept in river-floodplain systems. *Canadian Special Publications in Fisheries and Aquatic Sciences* **106:**110–127.

Kus, B. E. 1998. Use of restored riparian habitat by the endangered least Bell's vireo. *Restoration Ecology* **6:**75–82.

Mahoney, J. M., and S. B. Rood. 1998. Streamflow requirements for cottonwood seedling recruitment: An integrative model. *Wetlands* **18:**634–645.

Marti, E., S. G. Fisher, J. D. Schade, and N. B. Grimm. 1999. Flood frequency and stream-riparian linkages in arid lands. In *Streams and Ground Waters,* ed. by J. B. Jones and P. J. Mulholland, pp. 111–136. Academic Press, New York.

Milchunas, D. G., and W. K. Lauenroth. 1995. Inertia in plant community structure: State changes after cessation of nutrient-enrichment stress. *Ecological Applications* **5:**452–458.

Molles, M. C., C. S. Crawford, L. M. Ellis, H. M. Valett, and C. N. Dahm. 1998. Managed flooding for riparian ecosystem restoration: Managed flooding reorganizes riparian forest ecosystems along the middle Rio Grande in New Mexico. *BioScience* **48:**749–756.

Molles, M. C., Jr., C. N. Dahm, and M. T. Crocker. 1992. Climatic variability and streams and rivers in semi-arid regions. In *Aquatic Ecosystems in Semi-arid Regions: Implications for Resource Management,* ed. by R. D. Roberts and M. L. Bothwell, pp. 197–202. NHRI Symposium Series, Environment Canada, Saskatoon, Canada.

Naiman, R. J., H. DeCamps, and M. Pollock. 1993. The role of riparian corridors in maintaining regional biodiversity. *Ecological Applications* **3:**209–212.

Nilsson C., S. J. Xiong, M. E. Johanssonn, and L. B. M. Vought. 1999. Effects of leaf-litter accumulation on riparian plant diversity across Europe. *Ecology* **80:** 1770–1775.

Ohmart, R. D., B. W. Anderson, and W. C. Hunter. 1988. The ecology of the lower Colorado River from Davis Dam to the Mexico-United States international boundary: A community profile. *United States Fish and Wildlife Service Biological Report* **85(7.19):**1–296.

Ohmart, R. D., W. O. Deason and C. Burke. 1977. A riparian case history: The Colorado River. *United States Department of Agriculture Forest Service General Technical Report* **RM-43:**35–47.

Patten, D. T. 1998. Riparian ecosystems of semi-arid North America: Diversity and human impacts. *Wetlands* **18:**498–512.

Poff, N. L., J. D. Allan, M. B. Bain, J. R. Karr, K. L. Prestegaard, B. D. Richter, and J. C. Stromberg. 1997. The natural flow regime: A paradigm for river conservation and restoration. *BioScience* **47:**769–784.

Pollock, M. M., R. J. Naiman, and T. A. Hanley. 1998. Plant species richness in riparian wetlands: A test of biodiversity theory. *Ecology* **79:**94–105.

Rea, A. M. 1983. *Once a River: Bird Life and Habitat Changes on the Middle Gila.* University of Arizona Press, Tucson.

Roelle, J. E, and D. N. Gladwin. 1999. Establishment of woody riparian species from natural seedfall at a former gravel pit. *Restoration Ecology* **7:**183–192.

Rood, S. B., and J. M. Mahoney. 1990. Collapse of riparian poplar forests downstream from dams in western prairies: Probable causes and prospects for mitigation. *Environmental Management* **14:**451–464.

Schmidt, J. C., R. H. Webb, R. A. Valdez, G. R. Marzolf, and L. E. Stevens. 1998. Science and values in river restoration in the Grand Canyon. *BioScience* **48:**735–747.

Scott, M. L., G. T. Auble, and J. M. Friedman. 1997. Flood dependency of cottonwood establishment along the Missouri River, Montana, USA. *Ecological Applications* **7:**677–690.

Scott, M. L., J. M. Friedman, and G. T. Auble. 1996. Fluvial processes and the establishment of bottomland trees. *Geomorphology* **14:**327–339.

Scott, M. L., P. B. Shafroth, and G. T. Auble. 1999. Responses of riparian cottonwoods to alluvial water table declines. *Environmental Management* **23:**347–358.

Shafroth, P. B., J. M. Friedman, and L. S. Ischinger. 1995. Effects of salinity on establishment of *Populus fremontii* (cottonwood) and *Tamarix ramosissima* (saltcedar) in southwestern United States. *Great Basin Naturalist* **55:**58–65.

Shafroth, P. B., J. C. Stromberg, and D. T. Patten. In press. Riparian vegetation response to altered disturbance and stress regimes. *Ecological Applications.*

Shafroth, P. B., G. T. Auble, J. C. Stromberg, and D. T. Patten. 1998. Establishment of woody riparian vegetation in relation to annual patterns of streamflow, Bill Williams River, Arizona. *Wetlands* **18:**577–590.

Sher, A. A., D. L. Marshall, and S. A. Gilbert. 2000. Competition between native *Populus deltoides* and invasive *Tamarix ramosissima* and the implications for reestablishing flooding disturbance. *Conservation Biology* **14:**1744–1754.

Shields, F. D., A. Simon, and L. J. Steffen. 2000. Reservoir effects on downstream channel migration. *Environmental Conservation* **27:**54–66.

Smith, S. D., D. A. Devitt, A. Sala, J. R. Cleverly, and D. E. Busch. 1998. Water relations of riparian plants from warm desert regions. *Wetlands* **18:**687–695.

Springer, A. E., J. M. Wright, P. B. Shafroth, J. C. Stromberg, and D. T. Patten. 1999. Coupling ground-water and riparian vegetation models to simulate riparian vegetation changes due to a reservoir release. *Water Resources Research* **35:**3621–3630.

Stanford, J. A., J. V. Ward, W. J. Liss, C. A. Frissell, R. N. Williams, J. A. Lichatowich, and C. C. Coutant. 1996. A general protocol for restoration of regulated rivers. *Regulated Rivers: Research and Management* **12:**391–413.

Stein, B. A., L. S. Kutner, and J. S. Adams. 2000. *Precious Heritage: The Status of Biodiversity in the United States.* Oxford University Press, Oxford, U.K.

Stevens, L. E. 1989. Mechanisms of Riparian Plant Community Organization and Succession in the Grand Canyon, Arizona. Ph.D. diss., Northern Arizona University, Flagstaff, AZ.

Stevens, L. E., J. C. Schmidt, T. J. Ayers, and B. T. Brown. 1995. Flow regulation, geomorphology, and Colorado River marsh development in the Grand Canyon, Arizona. *Ecological Applications* **5:**1025–1039.

Stromberg, J. C. 1997. Growth and survivorship of Fremont cottonwood, Goodding willow, and salt cedar seedlings after large floods in central Arizona. *Great Basin Naturalist* **57:**198–208.

Stromberg, J. 1998a. Dynamics of Fremont cottonwood (*Populus fremontii*) and saltcedar (*Tamarix chinensis*) populations along the San Pedro River, Arizona. *Journal of Arid Environments* **40:**133–155.

Stromberg, J. C. 1998b. Functional equivalency of saltcedar (*Tamarix chinensis*) and Fremont cottonwood (*Populus fremontii*) along a free-flowing river. *Wetlands* **18:**675–686.

Stromberg, J. C. 2001a. Biotic integrity of *Platanus wrightii* riparian forests in Arizona: First approximation. *Forest Ecology & Management* **142:**249–264.

Stromberg, J. C. 2001b. Influence of stream flow regime and temperature on growth rate of the riparian tree, *Platanus wrightii,* in Arizona. *Freshwater Biology* **46:**227–240.

Stromberg, J. C., and D. T. Patten. 1990. Seed production and seedling establishment of a Southwest riparian tree, Arizona walnut (*Juglans major*). *Great Basin Naturalist* **50:**47–56.

Stromberg, J. C., J. Fry, and D. T. Patten. 1997. Marsh development after large floods in an alluvial, arid-land river. *Wetlands* **17:**292–300.

Stromberg, J. C., D. T. Patten, and B. D. Richter. 1991. Flood flows and dynamics of Sonoran riparian forests. *Rivers* **2:**221–235.

Stromberg, J. C., R. Tiller, and B. Richter. 1996. Effects of groundwater decline on riparian vegetation of semiarid regions: The San Pedro River, Arizona, USA. *Ecological Applications* **6:**113–131.

Stromberg, J. C., S. D. Wilkins, and J. A. Tress. 1993a. Vegetation-hydrology models as management tools for velvet mesquite (*Prosopis velutina*) riparian ecosystems. *Ecological Applications* **3:**307–314.

Stromberg, J. C., B. D. Richter, D. T. Patten, and L. G. Wolden. 1993b. Response of a Sonoran riparian forest to a 10-year return flood. *Great Basin Naturalist* **53:**118–130.

Stromberg, J. C., J. A. Tress, S. D. Wilkins, and S. Clark. 1992. Response of velvet mesquite to groundwater decline. *Journal of Arid Environments* **23:**45–58.

Stromberg, J. C., M. R. Sommerfeld, D. T. Patten, J. Fry, C. Kramer, F. Amalfi, and C. Christian. 1993c. Release of effluent into the Upper Santa Cruz River, Southern Arizona: Ecological considerations. In *Proceedings of the Symposium on Effluent Use Management,* ed. by M. G. Wallace, pp. 81–92. American Water Resources Association, Bethesda, MD.

Swetnam, T. W., and J. L. Betancourt. 1998. Mesoscale disturbance and ecological response to decadal climatic variability in the American Southwest. *Journal of Climate* **11:**3128–3147.

Taylor J. P., and K. C. McDaniel. 1998. Restoration of saltcedar (*Tamarix* sp.)-infested floodplains on the Bosque del Apache National Wildlife Refuge. *Weed Technology* **12:**345–352.

Taylor, J. P., D. B. Wester, and L. M. Smith. 1999. Soil disturbance, flood management, and riparian woody plant establishment in the Rio Grande floodplain. *Wetlands* **19:**372–382.

Tellman, B., R. Yarde, and M. G. Wallace. 1997. Arizona's changing rivers: How people have affected the rivers. *University of Arizona Water Resources Research Center Issue Paper* **19:**1–198.

Trimble, S. W., and A. C. Mendel. 1995. The cow as a geomorphic agent: A critical review. *Geomorphology* **13:**233–253.

Turner, R. M. 1974. Quantitative and historical evidence of vegetation changes along the Upper Gila River, Arizona. *U.S. Geological Survey Professional Paper 655-H,* U.S. Government Printing Office, Washington, DC.

Unitt, P. 1987. *Empidonax traillii extimus:* An endangered subspecies. *Western Birds* **18:**137–162.

U. S. Army Corps of Engineers. 1998. *Final Environmental Impact Statement, Rio Salado Environmental Restoration,* U.S. Army Corps of Engineers *Los Angeles, California.*

Webb, R. H., and J. L. Betancourt. 1992. Climatic variability and flood frequency of the Santa Cruz River, Pima County, Arizona. *U. S. Geological Survey Water Supply Paper* **2379:**1–40.

Weintraub, J. 1993. Effects of tamarisk removal on avian distributions at Camp Cady Wildlife Area in the California Mojave Desert. Ph.D. diss., California State University, Fullerton, CA.

Whittaker, D., and B. Shelby. 2000. Managed flow regimes and resource values: Traditional versus alternative strategies. *Rivers* **7:**233–244.

Wolden, L. G., and J. C. Stromberg. 1997. Experimental treatments (and unplanned natural events) for restoration of the herbaceous understory in an arid-region riparian ecosystem. *Restoration & Management Notes* **15:**161–167.

Wolden, L. G., J. C. Stromberg, and D. T. Patten. 1994. Flora and vegetation of the Hassayampa River Preserve. *Journal of the Arizona-Nevada Academy of Science* **28:**76–111.

Workman, S. R., and S. E. Serrano. 1999. Recharge to alluvial valley aquifers from overbank flow and excess infiltration. *Journal of the American Water Resources Association* **35:**425–432.

Zamora-Arroyo, F., P. Nagler, M. Briggs, D. Radtke, H. Rodriquez, J. Garcia, C. Valdes, A. Huete, and E. Glenn. 2001. Regeneration of native trees in response to flood releases from the United States into the delta of the Colorado River, Mexico. *Journal of Arid Environments.* **49:**49–64.

3

The Role of the Flood Pulse in Ecosystem-Level Processes in Southwestern Riparian Forests: A Case Study from the Middle Rio Grande

Lisa M. Ellis, Clifford S. Crawford, and Manuel C. Molles Jr.
University of New Mexico, Albuquerque, New Mexico

Riparian forests in arid and semiarid regions around the world are especially heavily impacted by river regulation. In areas with naturally limited water availability, anthropogenic domination of water resources has had particularly severe consequences. In part, impacts result from the alteration of flow variability; regulation to meet agricultural demand for water during the growing season often leads to changes in the timing of water

Flood Pulsing in Wetlands: Restoring the Natural Hydrological Balance, edited by Beth A. Middleton.
ISBN 0 471-41807-2

availability. This typically reduces the seasonal flood pulse that characterizes large-river floodplains and orchestrates a number of population and ecosystem-level responses in both the river and the floodplain. In the southwestern United States, conflicting demands for water and streamside habitat have resulted in the decline, and in many cases disappearance, of riparian ecosystems, so that considerable public and governmental attention has been focused on riparian conservation.

This chapter presents a case study of the effects of river regulation on riparian ecosystems along the Middle Rio Grande in central New Mexico. It discusses the decline of Rio Grande cottonwoods, the endangerment of the Rio Grande silvery minnow, and the invasion of the floodplain by exotic plants. The focus is on the results of an experimental study designed to mimic the natural flood pulse, and particularly on the effects of flooding on decomposition and the buildup of organic debris within the forest. The significance of increasing accumulations of organic debris with respect to increasing fire severity in the valley is addressed, with a look at the specific effects of a riparian wildfire. Finally, the discussion turns to the changing attitude toward river and riparian management, highlighting several programs currently under way in the Middle Rio Grande Valley. These include public outreach and education, designed to increase awareness of and appreciation for the natural ecology of floodplain ecosystems, and a collaborative restoration project intended to mimic the flood pulse. Rapid growth and development along the Middle Rio Grande Valley continue to threaten the existence of river and riparian ecosystems, but a strong advocacy for the river and riparian forest offers hope for the future.

In the southwestern United States, as in arid and semiarid regions throughout the world, river flow regimes are characterized by high variability (Molles et al., 1992; Walker et al., 1995). Conditions appropriate for the establishment of the dominant riparian cottonwoods (*Populus* spp.), for example, do not occur every year, but reflect this variability in river flow (Stromberg et al., 1991). Many processes in these riparian ecosystems, such as forest floor respiration and decomposition, are limited by moisture availability and occur at relatively high rates when overbank flooding occurs (Molles et al., 1998; Ellis et al., 1999). In spite of this inherent variability, however, riparian ecosystems in the Southwest provide reliable resources for a great diversity of plants and animals, with both abundance and species richness typically higher than in the surrounding uplands (Ohmart and Anderson, 1982; Knopf et al., 1988; Szaro, 1991).

The Rio Grande/Rio Bravo travels some 3220 km from its headwaters in the rugged San Juan Mountains of southwestern Colorado to the Gulf of Mexico. The Rio Grande and its native biota face intense competition from humans for access to this water, as along its course the river passes through three states in the United States, four in the Republic of Mexico, and more than 20 Indian nations. With a drainage basin of about 470,000 km^2, the Rio Grande dominates a large portion of the Southwest, and ecological problems within the region's riparian systems dominate local conservation concerns.

The area in central New Mexico known as the Middle Rio Grande Valley extends from Cochiti Reservoir, approximately 73 km north of Albuquerque, 260 river km downstream to San Marcial at the head of Elephant Butte Reservoir (Figure 3-1). This represents about 14 percent of the river's drainage and 8 percent of its total length and includes three biotic regions: Great Basin Grassland, Semidesert Grassland, and Chihuahuan Desert. Particularly important to the hydrologic regime of the region is the influence of climatic variation resulting from the El Niño-Southern Oscillation phenomenon that affects semiarid regions throughout the world (Ropelewski and Halpert, 1987; Nicholls, 1988; Scurlock, 1998). Locally, El Niño events result in increased winter and spring precipitation, whereas La Niña episodes are characterized by lower-than-average precipitation or drought (Ropelewski and Halpert, 1986; Molles and Dahm, 1990). Stream flow in the region is tied to this variation in precipitation, with increased discharge occurring during wet El Niño periods and reduced flow during dry La Niña years (Molles and Dahm, 1990). The local flora and fauna are undoubtably adapted to this variation in river flow, but human inhabitants of the valley have often been adversely impacted by periodic floods and droughts (Scurlock, 1998).

AN ALTERED RIVER: THE CASE OF THE MIDDLE RIO GRANDE

For much of its length the Rio Grande is a floodplain river. It formerly pulsed with an active, moving channel that supported numerous wetlands within a patchwork of small stands of cottonwood and willow forests and open fields (Crawford et al., 1993). Although the prehistoric range of ri-

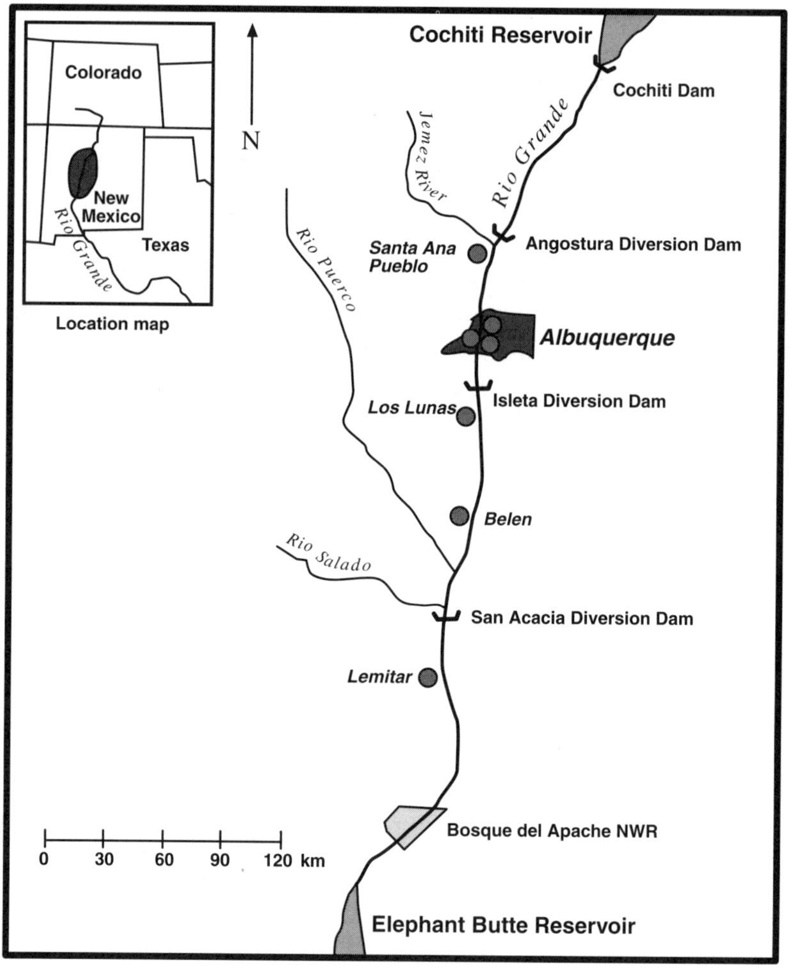

Figure 3-1. *Map of the Middle Rio Grande Valley, New Mexico. Cochiti Dam and Elephant Butte Reservoir define the northern and southern ends of the valley, respectively. Bosque del Apache NWR, site of much of our flooding research, is indicated, as are locations of the Bosque Ecosystem Monitoring Program (BEMP) sites (*circles*).*

parian forest dominated by Rio Grande cottonwoods (*Populus deltoides* ssp. *wislizenii*) was probably similar to its range today (Crawford et al., 1993), the actual prehuman extent of forest patches in what is now known as New Mexico's Middle Rio Grande Valley (MRGV) is unknown. A description in the 1600s noted an extensive stand of cottonwoods stretching from Alameda Pueblo to Albuquerque along the east side of the river (Adams and Chavez, 1956); this *Bosque Grande,* as it was known, remained a prominent feature of the valley at least until the early 1700s (Scurlock, 1998).

When the first written records were made by members of Don Francisco Vasquez de Coronado's expedition in 1540, the valley was already heavily impacted by Indian settlements, which resulted in clearing land for agriculture, cutting trees for firewood and construction, and altering the distribution and abundance of local plants and animals (Crawford et al., 1993; Crawford et al., 1996a; Tainter and Tainter, 1996). Hydrologic changes to the system were limited at that time, however, as the early Pueblo people practiced floodplain agriculture, using water conservation methods without the aid of water diversion structures (Wozniak, 1996). One of Coronado's chroniclers stated that "the valley of Our Lady—La Señora—had bountiful crops, and there were pleasant groves of cottonwood trees, that unfailing sign of water and fertility in a desert country" (Bolton, 1990).

The Spanish introduced irrigation techniques that actively diverted water from the river, including the *acequias,* or irrigation canals, that have remained an important part of local agriculture today, along with crops such as wheat and lettuce that required these new techniques (Wozniak, 1996; Scurlock, 1998). Between the late 1500s and late 1800s, the use of the floodplain for agriculture increased dramatically, first with the influx of Spanish settlers and later with Anglos moving west (Crawford et al., 1993; Scurlock, 1998). The extent of agriculture in the middle valley reached a peak around 1880, when about 50,607 ha were under cultivation (Burkholder, 1928 cited in Crawford et al. 1993). Extensive irrigation upstream, however, along with livestock grazing and timber harvests on adjacent uplands, took their toll, and by 1925 only about 18,220 ha were irrigated (Burkholder, 1928 cited in Crawford et al. 1993). This reduction was caused by many factors, including increasingly shallow groundwater levels (resulting in seepage and waterlogging), increasing soil salinity, a decreasing supply of water for irrigation, droughts alternating with floods, and an aggrading riverbed due to heavy sediment loads from the overgrazed and deforested watershed combined with an often decreased river flow (Crawford et al., 1993; Crawford et al., 1996a; Wozniak, 1996).

Hydrological imbalances and decreased agricultural productivity required technological intervention. By the early 1900s, increasing numbers of water management facilities designed to support irrigated agriculture and floodplain development were constructed along the Rio Grande in southern Colorado and New Mexico. These included Elephant Butte Dam at the lower end of the MRGV, which was completed in 1916 (Crawford et al., 1993). In 1925, the Middle Rio Grande Conservancy District (MRGCD) was created to improve irrigation supplies, to improve drainage, and to provide flood control facilities (Crawford et al., 1993; Crawford et al., 1996a; Wozniak, 1996). The MRGCD built an extensive system of drainage and conveyance canals to lower the water table and increase the efficiency of water transport, and irrigation ditches to provide irrigation water. The MRGCD also created a floodway approximately 460 m wide with the construction of earthen levees, added six diversion dams or headings along the middle valley, and built a dam and storage reservoir upstream on a primary tributary, the Rio Chama (Bullard and Wells, 1992).

These developments by the MRGCD greatly reduced waterlogging problems in the valley and improved the distribution of irrigation water, allowing a resurgence of agriculture in the valley that continues to the present time (Crawford et al., 1993). A major flood in 1941, however, overtopped the levees and inundated communities along the river, including parts of downtown Albuquerque. This event, combined with continuing problems such as water shortages and streambed aggradation, resulted in the preparation of the Rio Grande Comprehensive Plan, produced jointly by the U.S. Army Corps of Engineers and the U.S. Bureau of Reclamation (Bullard and Wells, 1992). This effort led to the creation of four flood control reservoirs on the main stem of the Rio Grande and its tributaries, extensive modification, improvement, and expansion of the levee system, clearance of the floodway, straightening and stabilization of the bank, and rehabilitation of existing drains and irrigation facilities (Crawford et al., 1993). Cochiti Dam, the most recent dam affecting the middle valley, was completed in 1975. An additional reservoir located upstream on the Rio Chama was created in 1971 to collect water, brought through an underground tunnel from the San Juan River in the Colorado River Basin, and deliver it to the Rio Grande.

The allocation of Rio Grande waters is mediated through a complex assortment of state, interstate, federal, and international water laws, agreements, and regulations (Crawford et al., 1993). These include the Treaty of 1906 between the United States and Mexico to provide a minimum water delivery to Mexico, and the Rio Grande Compact, signed in 1938, which allocates water between the states of Colorado, New Mexico, and Texas (Bullard and Wells, 1992). In effect, every drop of the Rio Grande is allocated, with certain restrictions concerning its use.

CONSEQUENCES OF THE ALTERED RIVER: SOME OBVIOUS PROBLEMS

The results of the comprehensive water management strategy were predictable. Wetlands in the valley decreased drastically after installation of the drainage system (Van Cleave, 1935). Initially, the stabilized banks provided appropriate habitat for cottonwood trees that had been unable to grow in waterlogged soils, but the paralysis of the meandering channel has also led to the forest's decline (Figure 3-2). The flood pulse, on which many of the native riparian plants depend for establishment, has largely been eliminated (Figure 3-3), leading not only to decreased regeneration (Howe and Knopf, 1991) but to ecosystem-level consequences as well (Molles et al., 1998; Ellis et al., 1999). Recent increases in the frequency and severity of fire along the riparian forest have probably resulted, at least in part, from decreased decomposition and the subsequent large accumulations of organic debris (Stuever et al., 1997; Ellis et al.,1998; Ellis et al.,

Figure 3-2. An aerial view along the Middle Rio Grande showing a riparian forest and river channel constrained within riverside levees, with agricultural and urban development in the floodplain outside the levees. (Photograph by Clifford Crawford.)

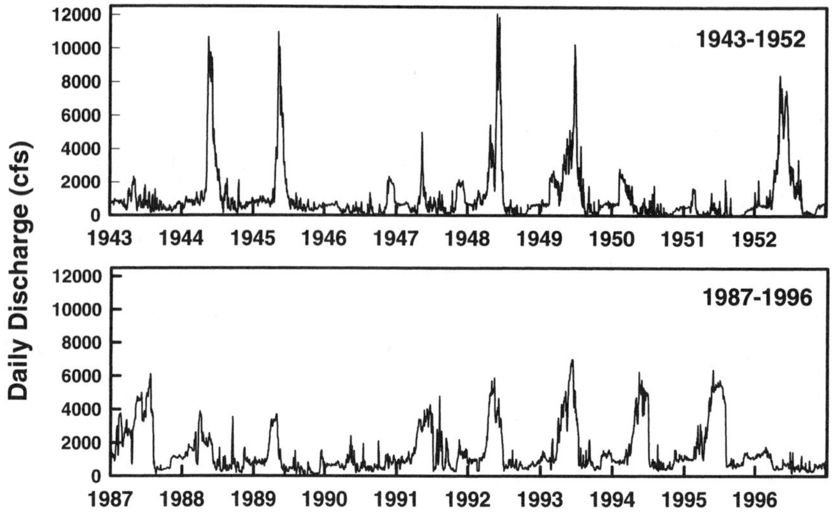

Figure 3-3. *Hydrograph recording Rio Grande discharge (in cubic feet second^{-1}, cfs) through Albuquerque* before *(1943–1952) and* after *(1987–1996) the installation of Cochiti Dam, located upstream from Albuquerque. Note both the decrease in maximum discharge and protraction of spring peaks, as well as the increase in minimum flows. (Data from U.S. Geological Survey, 2000.)*

1999). Three associated problems, the decline of Rio Grande cottonwoods, the endangerment of the Rio Grande silvery minnow, and the invasion of exotic plants on the floodplain, are discussed briefly in the following paragraphs. A more extensive discussion follows on the interrupted decomposition and the increased incidence and impact of fire related to the cessation of the flood pulse.

Decline of Rio Grande Cottonwoods

Lowland riparian areas in the Middle Rio Grande were historically dominated by the Rio Grande cottonwood (*Populus deltoides* ssp. *wislizenii*), which, together with tree willows (*Salix gooddingii, S. amygdaloides*) and a variety of native shrubs (e.g., *Amorpha fruticosa, Baccharis glutinosa, Forestiera neomexicana, Salix exigua*), formed the riparian ecosystem known locally as the *bosque,* a Spanish word for "forest." The reproductive biology of these floodplain trees is adapted to the annual flood pulse, as well as to the climatic variability in the region.

The release of cottonwood seeds coincides with the spring flood pulse. The wind- and water-dispersed seeds must land on flood-scoured banks or sand bars that are free of existing vegetation and shade from larger trees, such as those created by a naturally meandering river (Scott et al., 1993, 1997; Braatne et al., 1996). The growing roots from newly germinated seeds must remain moist, with a soil moisture subsidence of no more than about 5 mm day^{-1} over the first one to two weeks (Mahoney and Rood, 1991; Mahoney and Rood, 1992; Segelquist et al., 1993; Braatne et al., 1996; Crawford, et al., 1996a). Seedlings are susceptible to mortality resulting from desiccation due to soil moisture subsidence that is too rapid, from oxygen deprivation due to prolonged inundation, and from mechanical removal caused by recurring heavy floods (Stromberg et al., 1991; Crawford et al., 1996a). Above all, successful seedling establishment in the Southwest reflects climatic variation; appropriate conditions are not present every year. Years of high precipitation and, subsequently, high river flows resulting in overbank flooding in fall, winter, or spring favor recruitment, but successful establishment requires that subsequent late summer or fall flooding be minimal to reduce seedling mortality (Stromberg et al., 1991).

The regulation of river flows has contributed to the decline of cottonwood forests throughout western North America (e.g., Rood and Heinze-Milne, 1989; Rood and Mahoney, 1990; Howe and Knopf, 1991; Stromberg and Patten, 1992; Busch and Smith, 1995; Braatne et al., 1996; Scott et al., 1997). In New Mexico, the regulation of flows along the Rio Grande has greatly altered the cottonwood forest. The absence of scouring floods has limited most germination to sandbars within the active channel, where subsequent high flows wash seedlings away (Howe and Knopf, 1991; Crawford et al., 1993). This has restricted regeneration through much of the MRGV during the past 40 to 50 years (Howe and Knopf, 1991). Thus, within the stabilized floodplain, many stands of cottonwoods are senescing without replacement.

As phreatophytic plants typically exposed to mesic conditions, cottonwoods are not tolerant of water stress (Leffler et al., 2000); however, reduced water availability, attributable to regulation, has left adult cottonwoods susceptible to desiccating conditions. Several studies have shown reduced growth or increased mortality of cottonwoods downstream from dams or diversions, in part due to direct links between the soil water table and river water levels (Reily and Johnson, 1982; Stromberg and Patten,

1991; Stromberg and Patten, 1992; Tyree et al., 1994; Stromberg and Patten, 1996). Physiological studies have shown that *Populus* species in general, and particularly *P. deltoides,* are especially susceptible to drought-induced cavitation (Tyree and Ewers, 1991; Tyree et al., 1992; Tyree et al., 1994; Pockman et al., 1995), which may explain diebacks following flow regulation (Blake et al., 1996). Cavitation is the vaporization of gases from tissues surrounding xylem elements, resulting in embolism, or blockage of the water transport mechanism by air in the conduits (Tyree and Dixon, 1986; Blake et al., 1996). Extensive cavitation may result in the death of the shoot and thus may contribute to the decline of riparian cottonwoods (Tyree et al., 1994; Blake et al., 1996).

Recent work on cottonwoods in the MRGV found the local population extremely vulnerable to cavitation, with a particularly small safety margin between the measured and the critical xylem pressure that can be maintained without catastrophic cavitation (Leffler et al., 2000). Our observations during more than a decade of work in the Albuquerque *bosque* suggest that branch loss in this population has been considerable. Thus, although the cottonwood *bosque* along the MRGV has been touted as the longest remaining stretch of native riparian forest left in the Southwest (Howe and Knopf, 1991; Crawford et al., 1993), this ecosystem is at extreme risk of both reduced regeneration and potentially high adult mortality as the result of river regulation.

Studies of herbivory in local cottonwoods suggest an additional threat to the native forest along the Middle Rio Grande, which may reflect a combination of drought, river regulation, and increased ozone levels resulting from local air pollution. Herbivory by the cottonwood leaf beetle (*Chrysomela scripta*) was reported at outbreak proportions for two sites near Albuquerque in 1996 and 1997, and premature leaf fall due to the beetles was particularly high at those urban sites, as compared with rural sites south of Albuquerque where herbivory levels were more typical (Eichhorst, 1999). Ozone levels during 1995–1997 were much higher at the urban sites, where they exceeded levels at which cellular damage occurs in plants (Eichhorst, 1999), and this may be related to the beetle outbreak there. Beetle predation on cottonwood leaves has been shown to increase after treatment with ozone (Coleman and Jones, 1988; Jones and Coleman, 1988; Jones et al., 1994), as well as when plants are exposed to drought and chronic water stress (Waring and Cobb, 1992). Extensive loss of photosynthetic tissue due to herbivory can decrease net photosynthesis levels and plant carbon gain (Mooney and Gulmon, 1982; Proctor et al., 1982; Chapin et al., 1987; Morrison and Reekie, 1995), thereby affecting plant growth, reproduction, and survivorship (Bassman et al., 1982; Heinrichs, 1988; Quesada et al., 1995). Thus, high levels of beetle herbivory near Albuquerque, possibly reflecting both reduced river flow and increased ozone levels in the urban area, may further contribute to the decline of this species (Eichhorst, 1999).

Endangerment of the Rio Grande Silvery Minnow

To local farmers, the equivalent of the spotted owl is the Rio Grande silvery minnow (*Hybognathus amarus,* Cyprinidae), a tiny fish endemic to the Rio Grande drainage and now listed as a federally endangered species (U.S. Department of the Interior, 1994). Once relatively abundant and widespread in mainstream habitats, the silvery minnow historically occurred along the Rio Grande from Española to the Gulf of Mexico, as well as along the Pecos River through eastern New Mexico (Bestgen and Platania, 1991). It is now restricted to about 5 percent of its known historic range, having been extirpated from the Pecos River, and is currently found only along a 274 km reach of the Middle Rio Grande (Bestgen and Platania, 1991; Platania, 1991). This 90 mm minnow is the only surviving species of five small, mainstream cyprinids, four of which have been extirpated from the Rio Grande in New Mexico largely as a result of flow regulation (Bestgen and Platania, 1991).

A number of factors have contributed to the decline of the silvery minnow, including the introduction of competing or predatory nonnative fish, hybridization with the plains minnow (*H. placitus*), and degraded water quality (Bestgen and Platania, 1991). However, habitat modification, including channelization, water diversion resulting in channel desiccation, and impoundment, is probably the primary factor contributing to the decline of this species (Bestgen and Platania, 1990, 1991). Silvery minnows spawn in response to increased stream flow resulting from spring runoff or summer thunderstorms (U.S. Fish and Wildlife Service, 1999), after which the eggs drift in the water column (Platania and Altenbach, 1998). Eggs and larvae are typically transported downstream, where three diversion dams along the middle river now prevent the return of adults upstream. This has resulted in the minnow being concentrated in the lowermost portion of its current range (Platania and Altenbach, 1998).

Particularly detrimental for the silvery minnow, and a factor that brings it into direct conflict with local farmers, is the dewatering of the river during the irrigation season. During drought years, the Rio Grande, particularly in the lower third of the minnow's current range south of the San Acacia Diversion Dam, often consists of nothing more than extensive sandbars, with essentially all river water diverted for irrigation or for more efficient transport through riverside channels (Bestgen and Platania, 1990). Although historically the minnows were able to withstand periods of drought by swimming upstream and retreating to pools and backwater refugia, the combination of dams preventing upstream movement and the relatively rapid removal of water for irrigation now typically results in minnows being trapped in the dewatered zones. The outcome of this is obvious; in 1996, 40 percent of the population was lost as the combined drought and water withdrawals for irrigation left 60 miles of the river dry (Harris, 1998). Efforts by the City of Albuquerque, the Bureau of Reclamation, and the U.S. Fish and Wildlife Service to rewater the river averted what would have been certain extinction for the species (Harris, 1998). In areas in which complete desiccation is avoided, minnows are still concentrated into smaller pools of water, where predation by native and exotic fish, as well as birds and mammals, is high and pathogenic disease outbreaks spread quickly.

Critical habitat for the silvery minnow was designated along the active channel of the main stem of the Rio Grande through the MRGV, with primary components including stream morphology, water quality and quantity (U.S. Department of the Interior, 1999). With this provision in place, silvery minnows were raised by the U.S. Fish and Wildlife Service and the Albuquerque Aquarium and released into upstream reaches (S. Platania, personal communication), but these measures have not ensured the survival of the species. Drought conditions during the summer of 2000 again resulted in the river going dry below San Acacia, leaving perhaps thousands of minnows to die. A lawsuit brought by environmentalists demanding that federal agencies comply with the Endangered Species Act by ensuring suitable minnow habitat (a wet river) resulted in unprecedented cooperation among the MRGCD, representing the farmers, the City of Albuquerque, and the U.S. Bureau of Reclamation to protect both the minnow and irrigation. This federal court-mediated agreement, however, raised further questions over the rights of the federal government versus local (New Mexico) water law, and the debate over who has rights to Rio

Grande water continues. In late November 2000, a U.S. district court ruling threw out the critical habitat designation, noting that Rio Grande water is a limited commodity and cannot be allocated to fish at the expense of people (Linthicum, 2000). A three-year agreement was signed in June 2001 between the state of New Mexico and the federal government to provide water for the minnow, and a "biological opinion" issued by the Fish and Wildlife Service at the same time stated that federal water managers were sufficiently protecting the species and thus were in compliance with the Endangered Species Act. Environmentalists continue to challenge these rulings, however, to seek more protection for the fish. The ultimate outcome remains uncertain, but it is certain that that cooperation among farmers, environmentalists, and resource managers, and between local and federal governments, will be essential for the survival of both this species and the river.

Invasion by Exotic Plants

A walk through the *bosque* in Albuquerque yields a plant list dominated by exotics: Siberian elm (*Ulmus pumila*), Russian olive (*Elaeagnus angustifolia*), saltcedar (*Tamarix ramosissima*), white mulberry (*Morus alba*), tree of heaven (*Ailanthus altissima*), sweet clover (*Melilotus alba*), Johnson grass (*Sorghum halapense*). All of these species are well established, and it is easy to believe predictions by Howe and Knopf (1991) that exotic trees and shrubs will dominate the Rio Grande riparian community over the next 50 to 100 years. Indeed, in many areas they already do. In New Mexico and throughout the western United States, the replacement of native riparian plants with exotics represents yet another effect of river flow regulation and the loss of the flood pulse.

The greatest botanical bane in the Southwest is saltcedar, several species of *Tamarix* introduced to the western United States from southern Europe or the eastern Mediterranean in the mid-1800s for use as ornamentals, windbreaks, and stream bank stabilization (Horton, 1964; Brock, 1994). Saltcedar had become naturalized by the late 1800s, most major river systems in the Southwest had been invaded by the 1920s, and 1.5 million acres of riparian habitat were dominated by saltcedar by the late 1980s (DiTomaso, 1998; Cox, 1999). Saltcedar (*T. ramosissima*) dominates about 64 percent of the Middle Rio Grande (Cox, 1999), forming vast, monotypic stands, particularly south of Bernardo, New Mexico.

Excellent reviews of saltcedar biology and ecology are provided by Brock (1994), DiTomaso (1998), and Cox (1999). Here we simply highlight some of the many characteristics that make saltcedar such a successful invader, particularly as they relate to altered hydrologic regimes. A critical aspect of saltcedar biology is its reproductive phenology. Unlike native cottonwoods and willows, which produce short-lived seeds for only the brief period in early summer corresponding to the historic flood pulse, saltcedars produce seeds from early spring until late fall (Brock, 1994). These wind- and water-dispersed seeds, which can be produced as early as the plant's first year (Brock, 1994; DiTomaso, 1998), can utilize moisture from various sources, including late summer thunderstorms and flooding (Cox, 1999). Like cottonwood, saltcedar seedlings are intolerant of shade and need moist, bare soil to germinate. Cottonwood seedlings can outcompete these exotics if both germinate at the same time (Sher et al., 2000), but because saltcedar is not tied to the now-reduced spring flood pulse, its window of establishment is much longer.

Saltcedars recover rapidly from many types of disturbance, including cutting and fire (Crins, 1989). Once established, they tolerate drought, flooding, heat and cold, and high salinity. By excreting salts from their leaves, they tend to make soils more saline and thus inhibit the growth of other plants (DiTomaso, 1998). With higher water-use efficiency (Busch and Smith, 1995) and drought tolerance (Cleverly et al., 1997) than native riparian species, they are especially well adapted to reductions in river flow. Saltcedars also exhibit very rapid growth and develop an extensive root system that produces adventitious shoots as well as roots that originate from stems buried by sand and silt during flooding (DiTomaso, 1998; Cox, 1999). Thus, they can quickly form dense stands that stabilize the stream banks (ironically, one of the reasons they were initially introduced), and this contributes to downcutting and, ultimately, the inhibition of the natural meandering of the river channel and hydrologic processes such as the flood pulse.

The northern MRGV is more heavily impacted by Russian olive than by saltcedar. This drought-tolerant Eurasian native is more resistant to shade than other riparian dominants (Shafroth et al., 1995) and can therefore spread through a forest even without removal of native trees (Cox, 1999). Russian olive was first planted to create windbreaks in the Great Plains in the late 1800s, and subsequently throughout the western United States and Canada (Knopf and Olson, 1984). By 1986 it had colonized all 17 western states (Olson and Knopf, 1986). Initially this species was favored for erosion control, windbreaks, wildlife habitat improvement, and highway landscaping and today is still widely available from nurseries (Cox, 1999). However, like saltcedar, Russian olive can form dense thickets that stabilize floodplain terraces, causing relatively deep, narrow channels that reduce flooding on the floodplain (Cox, 1999). Russian olive also promotes an understory of exotic forbs and grasses (Knopf and Olson, 1984), further disrupting the ecological balance of floodplain ecosystems.

Animal responses to saltcedar and Russian olive are mixed. A number of studies have considered birds' use of saltcedar (see DiTomaso, 1998; Cox, 1999, for reviews). The general picture that emerges is that many species of birds use saltcedar, particularly when native riparian vegetation is absent, as along the Pecos River of New Mexico (Hunter et al., 1988) and the Colorado River in the Grand Canyon (Brown and Trosset, 1989). However, in some areas where saltcedar is replacing native vegetation—for example, along the lower Colorado River or the lower Rio Grande—the diversity and density of birds have decreased (Hunter et al., 1988). Many bird species use saltcedar in the MRGV, including some riparian obligates, but others, such as summer tanagers and white-breasted nuthatches, clearly prefer mature cottonwood stands (Raitt et al., 1980; Ellis, 1995). For these species, preservation of the native *bosque* will be essential. Rodents, particularly the white-footed mouse (*Peromyscus leucopus*), which dominate Middle Rio Grande floodplain forests, reach high numbers in saltcedar as well as cottonwood (Ellis et al., 1997a). Along the lower Colorado River, abundances of several species of rodents were similar in saltcedar and native-dominated communities, although rodent abundance increased following saltcedar removal (Anderson and Ohmart, 1985).

Russian olive has been considered desirable by wildlife management personnel because it produces fruits eaten by many birds and mammals, in addition to providing winter cover and summer nesting sites. Few studies have looked at the effects of Russian olive on riparian fauna. One study showed that populations of mice were higher in Russian olive than in native-dominated forests, whereas mammal species richness was similar in the two forest types (Knopf and Olson, 1984). In contrast, fewer species of birds used Russian olive as compared with native riparian vegetation, with hole-nesting species notably absent (Knopf and Olson, 1984). From the point of view considering birds and mammals, Russian olive may be acceptable as an understory shrub in low densities, but not when it becomes the dominant vegetation to the exclusion of native cottonwoods and other trees and shrubs (Cox, 1999).

Arthropod populations are abundant in both saltcedar and Russian olive. Species richness of ground-dwelling arthropods was similar or higher under saltcedar than under cottonwood, with abundance of most taxa similar under the two vegetation types (Ellis et al., 2000). Although the abundance of exotic isopods (*Armadillidium vulgare* and *Porcellio laevis*) was greater at cottonwood sites, spiders were more abundant in saltcedar. Our observations indicate that terrestrial arthropods are also quite abundant and diverse in stands of Russian olive. Arboreal arthropods also readily use both saltcedar and Russian olive. Mund-Meyerson (1998) found no differences in the numbers of orders or families of arboreal arthropods, or in their average body size, in cottonwood, saltcedar, and Russian olive during the avian breeding season. Saltcedar had more individual arthropods per unit area than either cottonwood or Russian olive late in the breeding season, although cottonwoods generally had more arthropods per tree because of their larger size. Thus, saltcedar and Russian olive may support an abundant source of arboreal arthropods for breeding birds (Mund-Meyerson, 1998).

That saltcedar and Russian olive can be utilized successfully by a variety of wildlife does not mean that these plants should be allowed to continue their spread through the valley unchecked. In low densities, both species probably will not adversely affect wildlife communities. However, the extent to which they alter riparian ecosystems when densities become high, and particularly their effect on river hydrology through stream bank stabilization, mean that control measures are necessary (Cox, 1999). Although the cost of controlling exotics can be quite high, native plants can be restored successfully (Taylor and McDaniel, 1998). This will be an important component of restoring the natural flow regime.

RESEARCH AT BOSQUE DEL APACHE NATIONAL WILDLIFE REFUGE: FLOODS, FIRE, AND THE LITTER CONNECTION

Much of our research during the 1990s was centered on an experimental study designed to restore the natural flood pulse to a cottonwood forest site that had been isolated from flooding for more than 50 years. Although the role of flooding in the establishment of young cottonwoods was well established (see the earlier section "Decline of Rio Grande Cottonwoods"), we were interested in understanding the influence of flooding on ecosystem-level processes and other components of the forest. Therefore, our study addressed a number questions at the population and ecosystem levels, ranging from the effects of flooding on rodent and arthropod populations (Ellis et al., 1997b; Ellis et al., 2001) to litter storage and nutrient cycling (Molles et al., 1998; Ellis et al., 1999). The picture that emerged with overriding importance is that flooding plays a key role in litter dynamics, particularly in the breakdown of leaf litter and woody debris, and subsequently in the management of fire.

Bosque del Apache Study Sites and Experimental Flooding

The Bosque del Apache National Wildlife Refuge is located approximately 160 km south of Albuquerque, New Mexico (Figure 3-1), at an elevation of about 1400 m, and contains roughly 14.5 km of the Rio Grande and its associated riparian vegetation. We established two study sites in mixed cottonwood forest at the Refuge during the summer of 1991. These primary study sites were located 3.7 km apart in a strip of continuous forest, 200 to 300 m wide, with each intensive study area of approximately 3.1 ha. The canopy of each was dominated by Rio Grande cottonwood ranging from 8 to 15 m in height, with a subcanopy of *Salix gooddingii* (Goodding willow) and *Tamarix ramosissima* (saltcedar). The most common understory shrubs were *Baccharis glutinosa* (seepwillow), *Forestiera neomexicana* (New Mexico olive), and *Amorpha fruticosa* (desert indigobush). The composition and density of the herbaceous understory varied among years, but generally included *Conyza canadensis, Chamaesyce serpyllifolia, Ratibida tagetes, Solanum elaeagnifolium,* and *Sphaeralcea angustifolia.*

The sites were approximately 0.5 km west of the Rio Grande and separated from it by the low-flow conveyance channel and riverside levee that parallel the river. After we had collected baseline data at both sites for two years, one site, designated the "Experimental Flood" site, was inundated for approximately one month during each of the following three years: 1993 (May 17 to June 12), 1994 (May 19 to June 19), and 1995 (May 17 to June 17). Floods were timed to match the historical peak flow for the upper Rio Grande, based on the mean annual hydrograph for 1889–1990 at the U.S. Geological Survey gauging station at Embudo (Slack et al., 1993), located upstream from major flow regulation. Water was taken from a riverside canal via a water diversion structure, which included a combination of water diverted directly from the Rio Grande, irrigation return flows from agricultural fields, and groundwater recharge accumulated in the nearby low-flow channel (J. Taylor, personal communication). Experimental floods inundated approximately 10 ha of riparian forest floor at the Experimental Flood site during 1993, 1994, and 1995 (Figure 3-4). Floodwater depth varied from nearly 20 to 200 cm across the site because of topographic variation; average depth was about 50 cm. Throughout the study the second site, "Reference-1," remained unflooded.

In 1994 we found a section of forest within the refuge that flooded directly from the Rio Grande when flows exceeded approximately 4500 cubic feet second^{-1} (cfs). This riparian forest was approximately 200 m east of the primary sites, bounded on the east by the Rio Grande and on the west by the levee. We termed it the "Natural Flood" site and established a similar sampling regime there in August 1994. At that time an additional

Figure 3-4. *Experimental flooding at Bosque del Apache National Wildlife Refuge (NWR) inundated approximately 10 ha of riparian forest for one month in May to June during each of 1993, 1994, and 1995. (Photograph by Lisa Ellis.)*

dry reference site ("Reference-2") was located south of the Natural Flood site in an area largely protected from flooding by a groin. These two riverside sites covered approximately 1.1 ha each and contained a primarily cottonwood canopy, with understory vegetation including saltcedar, seepwillow, New Mexico olive, and Russian olive. Herbaceous understory vegetation was sparse, particularly at the Natural Flood site, where the overstory was denser and light reaching the forest floor was limited. In 1995, flooding at the Natural Flood site lasted for approximately 2.5 months from mid-May through late July and surface water height measured at one typical location averaged about 20 cm throughout the flood.

Rainfall varied considerably during the study (Figure 3-5), reflecting the El Niño-Southern Oscillation phenomenon that affects the southwestern United States (Nicholls, 1988). Locally, El Niño events result in increased fall, winter, and spring precipitation, as experienced in 1991–92 and 1992–93 (Dahm and Moore, 1994; Figure 3-5). Winter precipitation was relatively high during 1994–95, but spring precipitation was low. Both

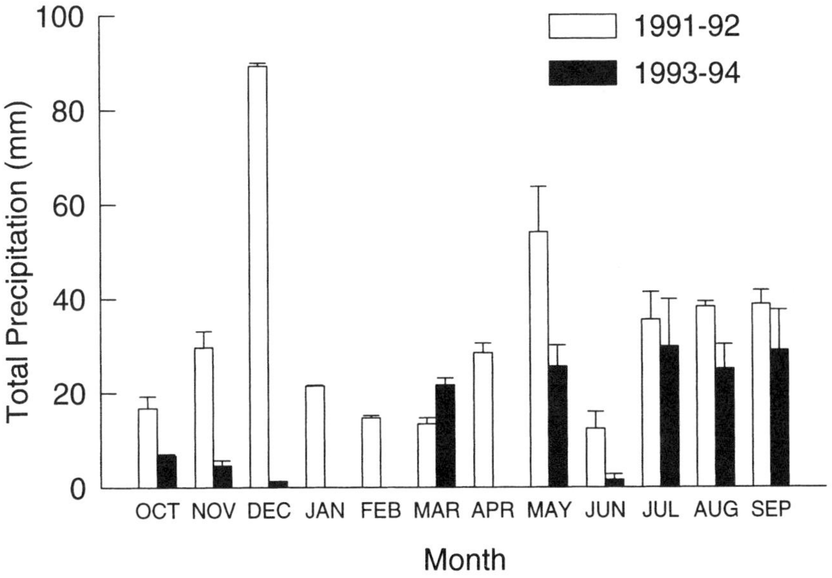

Figure 3-5. *Precipitation at Bosque del Apache NWR during 1991–92, illustrating a year with increased fall, winter, and spring precipitation, and during 1993–94, illustrating a year with relatively low winter precipitation. Data are presented by water year and are averaged for two locations within the refuge. (Reprinted from Ellis et al., 1999, copyright © Society for Ecological Restoration, by permission.)*

1990–91 and 1993–94 experienced relatively low (although normal) winter precipitation, with more of the annual precipitation arriving during the summer monsoon season.

Forest Floor Organic Matter—Accumulations Reflecting Lack of Flooding

One of the most obvious effects of removing the flood pulse has been the accumulation of organic matter within floodplain forests, which has slowed the cycling of nutrients and contributed to the increase in catastrophic fires in the region. Complicated scientific techniques to determine this effect are not required—a short visit to the *bosque* will provide ample evidence, with piles of downed branches, fallen trees, and leaf litter in abundance. Measurements of woody debris confirm that the buildup is considerable, ranging from 33.7 to 38.8 Mg ha^{-1} (Megagrams per hectare) at dry forest sites at Bosque del Apache (Ellis et al., 1999). The high level of drought-induced cavitation discussed earlier may contribute to the great quantity of woody debris caused by branch dieback. We were unable to quantify the woody debris on the forest floor before our experimental flooding began, but measurements taken in 1995 revealed significantly less total woody debris at the Natural Flood site as compared with the other three sites (Figure 3-6). Three consecutive years of experimental flooding was not enough to reduce accumulations of woody debris (for reasons discussed in "Wood and Leaf Decomposition" which follows), but comparison with the Natural Flood site suggests that continued flooding would lead to reductions in such debris.

Of particular interest to us with respect to woody debris was the considerable patchiness in wood accumulations at our dry (i.e., nonflooded) sites. Estimates of total woody debris biomass along 20 m transects at Reference-2 ranged from 4.9 to 106.9 Mg ha^{-1} (Ellis et al., 1999). In contrast, the distribution of woody debris at the Natural Flood site was uniformly lower, ranging from 3.8 to 22.5 Mg ha^{-1} along similar transects. These large piles of debris at dry sites can have important consequences for the spread of wildfire, as discussed later.

Published estimates of woody biomass in riparian forests along perennial streams in arid regions are limited. In contrast to our results, Robinson (1997, cited in Robertson et al., 1999) reported large pools of coarse woody debris along nonharvested, frequently flooded portions of flood-

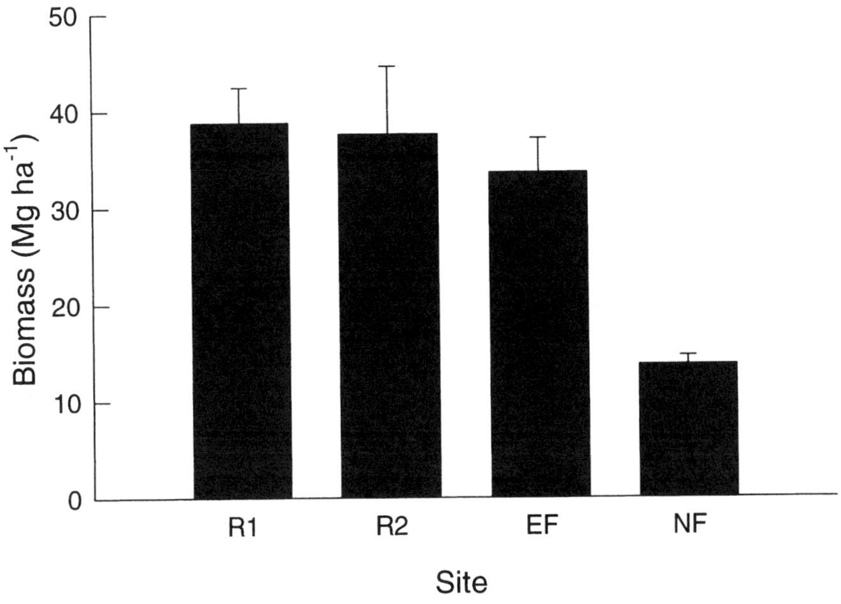

Figure 3-6. *Biomass of total woody debris for four study sites at Bosque del Apache NWR, measured in 1995. Values are the mean estimated total biomass (Mg ha^{-1}) across 20 transects for each site; vertical bars are standard errors. R1 = Reference-1, R2 = Reference-2, EF = Experimental Flood, NF = Natural Flood. (Adapted from Ellis et al., 1999, copyright © Society for Ecological Restoration, by permission.)*

plain rivers in Australia (6.3 kg C m^{-2}), with lower pools in infrequently flooded sites (1.2 kg C m^{-2}). The values in our study are comparable to the range recorded for temperate deciduous forests (Harmon et al., 1986). Estimates of coarse woody debris (diameter > 2.0 cm) for our nonflooded sites were greater than for a *Populus tremuloides* (aspen) forest in northern New Mexico that experienced over three times more annual precipitation than Bosque del Apache (Gosz, 1980; Dahm and Molles, 1992), whereas the biomass of coarse woody debris at our Natural Flood site was much less than that measured in the aspen forest.

In addition to woody debris, riparian forests along the Middle Rio Grande contain substantial accumulations of leaf litter on the forest floor. Measurements across five years at Bosque del Apache revealed considerable temporal variability in forest floor litter biomass, which probably reflected its patchy distribution (Figure 3-7). Litter within the cottonwood forest seems to reflect local microtopography, with leaves accumulating in

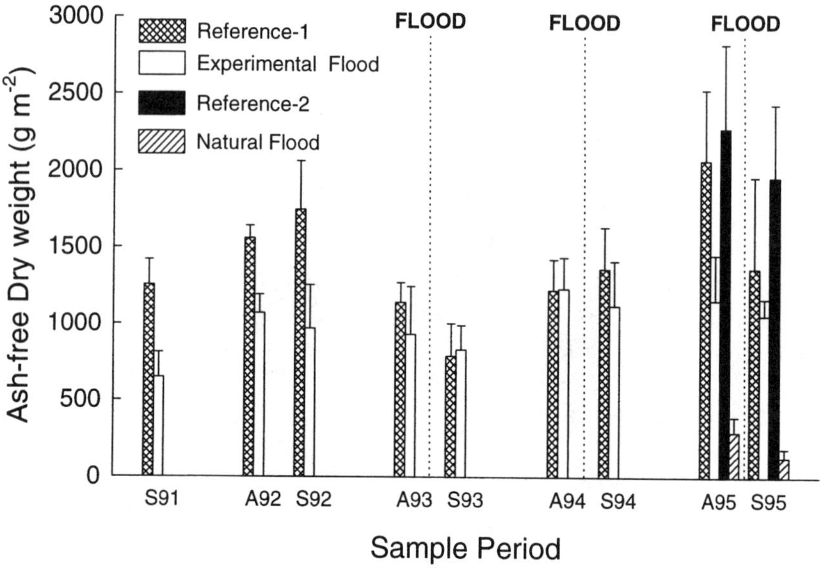

Figure 3-7. *Estimates of forest floor litter accumulations across five years of study in riparian forest sites at Bosque del Apache NWR. Values are the mean ash-free dry weights of all organic matter above mineral soil (n = 10 samples per collection) for April (A) and September (S); vertical bars are standard errors. (Reprinted from Ellis et al., 1999, copyright © Society for Ecological Restoration, by permission.)*

swales but being blown clear of high points. Such variation in litter accumulation was quantified in the cottonwood forest at the Rio Grande Nature Center in Albuquerque (C. Finnance, personal communication).

Comparisons with the Natural Flood site, however, indicated significantly less forest floor litter at that site as compared with nonflooded sites, even considering the range of variation in measurements at dry sites (Ellis et al., 1999; Figure 3-7). Again, three years of flooding did not reduce the buildup of forest floor litter at the experimental site, but lower quantities of leaf litter at the Natural Flood site suggested that repeated flooding does help to reduce the quantity of organic litter. Other studies also reported reduced forest floor litter at flooded riparian sites (Bell and Sipp, 1975; Peterson and Rolfe, 1982).

Wood and Leaf Decomposition—Positive Effects of Flooding

The large accumulations of woody debris in the forest suggested that wood decomposition rates were very slow in this system. Therefore, we designed an experiment to measure rates of wood decomposition in both flooded and dry forests, with preweighed and measured dry logs placed at each site during the first summer of the study. Subsequently, we collected subsets of these logs, first after two years, prior to experimental flooding, then after two and three floods, respectively.

Our measurements confirmed that in the absence of flooding, wood decomposition rates are indeed very slow (Ellis et al., 1999). We calculated log decay rates before and after flooding, using the exponential model $y_t = y_0 e^{-kt}$, where y_0 is the initial log mass, y_t is the log mass at time t, and k is the decay rate constant. Prior to flooding, wood decay rates (k) for 1991 to 1993 averaged 0.003 y^{-1} at the experimental site and 0.004 y^{-1} at Reference-1. Using these decay rates, we calculated log half-lives using the equation $-\ln(\frac{1}{2})k^{-1} = t$, where t is the log half-life and k is again the decay rate constant . The predicted half-life for wood lying in the forest prior to flooding at the experimental site was more than 200 years. There was some variation in the decay rate for the reference site over the course of the study, reflecting differences in annual precipitation; the overall decay rate for logs at Reference-1 during the entire five-year period was 0.010 y^{-1}, predicting a half-life of 69.3 years (Figure 3-8).

Three seasons of flooding slightly increased wood decay rates, with an average rate of 0.065 y^{-1} at the experimental site during the years with flooding (Ellis et al., 1999). Because of the small sample size of this study ($n = 5$ logs per collection), this difference in rates after flooding was only marginally significant (Wilcoxon Rank Sum test, $Z = -1.87$, $P = 0.06$; Figure 3-8), but the predicted half-life of 10.6 years was considerably less than that predicted for the dry site. Although this is a substantial reduction in the predicted half-life of downed wood in this system, it indicates that a decade or more of flooding will be needed to have a meaningful impact on the system.

Reported decay rates for log-bole mineralization (including respiration and leaching but not fragmentation) for temperate forest ecosystems range from 0.004 y^{-1}, predicting a half-life of 172 years, for *Pseudostuga menziesii* in the Pacific Northwest to 0.520 y^{-1}, predicting a half-life of 2.3 years, for *Liriodendron tulipera* in an eastern deciduous forest (Harmon et al., 1986). Although no decay rates are available for *P. deltoides,* rates for *P. tremuloides* range from 0.049 y^{-1} in northern Minnesota (Miller, 1983) to 0.070 y^{-1} in the moist mountains of northern New Mexico (Gosz, 1980). These rates predict half-lives of 14.0 and 9.8 years, respectively, which are

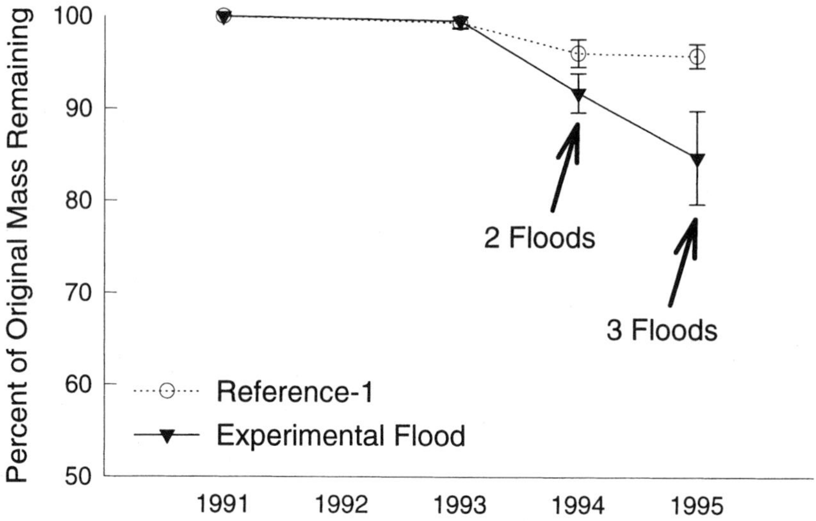

Figure 3-8. *Wood decomposition at experimentally flooded and dry riparian sites at Bosque del Apache NWR. Values are the percentages of the original mass of logs remaining, averaged across four logs for each collection at each site; vertical bars are standard errors. The 1994 and 1995 collections were made after two and three floods, respectively, at the experimental site. (Reprinted from Ellis et al., 1999, copyright © Society for Ecological Restoration, by permission.)*

similar to that predicted after flooding at our site. Martius (1997) reported that woody decay rates in the Central Amazon floodplain are enhanced by the seasonal oscillation between wetting and drying. Decay rates during the terrestrial phase of the flooding cycle were greatest for wood that had been immersed for one to four weeks, as compared with wood that had not been submerged or wood that had been submerged for six months (Martius, 1997).

Microbial and fungal activity influence the rates of wood decomposition (Harmon et al., 1986), and such activity in turn is affected by temperature, moisture, and aeration (Marra and Edmonds, 1994; Rayner and Todd, 1979). In our study, microbial and fungal activity generally increased with flooding (Molles et al., 1998). Responses included increased abundance of soil bacteria, fungi, and cellulose decomposers, higher mycorrhizal inoculum potential and root lengths colonized by mycorrhizal fungi, and increased dehydrogenase activity, indicating biological activity in the soil. We also used respiration chambers to measure forest floor respiration during the second two experimental floods, as well as during one natural flood.

Respiration rates at the experimental flood site were 300 times higher than at the unflooded control during the second experimental flood, and during the third flood, respiration rates at the experimental site increased to nearly 600 times that of the reference site (Molles et al., 1998). In contrast, respiration rates at the Natural Flood site were about one-tenth of the rates measured at the experimental site. This suggested that a process of ecosystem reorganization had been initiated within the experimental site, with large accumulations of organic matter being broken down at this site previously disconnected from the flood pulse (Molles et al., 1998).

Leaf decomposition rates varied among years in the absence of flooding, reflecting winter and summer precipitation. In leaf pack experiments, packs containing air-dried cottonwood leaves were placed at each site in early November of each year, with subsets of packs collected during the following April (prior to flooding), June (after flooding), and November. This allowed us to determine the timing of decomposition. These experiments revealed the greatest over-winter weight loss during the moist conditions of 1991–92 and the least during the dry winter of 1993–94 (Ellis et al., 1999; Figure 3-9). Subsequent decomposition at the reference site was relatively low after the moist winter, and later summer/fall decomposition was higher after the dry winters, likely reflecting monsoonal moisture input. Total annual mass loss for leaves at the reference site ranged from 18.6 to 32.8 percent.

In contrast, flooding significantly increased leaf decomposition during the summer months at the experimental site, with rapid weight loss coinciding with the period of flooding (Figure 3-9). Leaf packs exposed to flooding lost 30 to 50 percent of their initial weight during the month-long floods, whereas leaves at the reference site lost only 1 to 5 percent during the same period (Ellis et al., 1999). Total annual weight loss for leaf packs at the experimentally flooded site ranged from 46.4 to 56.4 percent, reflecting month-long floods, and leaf packs placed at the Natural Flood site during the final season lost an average 58.8 percent of their initial mass following two months of flooding. Flooding had the greatest influence on decomposition following the dry winter of 1993–94, when over-winter loss was lowest (Ellis et al., 1999).

Similar increases in leaf decomposition rates following flooding have been reported by other studies (Bell et al., 1978; Peterson and Rolfe, 1982; Shure et al., 1986). We suggest that summer flooding increases the break-

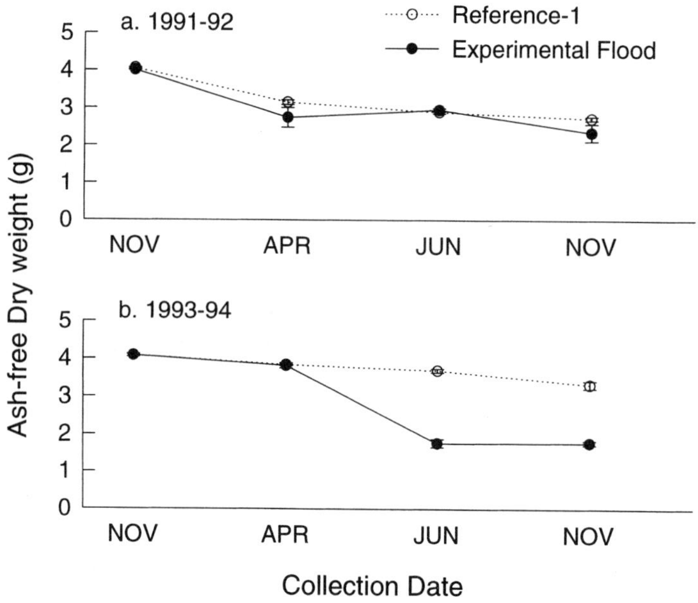

Figure 3-9. Decomposition of cottonwood leaves during 1991–92 and 1993–94 at two sites at Bosque del Apache NWR. The Experimental Flood site was inundated between April and June collections during 1993–94. Values are the mean ash-free dry weights of cottonwood leaves from five decomposition packs for each collection; vertical bars are standard errors. (Adapted from Ellis et al., 1999, copyright © Society for Ecological Restoration, by permission.)

down of leaves and, consequently, the release of essential nutrients within the ecosystem at a time critical for plant growth, and that this may benefit primary production (Ellis et al., 1999). In dry sites, and particularly after dry winters in this region, nutrients remain bound in dry, dead leaf tissue. Without flooding, much of the decomposition occurs following late summer thunderstorms, relatively late in the growing season. Restoring flooding early in the season may be important to promote this cycling of nutrients at a time more beneficial to plants. It may also help prepare leaves for handling by native macrodetritivores.

Much of the loss of mass from leaves during these one-year experiments was probably due to leaching. Leaves of deciduous trees may lose up to 30 percent of their weight during the first 24 to 48 hours after immersion (Cummins, 1974; Webster and Benfield, 1986; Bärlocher, 1992), with mass loss beyond about 35 percent probably reflecting fungal and microbial respiration (Harmon et al., 1986). Initial leaching increases subsequent rates of decomposition (Peterson and Rolfe, 1982; Day, 1983; Shure et al., 1986), suggesting that even short periods of flooding may help to decrease accumulations of leaf litter.

Whether nutrients leached from leaves and other organic debris are lost from a local site during flooding, or are instead deposited within that site, depends on a number of factors: the nature of the specific nutrients, the flooding cycle, and the vegetative cover (Junk et al., 1989). Although floodwaters often carry leached nutrients out of the forest, nutrient loads in groundwater passing through riparian sites are often reduced by vegetative demand (Gregory et al., 1991). For example, chemical analyses during our floods indicated a high rate of retention of nitrogen and phosphorus within the site (Grimm et al., 1997). Dissolved oxygen (DO) demand was also quite high during experimental flooding, reflecting the presence of large quantities of dissolved and particulate organic matter (Lieurance et al., 1994). Levels of about 8 mg l^{-1} of DO that entered the site in floodwaters were rapidly depleted, with undetectable levels at 15 cm above the submerged litter layer (Lieurance et al., 1994; Molles et al., 1998). In contrast, DO concentrations at the Natural Flood site remained above 4 mg l^{-1} within a few centimeters of the forest floor.

From a management point of view, there is some question as to whether short flood pulses would be more productive in terms of reestablishing ecosystem function than the longer floods used in our study. Excessive moisture may decrease the aeration that is essential to obligate aerobic fungi (Harmon and Chen, 1991), suggesting that fungi may be relatively inactive during periods of inundation. Residual moisture retained in wetted logs may provide conditions favoring bacterial and fungal growth, and thus intermittent wetting of riparian soils combined with water drawdowns to increase aeration may enhance potential decomposition even more than extended periods of inundation.

In addition to leaching and microbial and fungal activity, macroarthropods play a significant role in the breakdown of organic debris. The macrodetritivore probably most important in the prealteration system is the native cricket (*Gryllus alogus,* Gryllidae). Although the numbers of crickets were quite low at the dry reference sites, and prior to flooding at the experimental site, the crickets responded positively to flooding (Ellis et al., 2001). The response was not immediate—the increase in crickets was detected only after the second flood—but large numbers of early-instar crickets in postflood collections suggest that flooding promoted recruitment.

By far the most abundant macrodetritivore in riparian forests along the regulated Middle Rio Grande is the introduced terrestrial isopod (*Armadillidium vulgare,* Armadillidiidae). In contrast to the increase in abundance after flooding observed for crickets, isopods decreased after flooding (Ellis et al., 2000). This effect diminished, however, after subsequent floods, and it is possible that the local population was merely experiencing a temporary decline, reflecting the long absence of flooding. Isopods were observed clinging to floating wood or tree trunks during the floods, whereas crickets were seen fleeing in front of advancing floodwaters. Although the adaptive significance of each of these behaviors with respect to flooding is not known for this system, both vertical and horizontal migrations are reported in response to flooding for terrestrial invertebrates along the Amazon River (Adis, 1997). Thus, it seems likely that both behaviors could contribute to survival during the gradual inundations typical of the flood pulse along the Rio Grande.

Especially notable with respect to promoting decomposition are differences between crickets and isopods in their preferred food sources. In captivity, crickets prefer partially decayed leaves, such as those present after inundation, but isopods can consume relatively dry, intact leaves (C. S. Crawford, personal observation). This suggests that flood restoration, by the promotion of leaching and initial leaf decomposition, will promote the food resources preferred by crickets and thus increase cricket populations, and this will in turn result in the additional breakdown of leafy material on the forest floor. The ultimate effect of flooding on isopod populations remains to be determined. Although an exotic, *A. vulgare* seems to have little negative impact on the ecosystem, and if able to adjust to the reintroduction of flooding, it will probably continue to contribute to decomposition processes in this system.

Stored Organic Debris: The Time Factor in Restoration

Two salient conclusions from our research have broad implications for restoration. First is the recognition that restoring the *bosque* to something resembling its preregulation condition requires that the degraded ecosystem pass through what we call a "reorganization" phase, with certain processes such as forest floor respiration reaching levels unlike those of either the degraded or the naturally functioning states (Molles et al., 1998). It is particularly important for resource managers to recognize this requirement, as they may not see the results they expect immediately after restoration efforts are initiated. Only long-term monitoring of a number of population and ecosystem components after the restoration of natural flow conditions will enable recognition of some of the intermediate stages. It is also possible that different systems will experience different levels of reorganization and that geomorphic changes induced by channel migration will affect the trajectory of succession (Poiani et al., 2000).

The second conclusion from our work that we believe to be particularly relevant to restoration efforts is that ecosystem components reorganize at different rates (Molles et al., 1998). For example, bacterial and fungal populations and leaf decomposition all showed rapid responses to restored flooding (Molles et al., 1998; Ellis et al., 1999). As discussed earlier, cricket populations showed a time lag of one year, with detectable increases following the second flood. Changes in carabid beetle abundance showed a similar pattern to that of crickets, with concurrent changes in community composition, but comparisons with the naturally flooded site suggest that some compositional changes of carabids might continue after additional floods (Ellis et al., 2000). At the other extreme, wood decomposition rates predict that reducing the accumulations of woody debris will require a decade or more of annual flooding. Recognizing these temporal differences in reorganization will be crucial in determining the success of restoration projects.

FIRE: ITS RELATIONSHIP TO FLOODING AND LITTER BUILDUP

As indicated earlier, it is generally assumed that increasing accumulations of organic matter resulting from the absence of flooding have contributed greatly to the increase in fire in the Middle Rio Grande riparian forest (Stuever et al., 1997; Stuever, 1997; Ellis et al., 1998; Ellis et al., 1999). Although many ecosystems are adapted to fires and in many cases require them for regeneration, fire was probably not a significant component of this system prior to flow regulation. Upland areas were commonly burned by Indians (e.g., Sauer, 1950; Cooper, 1960; Dobyns, 1981; Pyne, 1982; Baisan and Swetnam, 1997; Scurlock, 1998), but although there is some suggestion that fires may have been used in riparian zones (DeBano and Neary, 1996), early accounts generally do not include lowland riparian habitats among reported fire locations (Dobyns, 1981; Bahre, 1985; Scurlock, 1998). The increasing role of fire in the decline of western riparian forests (Ohmart and Anderson, 1982; Busch and Smith, 1993; Busch, 1995; Stuever et al., 1997) necessitates a better understanding of the mechanisms involved in fire propagation and of the response of native plants to this new form of disturbance.

In 1996 we were presented with an ideal opportunity to test the relationship between fire severity and the accumulation of organic debris in riparian forests, as a wildfire burned our two Bosque del Apache flood sites. As discussed earlier, the accumulations of dead fuels differed considerably at the two sites. The biomass of woody debris at the Natural Flood site before the fire averaged 13.7 (± 0.9) Mg ha^{-1} across 20 transects, as compared with 33.7 (± 3.5 S. E.) Mg ha^{-1} at the Experimental Flood site (Ellis et al., 1999). In addition to differences in the horizontal distribution of wood at the two sites, there was also a much greater vertical distribution of woody debris at the Experimental Flood site. This debris included partially fallen trees as well as a number of dead snags that were not included in the woody biomass estimates. Forest floor litter storage estimates measured in the fall, nine months before the fire, averaged 1056.0 ± 108.7 g m^{-2} at the experimental flood site, as compared with 129.7 (± 57.4 S. E.) g m^{-2} at the natural flood site (Ellis et al., 1999). The fire history at these sites is unknown, but major fires had been excluded since at least 1940.

At the time of the fire, more than three-quarters of the Rio Grande Basin was experiencing severe to extreme long-term drought (Palmer Drought Severity Index; Brown and Heim, 1996). Conditions were extremely dry at the Refuge; precipitation for the water-year to date (October through May) was 33 mm, as compared with an average of 125 mm during the preceding five years, with no precipitation recorded at the Refuge headquarters during March through May prior to the fire (Bosque del Apache, unpublished data; Ellis, 2001). The riverbed itself was nearly dry, with most water routed into the low-flow conveyance channel (L. M. Ellis, personal observation), and because of the shortage of water, neither of the previously flooded sites was inundated that year.

The human-caused wildfire began on June 8, 1996, near the community of San Pedro, approximately 4 km north of Bosque del Apache (Ellis, 2001). High temperatures, low humidity, and high winds favored the spread of the fire, which initially moved north, but then south and into the Refuge. The Natural Flood site (referred to here as Burn-2) burned in the afternoon and probably partly at night as a backfire. In contrast, the Experimental Flood site (referred to as Burn-1) burned more quickly as a crowning headfire that same evening. After burning 1640 ha of the Refuge, including nearly 880 ha of cottonwood-willow forest, the fire was fully contained by June 11 (Bosque del Apache, unpublished report; P. Norton and J. Taylor, personal communications).

Fire Severity: Effect of Flood Histories

Fire severity was measured at each of the burn sites approximately two months after the fire. Severity was estimated around cottonwood trees that had been tagged and monitored before the fire and were monitored for survival after the fire. Fire severity estimates were based on the amount of leaf litter consumed by the fire. In "light" severity areas, leaf litter was burned in patches but not entirely consumed, in "moderate" severity areas most leaf litter was consumed, but some patches of unburned material remained, and in "high" severity areas all leaf litter was consumed (Stuever, 1997; Ellis, 2001). Fire severity was high around all 127 monitored trees at Burn-1; around 50 trees at Burn-2, 20 showed light fire severity, 27 moderate severity, and only 3 high severity (Ellis, 2001). As an additional estimate, we remeasured woody debris levels at all sites following the fire. Approximately 33 Mg ha^{-1} of woody debris were consumed from the forest floor at Burn-1, leaving only 1 percent of the preburn biomass present, whereas approximately 29 percent of the prefire wood biomass remained at Burn-2.

Observations at the two sites corroborated these measurements. At Burn-1, only ashen trails remained on the forest floor where fallen logs had been before the fire, and all remaining wood, which was probably input from previously standing trees, was heavily charred. Fire severity was unquestionably greater at that site. At Burn-2, considerable forest floor litter remained in patches throughout the site. Most important, areas of light fire severity at Burn-2 were located along a wide channel that passed through the site, where the most extensive flooding had occurred in past years and where prefire accumulations of organic debris were lowest. The area of high fire severity was located on a slight rise, where accumulations of organic debris were greater and densities of saltcedar greater than in the channel (Ellis, 2001). Thus, observed burn patterns at this site support the contention that organic debris accumulations increase fire severity.

Cottonwood Survival Relative to Fire Severity

The direct effects of the fire on the monitored cottonwood trees were estimated based on the amount of char in three regions of the tree: the root crown, the trunk, and the canopy (Stuever, 1997; Ellis, 2001). As expected, all cottonwoods at Burn-1 were heavily charred in all recorded variables, with no trees retaining green leaves in the canopy immediately after the fire, or producing canopy leaves during the subsequent two years (Table 3-1). Thus, top-kill was complete at this high-severity fire, and this supports similar findings by Stuever (1997) in the Albuquerque area.

In contrast, although the three cottonwoods at Burn-2 in high-severity areas were also completely charred, trees in the moderate to light fire severity categories were less extensively burned (Table 3-1). Further, six of the monitored trees (12 percent) at Burn-2 retained green leaves in the canopy after the fire; these trees were all located along the channel where fire severity was lowest and where litter accumulations prior to the fire had also been lowest. By two years after the fire, two of these trees no longer produced green leaves in the canopy, so that top-kill at Burn-2 was 94 percent for all trees; however, for trees in the light-severity category, top-kill

TABLE 3-1 Postfire condition of cottonwoods at Bosque del Apache

Data were recorded in August 1996 following the fire of early June. Values are the percentage of individuals in each category. Values at Burn-2 are given for all trees combined (total), as well as each of three fire severity categories (light, moderate, and high severity). All trees at Burn-1 were classified in the high-severity category, so only total values are given.

	Burn 1	Burn 2			
	Total	Light	Moderate	High	Total
% of trees green in canopy	0	30	0	0	12
% of trees with complete basal char	100	10	93	100	60
% of trees with root crown burned to wood	100	25	89	100	64
% of trees with trunk burned to wood	100	10	93	100	60
% of trees with crown charred	100	45	89	100	72
Average % of trunk height charred	100	14	77	100	53
Sample Size (*n*)	127	20	27	3	50

Source: Reprinted from *Biological Conservation,* vol. 97, L. M. Ellis, "Short-term response of woody plants to fire in a Rio Grande riparian forest, Central New Mexico, U.S.A.", pages 159–170, copyright 2001, with permission from Elsevier Science.

was 85 percent. This suggests an immediate benefit to lower litter accumulations, and correspondingly lower fire severity.

We were particularly interested in the postfire sprouting ability of Rio Grande cottonwoods, in that it has been suggested that some woody species along southwestern riparian ecosystems lack efficient postfire sprouting mechanisms (Busch and Smith, 1993). *Populus* species in general are known to sprout readily after disturbances, including flooding (Fowells, 1965; Rood et al., 1994; Braatne et al., 1996; Naiman and Décamps, 1997), with suckers arising from preexisting, suppressed peridermal buds that often produce clusters of shoots (Schier and Campbell, 1976). Several studies have reported asexual reproduction among *Populus deltoides* (Fuller, 1912; Everitt, 1968; Wilson, 1970; Barnes, 1985). Thus, although this species may not be adapted to fire per se, its physiology provides a mechanism to respond to fire disturbance.

In our study, burned trees began producing sprouts within the first month after the early June fire; they probably benefited from the rains that began in late June and continued through August. Although sucker mortality increased somewhat at both sites during the following two years, approximately 40 percent of individuals at each site had shoot suckers (suckers arising within 30 cm of the trunk) two years after the fire (Ellis, 2001; Figure 3-10). The percentage of trees at each site with root suckers (those within 30 to 90 cm of the trunk) was lower, with less than 15 percent at Burn-1 and just over 20 percent at Burn-2. Both shoot and root sprouts were recorded for 13 and 14 percent of the trees at Burn-1 and Burn-2, respectively. Overall, more of the trees at the site of lower fire severity (Burn-2) produced suckers, and individual trees at that site produced more suckers than trees at Burn-1. At both sites shoot suckers generally increased in size, but decreased in number per tree, during the two years after the fire (Figure 3-11). Unfortunately, because of other activities at the Refuge, we were able to monitor the trees for only two years after the fire. Additional studies are needed to determine the long-term survival of Rio Grande cottonwoods after fire, but our data suggest that survival of adult trees and sprouting ability were higher where fire severity was lower.

The timing of fire relative to the developmental condition of a population of trees may have important survival consequences. Generally, older, and typically larger, trees have better above-ground survival following fires, with bark thickness, stem diameter, and, generally, age all positively related and correlated with fire resistance (Whelan, 1995; Bond and van

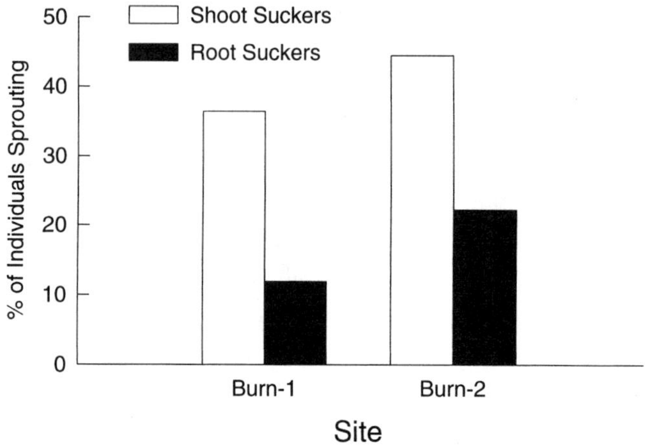

Figure 3-10. *Percentages of individual cottonwoods producing shoot and root suckers at two burned sites after the San Pedro fire at Bosque del Apache NWR, measured in August 1998, two years after the fire. (Reprinted from* Biological Conservation, *vol. 97, L. M. Ellis, "Short-term response of woody plants to fire in a Rio Grande riparian forest, Central New Mexico, U.S.A.," pages 159–170, Copyright (2001), with permission from Elsevier Science.)*

Wilgen, 1996). Sprouting ability often decreases with stem diameter or age (Blaisdell and Mueggler, 1956; Kayll and Gimingham, 1965; Burrows, 1985; Hobbs and Mooney, 1985), with smaller trees having more tissues capable of growth. In our study, smaller trees were more likely to produce shoot suckers and produced more suckers overall than large trees. Thus, whether a population persists by the survival of adult trees or by the vegetative production of new stems will depend in part on the initial age/size structure of the stand. Where very severe fires kill most adult trees, the loss of sexual reproduction may affect the genetic composition of the population.

Fire in this altered system, however, may provide an opportunity for sexual reproduction that was eliminated with the reduction of flooding. Where stabilized banks prevent a meandering river from creating appropriate habitat for seedling establishment, cottonwood populations senesce without replacement, because their seedlings cannot grow under the shade of a mature forest. Even in sites that are flooded, canopy shade precludes seedling establishment and populations eventually die out. However, we observed numerous seedlings at both burn sites after the fire; these were absent during prefire site monitoring, in spite of flooding. By opening the

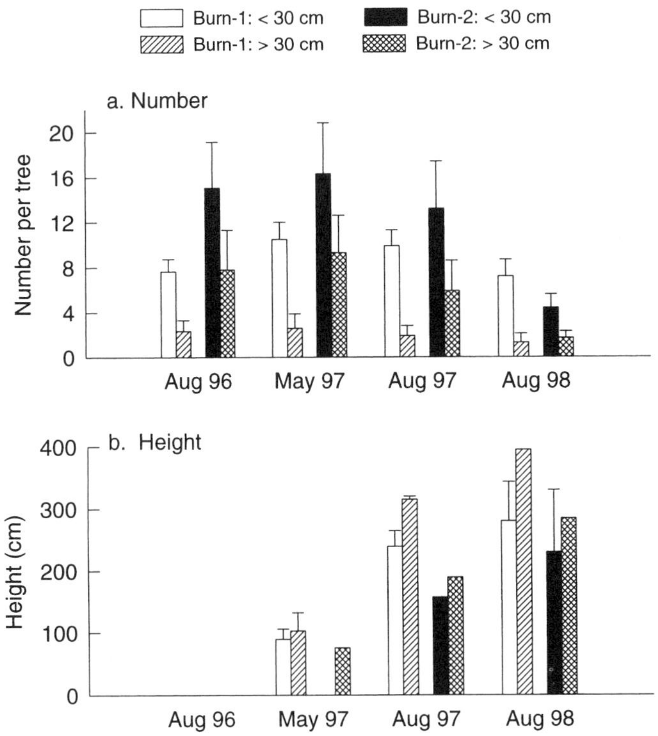

Figure 3-11. *Mean number* (a) *and height* (b) *of shoot sprouts produced by individual cottonwoods during two years following the San Pedro fire at Bosque del Apache NWR; vertical bars are standard error. ([a] Reprinted from* Biological Conservation, *vol. 97, L. M. Ellis, "Short-term response of woody plants to fire in a Rio Grande riparian forest, Central New Mexico, U.S.A.," pages 159–170, copyright (2001), with permission from Elsevier Science.)*

canopy, fire created ideal locations for seedling establishment, with enough adult trees nearby to produce the necessary seeds. Successful establishment resulted from flooding at some time during the two summers after the fire, either naturally (Burn-2) or artificially (Burn-1). Although we do not advocate controlled burns as a management tool in this ecosystem, our observations suggest that where fires occur, new forest patches can be promoted if irrigation is applied at the appropriate time. If seed trees remain in the vicinity, no artificial reseeding will be necessary, but in cases where all adult trees have been killed, seeds will have to be introduced artificially. Critical in this restoration strategy is the application of artificial flooding as a surrogate for the absent flood pulse.

Survival of Other Woody Species

Fire has been implicated in the spread of saltcedar, which sprouts readily after burning and may promote later fires through the accumulation of its deciduous litter (Mutch, 1970; Ohmart and Anderson, 1982). In comparing postfire survival of saltcedar to that of native plants, we measured sprouting rates for all woody species at both burn sites, using density plots that had been established before the fire. What we found surprised us.

Cottonwood was the only woody species in density plots to survive the San Pedro fire as a mature plant (i.e., not top-killed). This was true only at Burn-2 and included only 2.3 percent of individuals. We measured higher rates of shoot suckering among cottonwoods in density plots than in the individually monitored trees just discussed, with 77 percent of individuals sprouting in density plots at Burn-1 and 81 percent of individuals sprouting in plots at Burn-2 (Figure 3-12; Ellis, 2001). Root suckering was detected in 17 percent of individuals in density plots at Burn-1 and 5 percent of individuals in plots at Burn-2. Goodding willow, the main native subcanopy tree, was present in density plots only at Burn-1, where 73 percent of individuals produced shoot suckers and 18 percent produced root suckers (Figure 3-12). Among native shrubs (desert indigobush, seepwillow, and New Mexico olive) 12 to 14 percent produced shoot suckers at Burn-1, and sprouting rates ranged from 35 to 63 percent at Burn-2. Root suckering was lower for all three species at Burn-1, with less than 10 percent of individuals producing suckers, except for desert indigobush, 50 percent of which produced root suckers at Burn-2.

In contrast, our measures of sprouting in saltcedar were lower than we expected, based on suggestions in the literature that saltcedar recovers better than cottonwood after fires (Ohmart and Anderson, 1982; Crins, 1989; Busch and Smith, 1993; Busch, 1995). All saltcedars were top-killed, and root suckering was nearly absent. Sprouting from existing root crowns occurred in 53 percent of individuals at Burn-1 and 55 percent of individuals at Burn-2, lower than that measured for cottonwood and willow. Although it seemed that most of the saltcedar clumps remaining after the fire were sprouting, prefire density counts indicated that many clumps were completely consumed in the fire (Ellis, 2001). Thus, estimating saltcedar survivorship based only on postfire monitoring may inflate survival values. Certainly, more studies with preburn data are needed to confirm this inference. Particularly important may be the impact of repeated fires. Although

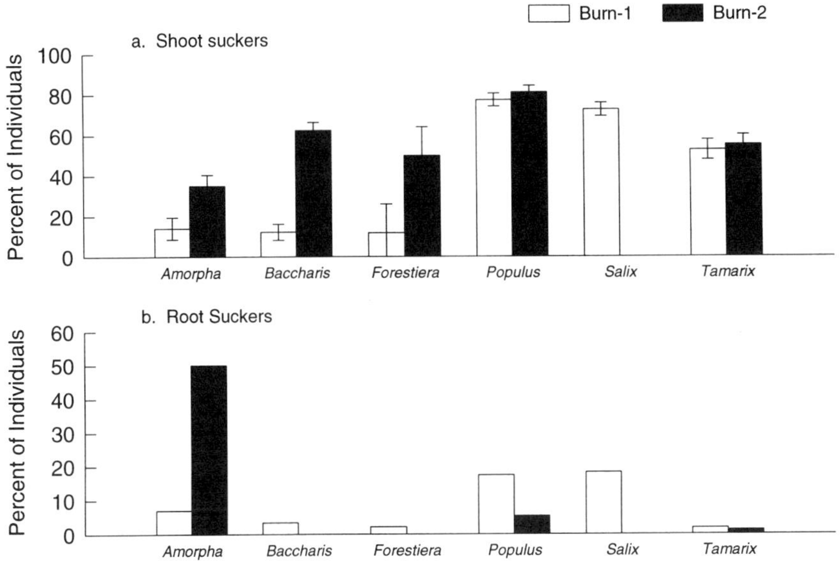

Figure 3-12. *Percentages of individuals of woody species with shoot suckers* (a) *and root suckers* (b) *four months after the San Pedro fire, based on prefire densities at each burned site. Error bars represent the percent error calculated at reference sites (Reprinted from* Biological Conservation, *vol. 97, L. M. Ellis, "Short-term response of woody plants to fire in a Rio Grande riparian forest, Central New Mexico, U.S.A.," pages 159–170, copyright (2001), with permission from Elsevier Science.)*

cottonwood may sprout well after a single fire, saltcedar may take over gradually after repeated fires. More studies are needed to determine how multiple burns affect survival and sprouting of both native and exotic species.

Other studies have found that the capacity of many plants to resprout after a fire depends in part on the severity of the fire (Bond and van Wilgen, 1996; Smith and Fischer, 1997), with higher survival typically recorded after burns of lower severity (Morgan and Neuenschwander, 1998). We found that for desert indigobush, seepwillow, and New Mexico olive, the three most common native shrubs, as well as for cottonwood, sprouting rates were higher at Burn-2. Thus, all of these species may recover better after moderate fires than after more severe ones. In cases of severe burning, resprouting may be completely prevented (Zasada, 1986). All of these data suggest that although completely eliminating fires in western riparian systems may be impossible, reducing fire *severity,* in part through the reduction of accumulated fuels, should be a goal for riparian ecosystem managers. Again, restoring the seasonal flood pulse will take us a long way toward that goal.

THE FUTURE: RESTORATION OF THE FLOOD PULSE

Need for the Flood Pulse: An Ecosystem-Level Problem

The importance of flow variability, and particularly the flood pulse, in river-floodplain ecosystems is now widely recognized (e.g., Junk et al., 1989; Sparks et al., 1990; Bayley, 1995; Walker et al., 1995; Poff et al., 1997; Richter et al., 1997). After more than a century of modifying and controlling the flow of rivers worldwide, experiments are now under way to restore the natural flow regime to a number of systems. Many of these efforts emphasize restoration of wetland or riparian vegetation, or the protection or recovery of native fishes, with a particular goal of establishing minimum flow requirements (see Poff et al., 1997, for review). However, there is increasing recognition that successful restoration of large river-floodplain systems will require a more holistic, ecosystem-level focus, with an emphasis on restoring the entire natural flow regime (Stromberg, 1993; Sparks, 1995; Poff et al., 1997; Haeuber and Michener, 1998).

There is an emerging consensus that no single management strategy will work for all ecosystems or organisms, even with the common goal of restoring flooding as a natural disturbance (Galat et al., 1998; Michener and Haeuber, 1998; Schmidt et al., 1998). For example, restoration of the ecological integrity of the Kissimmee River system in south central Florida focuses on methods to allow the river to regain its wandering course across the wide floodplain, including dechannelizing the river and removing flood control structures (Toth, 1995; Toth et al., 1998). Flooding along much of this system was historically a prolonged, low-intensity event. In contrast, a controlled flood along the Colorado River through the Grand Canyon of Arizona was intended to simulate the moderate-to high-intensity, short-duration disturbances that once scoured the channel, deposited sand bars, and removed streamside vegetation, maintaining a very different riparian ecosystem than that seen in the canyon after the installation of the Glen Canyon Dam (Collier et al., 1997; Schmidt et al., 1998). The Upper Mississippi River System (including the Illinois and the upper Mississippi Rivers) still experiences seasonal flood pulses, but navigation dams, levees, and wetland draining have altered the flow regime and interfere with the natural functioning of these floodplain ecosystems (Sparks et al., 1998). Several techniques are being used to restore, to some degree, the preregulation seasonal cycles, including the prevention of unnatural summer floods (Sparks et al., 1998). In all of these systems, restoration of their respective natural flow regimes will benefit a variety of processes and organisms more efficiently than restoration efforts aimed at target components. In the case of the Middle Rio Grande, restoring the spring flood pulse will have similar ecosystem-level benefits.

Riparian Restoration in the Middle Rio Grande Valley

Clearly, changes in water management are needed along the MRGV to ensure the survival of the riparian ecosystem and the many species of plants and animals that depend on that ecosystem. But just as clear is the realization that complete restoration of the Rio Grande's floodplain ecosystems is impossible. Agricultural use of the valley remains extensive, and urban sprawl is ever extending its grasp into areas once dominated by wetland and riparian vegetation. Along most of the MRGV, the *bosque* is confined to a narrow strip within the riverside levees (Figure 3-2). Through most of the northern valley, the river channel is downcut to the extent that overbank flooding no longer occurs. Meanwhile, the presence of a railroad bridge and development along parts of the aggraded channel in the lower valley constrain the release of high flows from Cochiti Reservoir north of Albuquerque. Although complete restoration of floodplain ecosystem structure and function to some presumed historical state is unrealistic, we believe that a more flexible concept of "partial restoration" is possible and would ensure the survival of this ecosystem (Crawford et al., 1996b; Crawford et al., 1999).

Partial restoration is possible through a combination of simulated flooding, manipulation of the current flow regime, and alteration of the existing bank structure (Crawford et al., 1999). These measures can promote the establishment and maintenance of native woody vegetation, as well as increase decomposition, mineralization, and nutrient cycling, and therefore decrease accumulated fuel loads and the risk of catastrophic fires. Although efforts are under way in a number of locations to replace the senescing cottonwoods, particularly via revegetation techniques such as pole planting, the current emphasis of hinging river restoration on the protection of endangered species will, we contend, fall short of the goal of ecosystem restoration.

We visualize two different types of sites being targeted for restoration via flooding (Crawford et al., 1996b). First, existing, mature *bosque* sites currently isolated from the river (either by levees or by incised channels that preclude overbank flooding) can be selected to receive flooding during the runoff season, either directly from the river or from ditches, as in our Bosque del Apache study. Such sites will be targeted for maintenance of existing *bosque,* in part to ensure a seed source for a future forest. Second, silt bars or treeless riverbanks, or areas cleared of existing exotic vegetation, can be flooded or at least kept moist to create germination sites for cottonwoods and willows. Clearly, the establishment of young forest patches along the river will be essential if senescing stands are to be replaced. The Albuquerque Overbank Project (discussed in the following section) is an example of this second objective, and shifting attitudes within management agencies such as the Bureau of Reclamation and MRGCD raise the possibility of its eventual realization along certain reaches, if not along the entire Middle Rio Grande.

We propose that the ultimate goal for partial restoration along the MRGV should be to establish and maintain a mosaic of riparian forest stands that can be easily flooded, with an emphasis on restoring ecosystem function (Crawford et al., 1996b). Ideally, this concept should apply to all reaches of the river to ensure an optimal array of habitats needed by migratory songbirds, waterfowl, and other native animals. For example, the Central Flyway follows the river valley and the *bosque* provides important habitat for both resident and visiting birds (Farley et al., 1994; Yong and Finch, 1997; Means and Finch, 1999). Maintaining a complex of different aged stands of mature and regenerating forests is essential to maintaining the diversity of birds that use the valley throughout the year (Farley et al., 1994).

Additional information is needed to fully understand the role of the flood pulse in this riparian forest. At the University of New Mexico, our multidisciplinary Hydrogeoecology Group, together with colleagues in other universities and a number of federal, state, and municipal agencies, is addressing this central question. Under study are ecosystem components, relating to flooding in native and exotic stands, that encompass decomposition and nutrient fluxes, primary production, and evapotranspiration. We anticipate that this continued research will lead to answers that will help conservation and restoration efforts, and that it will be useful to resource managers.

Changing Attitudes in Management: Application of the Flood Pulse

Management agencies operating along the Middle Rio Grande have a long-held mandate to deliver river water for human use. This has been accomplished by means that have decreased flooding and maintained a narrow channel. Currently, that directive, and the management attitudes that have developed from it, are changing. *The Middle Rio Grande Ecosystem: Bosque Biological Management Plan* (Crawford et al., 1993), developed by an interagency committee, provided recommendations for preserving and restoring the Middle Rio Grande floodplain ecosystem, including the restoration of its natural flow regime. We are now seeing cooperation among agencies that previously were thought to be "responsible" for the ecological problems in the valley. Despite a large amount of political contention over specific issues, shared goals to reestablish flooding and to promote the preservation and restoration of the *bosque* and other wetland ecosystems are beginning to emerge. These goals are manifested in a number of joint restoration programs, such as the Albuquerque Overbank Project (AOP).

The AOP is a collaborative effort, initiated in 1998, with participants from the U.S. Bureau of Reclamation, the University of New Mexico Department of Biology, the City of Albuquerque Open Space Division, the Middle Rio Grande Conservancy District, the New Mexico Natural Heritage Program, the U.S. Fish and Wildlife Service, the State of New Mexico Environment Department, and the U.S. Army Corps of Engineers (Crawford et al., 1999). The project has involved mechanically removing a dense 1.6 ha stand of Russian olive on an attached bar and then lowering part of the cleared area by about 0.6 m to promote flooding from the river. Channels were constructed on the newly cleared bar to foster cottonwood and willow establishment.

A series of parameters have been or currently are being monitored at the site, including groundwater levels, soil salinity, and herbaceous and woody plant diversity and growth, as well as terrestrial arthropod, bird, and beaver activity. Although data have not yet been formally presented, a couple of observations are noteworthy. Cottonwood and coyote willow establishment on the cleared and lowered area of the bar has been very successful; after three years, these native plants remain in far greater numbers than the exotics. Perhaps most interesting have been the changes in river channel morphology. Extensive erosional changes in the site's bank profile and in site topography have resulted in a channel with increased variability in depth, width, and velocity under variable discharges. These characteristics are in marked contrast to those of the channel prior to the alterations, when it maintained uniform depth, width, and velocity with variable discharges. Although monitoring will continue at the AOP site, after four growing seasons with very different flooding patterns we are able say that the experiment was successful in establishing a new stand of cottonwood and willow.

Need for Public Awareness: The Bosque Ecosystem Monitoring Program

Changing attitudes about the role of flooding along the Middle Rio Grande were evident within the management community by the early 1990s, but public understanding of the issue was still quite limited. An attempt to enhance that understanding had, however, been under way since 1986, when two of the authors of this chapter (CSC and MCM) began an applied ecology course at the University of New Mexico called "Bosque Biology." In many ways the course, which is still taught regularly, was the basis for much subsequent restoration-related *bosque* research and outreach.

In 1995 an educational program on the riparian forest, the *Bosque Education Guide,* was developed by a group of local educators, including the other author of this chapter (LME), whose goal was to translate for the general public a technical understanding of the *bosque* ecosystem as presented by the recently published *Bosque Biological Management Plan* (Crawford et al., 1993). Funding for the *Guide* was provided through the U.S. Fish and Wildlife Service from a special congressional appropriation, the "Bosque Initiative," which still funds it and other education and restoration activities in the Middle Rio Grande *bosque.* The *Guide,* originally developed for grades three to five, provides a background in basic ecological concepts associated with the flood pulse and riparian ecosystems, including topics such as cottonwood establishment and wildlife communities. It includes a trip to the river and *bosque* that is designed to help schoolchildren become aware of the values of this unique habitat so near their homes. For many children living in Albuquerque, this is their first realization that the river is there. The program is now used in a number of schools throughout the MRGV, as well as in other parts of the state.

In spite of the success of the Bosque Biology course and the *Bosque Education Guide,* we saw a need to inform even more of the public about the ecology of the riparian forest and changes in its management objectives. That, and an additional need for the collection of ongoing ecological data from many localities along the Middle Rio Grande, led to the creation of the Bosque Ecosystem Monitoring Program (BEMP). This long-term monitoring program has two objectives: first, to collect and make available to the management and research communities ongoing data about various riparian ecosystem components, and second, to provide educational opportunities to help citizens of the MRGV become more informed about the unique value of the *bosque* ecosystem and the specific conditions needed to maintain it.

Originated in 1996, BEMP was initiated through the University of New Mexico, with funding from the National Science Foundation's Informal Science Education Initiative. Subsequent annual support has come from the Bosque Initiative and the Sevilleta Long-Term Ecological Research (LTER) Schoolyard program. Valuable indirect support comes from the Bosque School, a private school in Albuquerque.

One of the management recommendations proposed by the *Bosque Biological Management Plan* was to establish long-term monitoring of populations and processes in the riparian forest (Crawford et al., 1993), a need that has been recognized elsewhere (National Resource Council, 1992; Christensen et al., 1996; Bricker and Ruggiero, 1998). Although several agencies monitor specific components of the ecosystem, such as shallow groundwater and stream flow rates, and a number of research projects have dealt with aspects of its vertebrate communities and wetland vegetation, there was no single coordinated effort to monitor key ecosystem features at sites throughout the valley on a long-term basis. BEMP does that, and currently includes seven sites: Lemitar (near Socorro), Belen, Los Lunas, Albuquerque (three sites), and Santa Ana Pueblo (listed south to north in Figure 3-1). Data collection methods and schedules, which are consistent throughout all sites, include recording shallow groundwater levels, precipitation, temperature, litter production, cottonwood density and growth, understory vegetation composition and cover, other tree and shrub densities, and terrestrial arthropod composition and activity. Data are managed through the University of New Mexico (UNM). Two of the sites are affected by flooding, while the others are not. The use of similar methods and a synchronous collecting schedule gives a unique opportunity for valleywide comparisons.

Equally important to the success of BEMP is public outreach. Citizen volunteers, currently ranging from primary and secondary students to adults, are involved in most of the data collection with help from BEMP staff. BEMP sponsors a Bosque Internship course for undergraduate and graduate students at the University of New Mexico. Interns learn about the *bosque* ecosystem and work with teachers and schoolchildren, who participate in data collection under careful supervision. These activities build a positive relationship between UNM students and an array of young and older citizens, working and learning together. At least two schools now involved with the program integrate the *bosque* fully into their science curriculum, with students ranging from second to sixth grade making at least monthly trips to the river and riparian forest. Some classes have invited students from other schools to visit the *bosque;* the host students shared with their guests their knowledge of the river and riparian forest and demonstrated sampling procedures. Additional outreach activities include an annual student congress, conferences with students and teachers representing different BEMP sites, and presentations to public and private organizations.

Participants in BEMP leave the program with a well-grounded knowledge of the ecology of the system and an appreciation for the scientific methods of data collection intended to meet the needs and goals of resource managers. Participants are also encouraged to inform their own communities about the riparian ecosystem. Most important, many gain a strong sense of ownership and stewardship of the river and the *bosque.* Generation of this attitude should help determine well-thought-out solutions to the pressing questions about the fate of this vulnerable ecosystem that has been, and more than ever remains, of vital importance to the region.

Outlook for the Future

The ever-growing human population along the MRGV continues to put increasing pressures on the river and its *bosque.* Rapid growth and development show no sign of diminishing, and there is little chance of increasing the region's water income without creating new supplies from existing sources (Middle Rio Grande Water Assembly, 2000). Albuquerque municipal water is currently taken exclusively from an underground aquifer, but plans are for the city to begin using river surface water for drinking water by the year 2005 (Anonymous, 2000). Although concern for the river and *bosque* ecosystem has been expressed during the water planning process, socioeconomic pressures threaten to downplay that concern. One impediment to *bosque* restoration involves legal constraints in regard to access to water. Water rights in New Mexico are complex (Shupe and Folk-Williams, 1988; Bokum et al., 1992), with "beneficial use" restrictions applied to water allotments. Until forest restoration and maintenance are designated legally as beneficial uses, and until regional growth and development are brought under more realistic control relative to available resources, efforts to restore the flood pulse to the *bosque* will be impeded. Changes in existing water laws may be needed to implement the most efficient restoration plans. Certainly, the economics of water use will be subject to revision (Brown et al., 1996).

Of utmost importance is continued education of the public, including young schoolchildren through top executives and politicians, about the changing dynamics of the ecosystem in question (Crawford et al., 1996b). Without public and political understanding and support, significant *bosque* restoration will have little hope for success. Continued, long-term monitoring of ecological variables along the river is essential to provide a scientific foundation for restoration efforts. Above all, for future ecosystem change to be positive, there will have to be open communication among all constituencies along the river. Sustaining quality of life along the valley, including the preservation of the river and the *bosque,* will depend on such information transfer. Today, people are talking, and there is a loud voice for the river and the *bosque,* but the future of this unique ecosystem is in no way ensured.

ACKNOWLEDGMENTS

Many people have contributed to the research and activities reviewed in this chapter. We thank the many undergraduate and graduate students and technicians at the University of New Mexico who have contributed work in the field and lab. Generous logistical support was provided by the staff of the Bosque del Apache National Wildlife Refuge. We especially thank M. Baker, C. Dahm, S. Dhillion, T. Kieft, M. Valett, and C. White for collaboration on various aspects of our research at the refuge. Support for our studies since the early 1990s has come from the Bosque Initiative (via the U.S. Fish and Wildlife Service) and the National Science Foundation (several sources). Preparation of this manuscript was partially funded by the National Science Foundation, Award DEB-9903973.

REFERENCES

Adams, E. B., and F. A. Chavez. 1956. *The Missions of New Mexico: A Description by Fray Francisco Atanasio Dominguez with Other Contemporary Documents.* University of New Mexico Press, Albuquerque.

Adis, J. 1997. Terrestrial invertebrates: Survival strategies, group spectrum, dominance and activity patterns. In *The Central Amazon Floodplain,* ed. by W. Junk, pp. 299–317. Springer, New York.

Anderson, B. W., and R. D. Ohmart. 1985. Riparian vegetation as a mitigating process in stream and river restoration. In *The Restoration of Rivers and Streams: Theories and Experience,* ed. by J. A. Gore, pp. 41–80. Butterworth Publishers, Boston.

Anonymous. 2000. *Sustainability.* A Quarterly Publication of the Water Resources Division Public Works Department. City of Albuquerque, NM. **June:**1.

Bahre, C. J. 1985. Wildfire in Southeastern Arizona between 1859 and 1890. *Desert Plants* **7:**190–194.

Baisan, C. H., and T. W. Swetnam. 1997. *Interactions of Fire Regimes and Land Use in the Central Rio Grande Valley.* USDA Forest Service Research Paper RM-RP-330. U.S. Department of Agriculture, Forest Service, Rocky Mountain Forest and Range Experiment Station, Fort Collins, CO.

Bärlocher, F. 1992. Effects of drying and freezing autumn leaves on leaching and colonization by aquatic hyphomycetes. *Freshwater Biology* **28:**1–7.

Barnes, W. 1985. Population dynamics of woody plants on a river island. *Canadian Journal of Botany* **63:**647–655.

Bassman, J., W. Myers, D. Dickmann, and L. Wilson. 1982. Effects of simulated insect damage on early growth of nursery-grown hybrid poplars in northern Wisconsin. *Canadian Journal of Forest Research* **12:**1–9.

Bayley, P. B. 1995. Understanding large river-floodplain ecosystems. *BioScience* **45:**153–158.

Bell, D. T., and S. K. Sipp. 1975. The litter stratum in a streamside forest ecosystem. *Oikos* **26:**391–397.

Bell, D. T., F. L. Johnson, and A. R. Gilmore. 1978. Dynamics of litter fall, decomposition, and incorporation in the streamside forest ecosystem. *Oikos* **30:**76–82.

Bestgen, K. R., and S. P. Platania. 1990. Extirpation of *Notropis simus simus* (Cope) and *Notropis orca* (Pisces: Cyprinidae) from the Rio Grande in New

Mexico, with notes on their life history. *Occasional Papers of the Museum of Southwestern Biology* **6:**1–8.

Bestgen, K. R., and S. P. Platania. 1991. Status and conservation of the Rio Grande silvery minnow, *Hybognathus amarus. Southwestern Naturalist* **36:**225–232.

Blaisdell, J. P., and W. F. Mueggler. 1956. Sprouting of bitterbush (*Purshia tridentata*) following burning or top removal. *Ecology* **37:**365–370.

Blake, T. J., J. S. Sperry, T. J. Tschaplinski, and S. S. Wang. 1996. Water relations. In *Biology of* Populus *and Its Implications for Management and Conservation,* ed. by R. F. Settler, H. D. Bradshaw, Jr., P. E. Heilman, and T. M. Hinckley, pp. 401–422. NRC Press, Ottawa.

Bokum, C., V. Gabin, and P. Morgan. 1992. *Living Within Our Means: A Water Management Policy for New Mexico in the 21st Century.* New Mexico Environmental Law Center, Santa Fe.

Bolton, H. E. 1990. *Coronado: Knight of Pueblos and Plains.* University of New Mexico Press, Albuquerque.

Bond, W. J., and B. W. van Wilgen. 1996. *Fire and Plants.* Chapman and Hall, New York.

Braatne, J. H., S. B. Rood, and P. E. Heilman. 1996. Life history, ecology, and conservation of riparian cottonwoods in North America. In *Biology of* Populus *and Its Implications for Management and Conservation,* ed. by R. F. Settler, H. D. Bradshaw Jr., P. E. Heilman, and T. M. Hinckley, pp. 57–85. NRC Press, Ottawa.

Bricker, O. P., and M. A. Ruggiero. 1998. Toward a national program for monitoring environmental resources. *Ecological Applications* **8:**326–329.

Brock, J. H. 1994. *Tamarix* spp. (salt cedar), an invasive exotic woody plant in arid and semi-arid riparian habitats of western USA. In *Ecology and Management of Invasive Riverside Plants,* ed. by L. C. de Waal, L. E. Child, P. M. Wade, and J. H. Brock, pp. 27–44. John Wiley & Sons, New York.

Brown, B. T., and M. W. Trosset. 1989. Nesting-habitat relationships of riparian birds along the Colorado River in Grand Canyon, Arizona. *Southwestern Naturalist* **34:**260–270.

Brown, F. L., S. C. Nunn, J. W. Shomaker, and G. Woodard. 1996. *The Value of Water: A Report to the City of Albuquerque in Response to RFP-95-010-SV.* Albuquerque. Unpublished report.

Brown, W. O., and R. H. Heim Jr. 1996. *United States June Climate in Historical Perspective.* Climate Variations Bulletin, Historical Climatology Series 4–7, vol. 8 (6). National Climate Data Center, NOAA, Asheville, NC.

Bullard, T. F., and S. G. Wells. 1992. *Hydrology of the Middle Rio Grande from Velarde to Elephant Butte Reservoir, New Mexico.* U.S. Fish and Wildlife Service, Resource Publication 179. U.S. Department of the Interior, Fish and Wildlife Service, Washington, DC.

Burrows, N. D. 1985. Reducing the abundance of *Banksia grandis* in the jarrah forest by the use of controlled fire. *Australian Forestry* **48:**63–70.

Busch, D. E. 1995. Effects of fire on southwestern riparian plant community structure. *Southwestern Naturalist* **40:**259–267.

Busch, D. E., and S. D. Smith. 1993. Effects of fire on water and salinity relations of riparian woody taxa. *Oecologia* **94:**186–194.

Busch, D. E., and S. D. Smith. 1995. Mechanisms associated with decline of woody species in riparian ecosystems of the southwestern U.S. *Ecological Monographs* **65:**347–370.

Chapin, F. S., A. J. Bloom, C. B. Field, and R. H. Waring. 1987. Plant responses to multiple environmental factors. *BioScience* **37:**49–57.

Christensen, N. L., A. M. Bartuska, J. H. Brown, S. Carpenter, C. D'Antonio, R. Francis, J. F. Franklin, J. A. MacMahon, R. F. Noss, D. J. Parsons, C. H. Peterson, M. G.Turner, and R. G. Woodmansee. 1996. The report of the Ecological Society of America Committee on the scientific basis for ecosystem management. *Ecological Applications* **6:**665–691.

Cleverly, J. R., S. D. Smith, A. Sala, and D. A. Devitt. 1997. Invasive capacity of *Tamarix ramosissima* in a Mojave Desert floodplain: The role of drought. *Oecologia* **111:**12–18.

Coleman, J. S., and C. G. Jones. 1988. Plant stress and insect performance: Cottonwood, ozone and a leaf beetle. *Oecologia* **76:**57–61.

Collier, M. P., R. H. Webb, and E. D. Andrews. 1997. Experimental flooding in Grand Canyon. *Scientific American* **276:**82–89.

Cooper, C. F. 1960. Changes in vegetation, structure, and growth of southwestern ponderosa pine since white settlement. *Ecological Monographs* **30:**129–164.

Cox, G. W. 1999. *Alien Species in North America and Hawaii: Impacts on Natural Ecosystems.* Island Press, Washington, DC.

Crawford, C. S., L. M. Ellis, and M. C. Molles Jr. 1996a. The Middle Rio Grande bosque: An endangered ecosystem. *New Mexico Journal of Science* **36:** 276–299.

Crawford, C. S., L. M. Ellis, M. C. Molles Jr., and H. M. Valett. 1996b. The potential for implementing partial restoration of the Middle Rio Grande ecosystem, In *Desired Future Conditions for Southwestern Riparian Ecosystems: Bringing Interests and Concerns Together,* D. W. Shaw and D. M. Finch, tech. coords., pp. 93–99. USDA Forest Service General Technical Report RM-GTR-272. U.S. Department of Agriculture, Forest Service, Rocky Mountain Forest and Range Experiment Station, Fort Collins, CO.

Crawford, C. S., L. M. Ellis, D. Shaw, and N. E. Umbreit. 1999. Restoration and monitoring in the Middle Rio Grande bosque: Current status of flood pulse related efforts. In *Rio Grande Ecosystems: Linking Land, Water and People, Toward a Sustainable Future for the Middle Rio Grande Basin.* D. M. Finch, J. C. Whitney, J. F. Kelly, and S. R. Loftin, tech. coords., pp. 158–163. USDA Forest Service Proceedings RMRS-P-7. U.S. Department of Agriculture, Forest Service, Rocky Mountain Forest and Range Experiment Station, Ogden, UT.

Crawford, C. S., A. C. Culley, R. Leutheuser, M. S. Sifuentes, L. H. White, and J. P. Wilber. 1993. *Middle Rio Grande Ecosystem: Bosque Biological Management Plan.* U.S. Fish and Wildlife Service, District 2, Albuquerque.

Crins, W. J. 1989. The Tamaricaceae in the southeastern United States. *Journal of the Arnold Arboretum* **70:**403–425.

Cummins, K. W. 1974. Structure and function of stream ecosystems. *BioScience* **24:**631–641.

Dahm, C. N., and M. C. Molles Jr. 1992. Streams in semiarid regions as sensitive indicators of global climate change. In *Troubled Waters of the Greenhouse Earth,* ed. by P. Firth and S. Fisher, pp. 250–260. Springer-Verlag, New York.

Dahm, C. N., and D. I. Moore. 1994. The El Niño/Southern Oscillation phenomenon and the Sevilleta Long-Term Ecological Research site. In *El Niño and Long-Term Ecological Research (LTER) Sites,* ed. by D. Greenland, pp. 12–30. Publication No. 18. LTER Network Office, University of Washington, Seattle.

Day, F. P. 1983. Effects of flooding on leaf litter decomposition in microcosms. *Oecologia* **57:**180–184.

DeBano, L. F., and D. G. Neary. 1996. Effects of fire on riparian systems. In *Effects of Fire on Madrean Province Ecosystems: A Symposium Proceedings,* P. F. Ffolliott, L. F. DeBano, M. B. Baker, et al., tech. coords., pp. 69–76. USDA Forest Service General Technical Report RM-GTR-289. U.S. Department of Interior, Forest Service, Rocky Mountain Forest and Range Experiment Station, Fort Collins, CO.

Di Tomaso, J. M. 1998. Impact, biology, and ecology of saltcedar (*Tamarix* spp.) in the southwestern United States. *Weed Technology* **12:**326–336.

Dobyns, H. F. 1981. *From Fire to Flood: Historic Human Destruction of Sonoran Desert Riverine Oases.* Ballena Press, Socorro, NM.

Eichhorst, K. D. 1999. Dynamics of Defoliator Activity Along an Urban to Rural Gradient in the Middle Rio Grande Riparian Corridor. Master's thesis, University of New Mexico, Albuquerque.

Ellis, L. M. 1995. Bird use of saltcedar and cottonwood vegetation in the Middle Rio Grande Valley of New Mexico, U.S.A. *Journal of Arid Environments* **30:**339–349.

Ellis, L. M. 2001. Short-term response of woody plants to fire in a Rio Grande riparian forest, Central New Mexico, U.S.A. *Biological Conservation* **97:** 159–170.

Ellis, L. M., C. S. Crawford, and M. C. Molles Jr. 1997a. Rodent communities in native and exotic riparian vegetation in the Middle Rio Grande Valley of Central New Mexico. *Southwestern Naturalist* **42:**13–19.

Ellis, L. M., C. S. Crawford, and M. C. Molles Jr. 1998. Comparison of litter dynamics in native and exotic riparian vegetation along the Middle Rio Grande of central New Mexico, U.S.A. *Journal of Arid Environments* **38:**283–296.

Ellis, L. M., C. S. Crawford, and M. C. Molles Jr. 2001. Influence of annual flooding on terrestrial arthropod assemblages of a Rio Grande riparian forest. *Regulated Rivers: Research & Management* **17:**1–20.

Ellis, L. M., M. C. Molles Jr., and C. S. Crawford. 1997b. Short-term effects of annual flooding on a population of *Peromyscus leucopus* in a Rio Grande riparian forest of central New Mexico. *American Midland Naturalist* **138:**260–267.

Ellis, L. M., M. C. Molles Jr., and C. S. Crawford. 1999. Influence of experimental flooding on litter dynamics in a Rio Grande riparian forest of central New Mexico. *Restoration Ecology* **7:**1–13.

Ellis, L. M., M. C. Molles Jr., C. S. Crawford, and F. Heinzelmann. 2000. Surface-active arthropod communities in native and exotic riparian vegetation in the Middle Rio Grande Valley, New Mexico, U.S.A. *Southwest Naturalist* **45:** 456–471.

Everitt, B. L. 1968. Use of the cottonwood in an investigation of the recent history of a flood plain. *American Journal of Science* **266:**417–439.

Farley, G. H., L. M. Ellis, J. N. Stuart, and N. J. Scott Jr. 1994. Avian diversity in different-aged stands of riparian forest along the Rio Grande, New Mexico. *Conservation Biology* **8:**1098–1108.

Fowells, H. A. 1965. *Silvics of Forest Trees of the United States.* United States Department of Agriculture, Agricultural Handbook No. 271. United States Government Printing Office, Washington DC.

Fuller, G. D. 1912. Soil moisture in the cottonwood dune association of Lake Michigan. *Botanical Gazette* **53:**512–514.

Galat, D. L., L. H. Fredrickson, D. D. Humburg, K. J. Bataille, J. R. Bodie, J. Dohrenwend, G. T. Gelwicks, J. E. Havel, D. L. Helmers, J. B. Hooker, J. R. Jones, M. F. Knowlton, J. Kinisiak, J. Mazourek, A. C. McColpin, R. B. Renkin, and R. D. Semlitsch. 1998. Flooding to restore connectivity of regulated, large-river wetlands. *BioScience* **48:**721–733.

Gosz, J. R. 1980. Biomass distribution and production budget for nonaggrading forest ecosystems. *Ecology* **61:**507–514.

Gregory, S. V., F. J. Swanson, W. A. McKee, and K. W. Cummins. 1991. An ecosystem perspective of riparian zones. *BioScience* **41:**540–551.

Grimm, N. B., S. G. Fisher, S. V. Gregory, G. R. Marzolf, D. M. McKnight, F. J. Triska, and H. M. Valett. 1997. Sustainability of western watersheds: Nutrients and productivity. In *Aquatic Ecosystems Symposium: A Report to the Western Water Policy Review Advisory Commission,* ed. by W. L. Minckley, pp. 31–43. Arizona State University, Tempe.

Haeuber, R. A., and W. K. Michener. 1998. Policy implications of recent natural and managed floods. *BioScience* **48:**765–772.

Harmon, M. E., and H. Chen. 1991. Coarse woody debris dynamics in two old-growth ecosystems. *BioScience* **41:**604–610.

Harmon, M. E., J. F. Franklin, F. J. Swanson, P. Sollins, S. V. Gregory, J. D. Lattin, N. H. Anderson, S. P. Cline, N. G. Aumen, J. R. Sedell, G. W. Lienkaemper, K. Cromack Jr., and K. W. Cummins. 1986. Ecology of coarse woody debris in temperate ecosystems. *Advances in Ecological Research* **15:** 133–302.

Harris, S. 1998. *The Truth About Silvery Minnows. http://www.coloradorivers.org/Status/Rio%20Grande/minnows.htm.* Colorado Rivers Alliance, Denver.

Heinrichs, E. A. 1988. Global food production and plant stress. In *Plant Stress–Insect Interactions,* ed. by E. A. Heinrichs, pp. 1–33. John Wiley & Sons, New York.

Hobbs, R. J., and H. A. Mooney. 1985. Vegetative regrowth following cutting in the shrub *Baccharis pilularis* spp. *consanguinea* (DC) C.B. Wolf. *American Journal of Botany* **72:**514–519.

Horton, J. S. 1964. *Notes on the Introduction of Deciduous Tamarisk.* USDA Forest Service Research Note RM-16. 7 pp. Fort Collins, CO. U.S. Department of Agriculture, Forest Service, Rocky Mountain Forest and Range Experiment Station.

Howe, W. H., and F. L. Knopf. 1991. On the imminent decline of the Rio Grande cottonwoods in central New Mexico. *Southwestern Naturalist* **36:**218–224.

Hunter, W. C., R. D. Ohmart, and B. W. Anderson. 1988. Use of exotic saltcedar (*Tamarix chinensis*) by birds in arid riparian systems. *Condor* **90:**113–123.

Jones, C. G., and J. S. Coleman. 1988. Plant stress and insect behavior: Cottonwood, ozone and the feeding and oviposition preference of a beetle. *Oecologia* **76:**51–56.

Jones, C. G., J. S. Coleman, and S. Findlay. 1994. Effects of ozone interactions among plants, consumers and decomposers. In *Plant Responses to the Gaseous Environment,* ed. by R. G. Alscher and A. Wellbum, pp. 339–363. Chapman and Hall, London.

Junk, W. J., P. B. Bayley, and R. E. Sparks. 1989. The flood pulse concept in river-floodplain systems. *Canadian Special Publication in Fisheries and Aquatic Sciences* **106:**110–127.

Kayll, A. J., and C. H. Gimingham. 1965. Vegetative regeneration of *Calluna vulgaris* after fire. *Journal of Ecology* **53:**729–734.

Knopf, F. L., and T. E. Olson. 1984. Naturalization of Russian olive: Implications to Rocky Mountain wildlife. *Wildlife Society Bulletin* **12:**289–298.

Knopf, F. L., R. R. Johnston, T. Rich, F. B. Samson, and R. C. Szaro. 1988. Conservation of riparian ecosystems in the United States. *Wilson Bulletin* **100:** 272–284.

Leffler, A. J., L. E. England, and J. Naito. 2000. Vulnerability of Fremont cottonwood (*Populus fremontii* Wats.) individuals to xylem cavitation. *Western North American Naturalist* **60:**204–210.

Lieurance, F. S., H. M. Valett, C. S. Crawford, and M. C. Molles Jr. 1994. Experimental flooding of a riparian forest: restoration of ecosystem functioning. In *Proceedings of the Second International Conference on Ground Water Ecology,* ed. by J. A. Stanford and H. M. Valett, pp. 365–374. American Water Resources Association, Herndon, VA.

Linthicum, L. 2000. Judge tosses safe zone for minnows back to agency. *The Albuquerque Journal,* 23 November 2000.

Mahoney, J. M., and S. B. Rood. 1991. A device for studying the influence of declining water table on poplar growth and survival. *Tree Physiology* **8:**305–314.

Mahoney, J. M., and S. B. Rood. 1992. Response of hybrid poplar to water table declines in different substrates. *Forest Ecology and Management* **54:**141–156.

Marra, J. L., and R. L. Edmonds. 1994. Coarse woody debris and forest floor respiration in an old-growth coniferous forest on the Olympic Peninsula, Washington, U.S.A. *Canadian Journal of Forest Research* **24:**1811–1817.

Martius, C. 1997. Decomposition of wood. In *The Central Amazon Floodplain,* ed. by W. Junk, pp. 267–276. Springer, New York.

Means, M. D., and D. M. Finch. 1999. Bird migration through Middle Rio Grande riparian forests, 1994 to 1997. In *Rio Grande Ecosystems: Linking Land, Water and People, Toward a Sustainable Future for the Middle Rio Grande Basin,* D. M. Finch, J. C. Whitney, J. F. Kelly, and S. R. Loftin, tech. coords., pp. 191–196. USDA Forest Service Proceedings RMRS-P-7. U.S. Department of Agriculture, Forest Service, Rocky Mountain Forest and Range Experiment Station, Ogden, UT.

Michener, W. K., and R. A. Haeuber. 1998. Flooding: Natural and managed disturbances. *BioScience* **48:**677–680.

Middle Rio Grande Water Assembly. 2000. Living within our water budget. *Middle Rio Grande Water Chronicle* **June:** 1–3.

Miller, W. E. 1983. Decomposition rates of aspen bole and branch litter. *Forest Science* **29:**351–356.

Molles, M. C., Jr., and C. N. Dahm. 1990. A perspective on El Niño and La Niña: Global implications for stream ecology. *Journal of the North American Benthological Society* **91:**68–76.

Molles, M. C., C. N. Dahm, and M. T. Crocker. 1992. Climatic variability and streams and rivers in semi-arid regions. In *Aquatic Ecosystems in Semi-Arid Regions: Implications for Resource Management,* ed. by R. D. Robarts and M. L. Bothwell, pp. 197–202. NHRI Symposium Series 7. Environment Canada, Saskatoon.

Molles, M. C., Jr., C. S. Crawford, L. M. Ellis, H. M. Valett, and C. N. Dahm. 1998. Managed floods: Restoration of riparian forest ecosystem structure and function along the Middle Rio Grande. *BioScience* **48:**749–756.

Mooney, H. A., and S. L. Gulmon. 1982. Constraints on leaf structure and function in reference to herbivory. *BioScience* **32:**198–206.

Morgan, P., and L. F. Neuenschwander. 1988. Shrub response to high and low severity burns following clearcutting in Northern Idaho. *Western Journal of Applied Forestry* **3:**5–9.

Morrison, K. D., and E. G. Reekie. 1995. Pattern of defoliation and its effect on photosynthetic capacity in *Oenothera biennis*. *Journal of Ecology* **83:** 759–767.

Mund-Meyerson, M. J. 1998. Arthropod Abundance and Composition on Native and Exotic Trees in the Middle Rio Grande Riparian Forest as Related to Avian Foraging. Master's thesis, University of New Mexico, Albuquerque.

Mutch, R. W. 1970. Wildland fires and ecosystems: A hypothesis. *Ecology* **51:**1046–1051.

Naiman, R. J., and H. Décamps. 1997. The ecology of interfaces: Riparian zones. *Annual Review of Ecology and Systematics* **28:**621–658.

National Resource Council. 1992. *Restoration of Aquatic Ecosystems: Science, Technology and Public Policy.* National Academy Press, Washington, DC.

Nicholls, N. 1988. El Niño-Southern Oscillation and rainfall variability. *Journal of Climate* **1:**418–421.

Ohmart, R. D., and B. W. Anderson. 1982. North American desert riparian ecosystems, In *Reference Handbook on the Deserts of North America,* ed. by G. L. Bender, pp. 433–479. Greenwood Press, Westport, CT.

Olson, T. E., and F. L. Knopf. 1986. Naturalization of Russian-olive in the western United States. *Western Journal of Applied Forestry* **1:**65–69.

Peterson, D. L., and Rolfe, G. L. 1982. Nutrient dynamics and decomposition of litterfall in floodplain and upland forests of Central Illinois. *Forest Science* **28:**667–681.

Platania, S. P. 1991. Fishes of the Rio Chama and Upper Rio Grande, New Mexico, with preliminary comments on their longitudinal distribution. *Southwestern Naturalist* **36:**186–193.

Platania, S. P., and C. S. Altenbach. 1998. Reproductive strategies and egg types of seven Rio Grande Basin cyprinids. *Copeia* **1998:**559–569.

Pockman, W. T., J. S. Sperry, and J. W. O'Leary. 1995. Sustained and significant negative water pressure in xylem. *Nature* **378:**715–716.

Poff, N. L., J. D. Allan, M. B. Bain, J. R. Karr, K. L. Prestegaard, B. D. Richter, R. E. Sparks, and J. C. Stromberg. 1997. The natural flow regime: A paradigm for river conservation and restoration. *BioScience* **47:**769–784.

Poiani, K. A., B. D. Richter, M. G. Anderson, and H. E. Richter. 2000. Biodiversity conservation at multiple scales: Functional sites, landscapes, and networks. *BioScience* **50:**133–146.

Proctor, J. T. A., J. M. Bodnar, W. J. Blackburn, and R. L. Watson. 1982. Analysis of the effects of the spotted tentiform leafminer (*Phyllonorycter blancardella*) on the photosynthetic charactersitics of apple leaves. *Canadian Journal of Botany* **60:**2734–2740.

Pyne, S. J. 1982. *Fire in America: A Cultural History of Wildland and Rural Fire.* Princeton University Press, Princeton, NJ.

Quesada, M., K. Bollman, and A. G. Stephenson. 1995. Leaf damage decreases pollen production and hinders pollen performance in *Cucurbita texana. Ecology* **76:**437–443.

Raitt, R. J., M. C. Delesantro, and T. G. Marr. 1980. *Avifauna Census, Elephant Butte and Caballo Reservoirs, New Mexico, Phase II.* Final report submitted to U.S. Bureau of Reclamation, Amarillo, TX. Unpublished report.

Rayner, A. D. M., and N. K. Todd. 1979. Population and community structure and dynamics of fungi in decaying wood. *Advances in Botanical Research* **7:**333–420.

Reily, P. W., and W. C. Johnson. 1982. The effects of altered hydrologic regime on tree growth along the Missouri River in North Dakota. *Canadian Journal of Botany* **60:**2410–2423.

Richter, B. D., J. V. Baumgartner, R. Wigington, and D. P. Braun. 1997. How much water does a river need? *Freshwater Biology* **37:**231–249.

Robertson, A. I., S. E. Bunn, P. I. Boon, and K. F. Walker. 1999. Sources, sinks and transformations of organic carbon in Australian floodplain rivers. *Marine and Freshwater Research* **50:**813–829.

Rood, S. B., and S. Hinze-Milne. 1989. Abrupt downstream forest decline following river damming in southern Alberta. *Canadian Journal of Botany* **67:** 1744–1749.

Rood, S. B., and J. M. Mahoney. 1990. Collapse of riparian poplar forests downstream from dams in western prairies: Probable causes and prospects for mitigation. *Environmental Management* **14:**451–464.

Rood, S. B., C. Hillman, T. Sanche, and J. M. Mahoney. 1994. Clonal reproduction of riparian cottonwoods in southern Alberta. *Canadian Journal of Botany* **72:**1766–1774.

Ropelewski, C. F., and M. S. Halpert. 1986. North American precipitation and temperature patterns associated with the El Niño / Southern Oscillation (ENSO). *Monthly Weather Review* **114:**2352–2362.

Ropelewski, C. F., and M. S. Halpert. 1987. Global and regional scale precipitation patterns associated with the El Niño / Southern Oscillation. *Monthly Weather Review* **115:** 1606–1626.

Sauer, C. O. 1950. Grassland climax, fire, and man. *Journal of Range Management* **3:**16–21.

Schier, G. A., and R. B. Campbell. 1976. Differences among *Populus* species in ability to form adventitious shoots and roots. *Canadian Journal of Forest Research* **6:**253–261.

Schmidt, J. C., R. H. Webb, R. A. Valdez, G. R. Marzolf, and L. W. Stevens. 1998. Science and values in river restoration in the Grand Canyon. *BioScience* **48:**735–747.

Scott, M. L., G. T. Aubel, and J. M. Friedman.1997. Flood dependency of cottonwood establishment along the Missouri River, Montana, U.S.A. *Ecological Applications* **7:**677–690.

Scott, M. L., M. A. Wondzell, and G. T. Auble. 1993. Hydrograph characteristics relevant to the establishment and growth of western riparian vegetation. In *Proceedings of the 13th Annual American Geophysical Union Hydrology Days,* ed. by H. J. Morel-Seytoux, pp. 237–246. Hydrology Days Publications, Atherton, CA.

Scurlock, D. 1998. *From the Rio to the Sierra: An Environmental History of the Middle Rio Grande Basin.* USDA Forest Service General Technical Report RMRS-GTR-5. U.S. Department of Agriculture, Forest Service, Rocky Mountain Research Station. Fort Collins, CO.

Segelquest, C. A., M. L. Scott, and G. T. Aubel. 1993. Establishment of *Populus deltoides* under simulated alluvial groundwater declines. *American Midland Naturalist* **130:**274–285.

Shafroth, P. B., G. T. Aubel, and M. L. Scott. 1995. Germination and establishment of the native plains cottonwood (*Populus deltoides* Marshall supsp. *monilifera*) and the exotic Russian-olive (*Eleagnus angustifolia* L.). *Conservation Biology* **9:**1169–1175.

Sher, A. A., D. L. Marshall, and S. A. Gilbert. 2000. Competition between native *Populus deltoides* and invasive *Tamarix ramosissima* and the implications for reestablishing flooding disturbance. *Conservation Biology* **14:**1744–1754.

Shupe, S. J., and J. Folk-Williams. 1988. *The Upper Rio Grande: A Guide to Decision Making.* Western Network, Santa Fe, NM.

Shure, D. J., M. R. Gottschalk, and K. A. Parsons. 1986. Litter decomposition processes in a floodplain forest. *American Midland Naturalist* **115:**314–327.

Slack, J. R., A. M. Lumb, and J. M. Landwehr. 1993. *Hydro-Climatic Data Network (HCDN) Streamflow Data Set, 1874–1988.* U.S. Geological Survey Water-Resources Investigations Report 93-4076. U.S. Geological Survey, Reston, VA.

Smith, J. K., and W. C. Fischer. 1997. *Fire Ecology of the Forest Habitat Types of Northern Idaho.* General Technical Report INT-GTR-363. U.S. Department of Agriculture Forest Service, Intermountain Research Station, Ogden, UT.

Sparks, R. E. 1995. Need for ecosystem management of large rivers and their floodplains. *BioScience* **45:**168–182.

Sparks, R. E., J. C. Nelson, and Y. Yin. 1998. Naturalization of the flood regime in regulated rivers. *BioScience* **48:**706–720.

Sparks, R. E., P. B. Bayley, S. L. Kohler, and L. L. Osborne. 1990. Disturbance and recovery of large floodplain rivers. *Environmental Management* **14:** 699–709.

Stromberg, J. C. 1993. Fremont cottonwood-Goodding willow riparian forests: A review of their ecology, threats, and recovery potential. *Journal of the Arizona-Nevada Academy of Science* **26:**97–110.

Stromberg, J. C., and D. T. Patten. 1991. Instream flow requirements for cottonwoods at Bishop Creek, Inyo County, California. *Rivers* **2:**1–11.

Stromberg, J. C., and D. T. Patten. 1992. Mortality and age of black cottonwood stands along diverted and undiverted streams in the eastern Sierra Nevada, California. *Madroño* **39:**205–223.

Stromberg, J. C., and D. T. Patten. 1996. Instream flow and cottonwood growth on the eastern Sierra Nevada of California, U.S.A. *Regulated Rivers: Research and Management* **12:**1–12.

Stromberg, J. C., D. T. Patten, and B. D. Richter. 1991. Flood flows and dynamics of Sonoran riparian forests. *Rivers* **2:**221–235.

Stuever, M. 1997. Fire induced mortality of Rio Grande Cottonwood. Master's thesis, University of New Mexico, Albuquerque.

Stuever, M. C., C. S. Crawford, M. C. Molles Jr., C. S. White, and E. Muldavin. 1997. Initial assessment of the role of fire in the Middle Rio Grande bosque. In *Proceedings: First Conference on Fire and Endangered Species and Habitats,* ed. by J. M. Greenlee, pp. 275–283. International Association of Wildland Fire, Fairfield, WA.

Szaro, R. C. 1991. Wildlife communities of south-western riparian ecosystems. In *Wildlife and Habitats in Managed Landscapes,* ed. by J. E. Rodiek and E. G. Bolen, pp. 174–201. Island Press, Washington, DC.

Tainter, J. A., and B. B. Tainter. 1996. Riverine settlement in the evolution of prehistoric land-use systems in the Middle Rio Grande Valley, New Mexico. In *Desired Future Conditions for Southwestern Riparian Ecosystems: Bringing Interests and Concerns Together,* D. W. Shaw and D. M. Finch, tech. coords. USDA Forest Service General Technical Report RM-GTR-272. U.S. Dept. of Agriculture, Forest Service, Rocky Mountain Forest and Range Experiment Station, Fort Collins, Colorado, pp. 22–32.

Taylor, J. P., and K. C. McDaniel. 1998. Restoration of saltcedar (*Tamarix* sp.)-infested floodplains on the Bosque del Apache National Wildlife Refuge. *Weed Technology* **12:**345–352.

Toth, L. A. 1995. Principles and guidelines for restoration of river/floodplain ecosystems—Kissimmee River, Florida. In *Rehabilitating Damaged Ecosystems* (2d ed), ed. by J. Cairns Jr., pp. 49–73. Lewis Publishers/CRC Press, Boca Raton, FL.

Toth, L. A., S. L. Melvin, D. A. Arrington, and J. Chamberlain. 1998. Hydrologic manipulations of the channelized Kissimmee River. *BioScience* **48:**757–764.

Tyree, M. T., and M. A. Dixon. 1986. Water stress induced cavitation and embolism in some woody plants. *Physiologia Plantarum* **66:**397–405.

Tyree, M. T., and F. W. Ewers. 1991. The hydraulic architecture of trees and other woody plants. *New Phytologist* **119:**345–360.

Tyree, M. T., J. Alexander, and J. Machado. 1992. Loss of hydraulic conductivity due to water stress in intact juveniles of *Quercus rubra* and *Populus deltoides. Tree Physiology* **10:**411–415.

Tyree, M. T., K. J. Kolb, S. B. Rood, and S. Patiño. 1994. Vulnerability to drought-induced cavitation of riparian cottonwoods in Alberta: A possible factor in the decline of the ecosystem? *Tree Physiology* **14:**455–466.

U.S. Department of the Interior. 1994. Endangered and threatened wildlife and plants: Final rule to list the Rio Grande silvery minnow as an endangered species. *Federal Register* **59:**36988–36995.

U.S. Department of the Interior. 1999. Endangered and threatened wildlife and plants: Final designation of critical habitat for the Rio Grande silvery minnow. *Federal Register* **64:**36274–36290.

U.S. Fish and Wildlife Service. 1999. *Rio Grande Silvery Minnow Recovery Plan.* Albuquerque, NM.

U.S. Geological Survey. 2000. *Rio Grande and Albuquerque, NM. http://waterdata.usgs.gov/nwis-w/NM/index.cgi?statnum=0830000.* United States Geological Survey, Albuquerque.

Van Cleave, M. 1935. Vegetative changes in the Middle Rio Grande Conservancy District. Master's thesis, University of New Mexico, Albuquerque.

Walker, K. F., F. Sheldon, and J. T. Puckridge. 1995. A perspective on dryland river ecosystems. *Regulated Rivers: Research & Management* **11:**85–104.

Waring, G. L., and N. S. Cobb. 1992. The impact of plant stress on herbivore population dynamics. In *Insect-Plant Interactions,* vol. 4, ed. by L. Bernays, pp. 167–226. Academic Press, New York.

Webster, J. R., and E. F. Benfield. 1986. Vascular plant breakdown in freshwater ecosystems. *Annual Review of Ecology and Systematics* **17:**567–594.

Whelan, R. J. 1995. *The Ecology of Fire.* Cambridge University Press, New York.

Wilson, R. 1970. Succession in stands of *Populus deltoides* along the Missouri River in southwestern South Dakota. *American Midland Naturalist* **83:** 330–342.

Wozniak, F. E. 1996. Human impacts on riparian ecosystems of the Middle Rio Grande Valley during historic times. In *Desired Future Conditions for Southwestern Riparian Ecosystems: Bringing Interests and Concerns Together,* D. W. Shaw and D. M. Finch, tech. coords., pp. 33–43. USDA Forest Service General Technical Report RM-GTR-272. U.S. Department of Interior, Forest Service, Rocky Mountain Forest and Range Experiment Station, Fort Collins, CO.

Yong, W., and D. M. Finch. 1997. Population trends of migratory landbirds along the middle Rio Grande. *Southwestern Naturalist* **42:**137–147.

Zasada, J. 1986. Natural regeneration of trees and tall shrubs on forest sites in interior Alaska. In *Forest Ecosystems in the Alaska Taiga: A Synthesis of Structure and Function.* ed. by K. Van Cleve, F. S. Chapin III, P. W. Flanagan, L. A. Viereck, and C. T. Dyrness, pp. 44–73. Springer-Verlag, New York.

4

The Role of the Flood Pulse in Maintaining Boltonia decurrens, *a Fugitive Plant Species of the Illinois River Floodplain: A Case History of a Threatened Species*

M. Smith

Southern Illinois University, Edwardsville, Illinois

P. Mettler

Southern Illinois University, Carbondale, Illinois

Over the past 100 years, the natural hydrologic regime of the river-floodplain system of the Illinois River Valley has been altered by human intervention. The diversion of water from Lake Michigan, the construction of

Flood Pulsing in Wetlands: Restoring the Natural Hydrological Balance, edited by Beth A. Middleton.
ISBN 0 471-41807-2

a series of navigation dams, and the development of agricultural levees have disrupted the annual flood pulse regime, threatening the biota that evolved in synchrony with the system.

Boltonia decurrens, a herbaceous species that is endemic to the Illinois River, is threatened with extinction. A fugitive plant that requires open areas for population establishment, it developed characteristics over evolutionary time that enabled it to colonize the vast areas of the floodplain that were available following the recession of the annual spring floods. Tolerance for hypoxic soils and long periods of inundation provided a critical advantage under the conditions present in the historical floodplain system. Formerly abundant in contiguous populations along the entire lower Illinois River Valley, it now occurs in disjunct populations separated by the three navigation dams on the river. Seed dispersal has been disrupted by the navigation dams, and suitable habitat has been reduced as a result of the construction of agricultural levees. Boltonia decurrens has survived by exploiting areas disturbed by human activity—moving from one area to another as a site becomes unsuitable through vegetative succession, agricultural use of the land, or industrial or urban development.

Although the species is listed as threatened by state and federal authorities, it is inadequately protected. A Federal Recovery Plan was developed for B. decurrens in 1990 to provide a strategy for managing populations on state and federal land; however, a lack of data concerning its life history requirements resulted in recommendations that were inappropriate. Conservation efforts have sought to preserve selected populations, but because of its fugitive nature, only habitat restoration will ensure its survival.

Economic and political factors combine to make the restoration of a natural flood pulse regime to the Illinois river-floodplain system unlikely under present conditions; however, because of the economic devastation following the Midwest flood of 1993, there is growing awareness of the futility of continuing the current policies regulating flood control on the Illinois River. Government agencies, private conservation organizations, and citizen groups are considering alternatives to the continued reliance on levee systems to provide protection for residential and commercial development in the floodplain. There is general consensus that human utilization of natural resources is a component of the current environment, and planned strategies must seek to improve functional attributes of the ecosystem within the constraints of continued development and use by humans. At present, there is little recognition of the value of native species, beyond those that are of economic or recreational interest. Despite some changes in attitude toward managing river systems, efforts to restore the natural flood pulse regime to the Illinois River, which would require extensive cooperation and concession on the part of government agencies, landholders, and private interest groups, are not to be expected in the near future. To date, private conservation organizations are the only groups targeting the preservation of threatened native species and planning for the restoration of the natural hydrology to portions of the Illinois River.

Flooding as an integral part of natural river-floodplain ecosystems is best expressed by the "flood pulse" concept, which identifies the predictable advance and retraction of water on the floodplain as the principal agent controlling the adaptations and composition of the biota (Junk et al., 1989; Bayley, 1991; Bayley, 1995; Sparks, 1995). The flood pulse maintains river-floodplain systems in dynamic equilibrium and should not be defined as either a "perturbance" or "disturbance," as described by Vogl (1980), Sousa (1984), and White and Pickett (1985), who refer to unpredictable, brief, and violent discrete events as disturbances. When systems and their biota have been dependent on repetitive events and have evolved in synchrony with them, the so-called perturbations are part of the natural system (Vogl 1980). A system characterized by a regular flood pulse, with species adapted to this regime, is disturbed when there are significant departures from the average hydrological regime (Bayley, 1995), such as the prevention of floods or a catastrophic flood event, as occurred in the Midwest in 1993. Predictable, systemwide flood pulses, therefore, are not to be considered disturbances to the regime, but part of the natural order (Sparks et al., 1990; Bayley, 1991, 1995). This chapter refers to discrete events, natural or man-made, that disrupt the ecosystem, community, or population structure (Sparks et al., 1990) as disturbances to the Illinois River-floodplain system. These include anthropogenic disturbances, whether or not they were initiated by conservation concerns, and amplified, reduced, or mistimed floods (Sparks et al., 1990, 1998; Sparks, 1995). An understanding of the relationship between the organisms of the Illinois River Valley and the historical flood pulse regime helps to explain the endangered status of many Illinois River endemics that have evolved to require the predictable, moderate flood events natural to the river system (Bellrose et al., 1983; Sparks, 1984).

Fugitive species are defined by Hutchinson (1951) as successional species that are always on the move, doomed to extinction in their present habitat as they succumb to competitive pressure and surviving by escaping and reestablishing in other areas as a new niche opens. These species depend on frequent fluctuations in habitat to provide refuges in which to establish new populations (Hutchinson, 1951; Harper, 1977). The natural water level fluctuations observed in large river-floodplain ecosystems create and maintain an early successional environment (Junk et al., 1989; Bayley, 1991) ideal for the establishment and persistence of fugitive species. The historical cycle of annual, late winter, and early spring flooding of the Illinois River provided the mechanism that created this habitat (Sparks, 1995). Established vegetation was removed by the inundation of floodwater, and subsequent drawdown of the floodplain wetlands renewed native herbaceous wetland vegetation. When the natural hydrologic regime is altered, as in the Illinois River Valley (Sparks, 1995; Sparks et al., 1998), this habitat can be fragmented and destroyed, resulting in significant effects on the floodplain flora and fauna. The modification of flooding and flood characteristics of the Illinois River has reduced habitat availability for fugitive species (Smith et al., 1998; Sparks et al., 1998; Sparks and Spink, 1998), which are particularly sensitive to habitat alteration and loss (Hutchinson, 1951).

THE FLOOD PULSE AND *BOLTONIA DECURRENS*

The case of *Boltonia decurrens,* a federally listed threatened species, is an excellent example of the consequences of disruption of the natural flooding regime for species adapted to dynamic, riverine habitats. *Boltonia decurrens,* a fugitive species, is endemic to the floodplains of the Illinois River and the area of its confluence with the Mississippi (Schwegman and Nyboer, 1985; U.S. Fish and Wildlife Service, 1990; Smith and Keevin, 1998). In spite of prolific seed production and the ability to reproduce vegetatively (Schwegman and Nyboer, 1985; Smith and Keevin, 1998; Smith et al., 1998), the number of naturally occurring populations, which fluctuates annually, has declined over the past 100 years (Schwegman and Nyboer, 1985; U.S. Fish and Wildlife Service, 1990). In 1988, the U.S. Fish and Wildlife Service placed *B. decurrens* on the federal list of threatened species. It is currently listed as endangered in Missouri (Missouri Department of Conservation, 1999) and threatened in Illinois (Herkert, 1991).

Boltonia decurrens is an early successional species that requires either natural or suitable human disturbance to create and maintain habitat (Schwegman and Nyboer, 1985; U.S. Fish and Wildlife Service, 1990; Smith et al., 1998). It appears that *B. decurrens* originally grew in wet prairies, in shallow marshes, and on the open shores of creeks and lakes of the Illinois River (Schwegman and Nyboer, 1985). Its habitat was described by its first collector, Charles Wilkins Short, as "wet prairies of Illinois" (Torrey and Gray, 1841). It was associated with bottomland prairies and lakeshores in the area of Illinois across the Mississippi from St. Louis (Hus, 1908), and labels from nineteenth-century herbarium collections indicate that it grew in contiguous populations on the shores of backwater lakes along the Illinois River (U.S. Fish and Wildlife Service, 1990). In 1934, Turner recorded the presence of *Boltonia asteroides* var. *decurrens* in low floodplain prairies in the lower Illinois River Valley. As recently as 1949, it was collected from marshland along the Illinois River (Schwegman and Nyboer, 1985). Collections since 1970 are limited to disturbed alluvial ground near the Illinois and its confluence with the Mississippi River (Morgan, 1980) and open, muddy edges of floodplain forests (Kurz, 1981). Populations now occur in three clusters (Figure 4-1) that are associated with pools created by navigation dams—Peoria and La Grange on the Illinois River and Melvin Price (Lock and Dam 26) on the Mississippi River (Figure 4-1) (U.S. Fish and Wildlife Service, 1990; Smith et al., 1998). It is the general consensus that the threatened status of *B. decurrens* is due to a reduction in suitable habitat (U.S. Fish and Wildlife Service, 1990; Smith et al., 1998).

Under ideal growing conditions, mature plants of *B. decurrens* can reach a height of c. 2.5 m (Figure 4-2); however, if light is limited or seed germination is delayed by late summer floods, plants often reproduce and senesce when less than 0.5 m. Plants flower from August through October, and seeds mature during October and November. If environmental conditions are satisfactory, basal rosettes form as plants senesce in October (Figure 4-3), and bolt and flower the following year (Schwegman and Nyboer, 1985; Smith and Keevin, 1998). Information collected in a series of field and laboratory studies (Smith, 1990, 1991, 1993, 1995; Smith and Keevin, 1998) has resulted in the development of a life cycle model for the species (Figure 4-4). Although *B. decurrens* is a perennial under optimal conditions, various life stages may behave as winter annuals (winter seedlings bolt, flower, and senesce the following summer), as summer an-

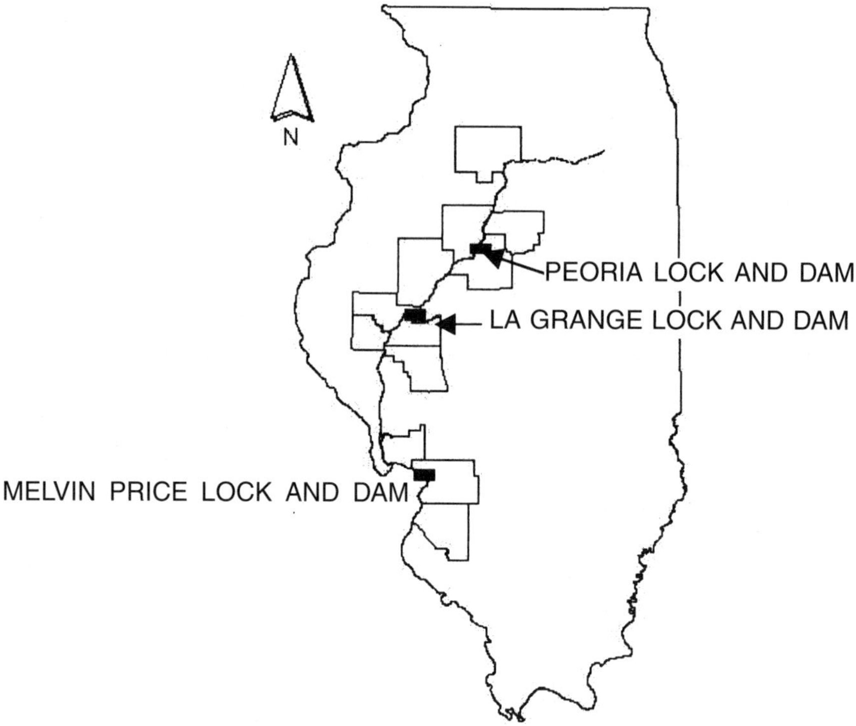

Figure 4-1. *Illinois counties containing populations of* Boltonia decurrens: *from North to South on the west side of the river (Bureau, Peoria, Fulton, Schuyler, and Brown) and from North to South on the east side of the river (Woodford, Tazewell, Scott, Morgan, Jersey, and St. Clair). Navigation dams that affect the distribution of populations of* B. decurrens: *Peoria and La Grange on the Illinois River and Lock and Dam 26 on the Mississippi River.*

nuals (summer seedlings bolt, flower, and senesce the same summer), or as biennials (ca. 25 percent of spring seedlings overwinter as rosettes that bolt, flower, and senesce the following summer). In any situation, an individual senesces after bolting and dies during the same year, leaving no persistent root stock; however, under favorable conditions, rosettes are formed at the stem base and become nutritionally independent from the dying parent before winter. Field observations (Schwegman and Nyboer, 1985; U.S. Fish and Wildlife Service, 1990) indicate that in an established population, seedling recruitment is rare and the species depends on vegetative reproduction for the population to persist.

Figure 4-2. *Marian Smith with a mature* Boltonia decurrens *plant in September 1992. (Photograph by Thomas Keevin; used by permission.)*

Boltonia decurrens is a prolific seed producer (ca. 30,000 viable seeds per mature individual) (Smith and Keevin, 1998) and can produce up to 15 vegetative ramets at the base of a large senescing flowering plant (Redmond, 1993; Baker, 1997; Smith and Keevin, 1998). Its reproductive potential, however, is seldom realized. Seedling establishment in the field is often low or lacking entirely (Smith and Keevin, 1998), and vegetative rosettes are not produced in extremely dry years (M. Smith, personal observation) or when succession has been uninterrupted for several years (U.S. Fish and Wildlife Service, 1990). *Boltonia decurrens* requires open wetland areas for population establishment. Immediately following a disturbance in a suitable floodplain site, *B. decurrens* colonizes and often

Figure 4-3. *Vegetatively produced rosette at the base of a senescent mother plant. (Photograph by Nancy Parker; used by permission.)*

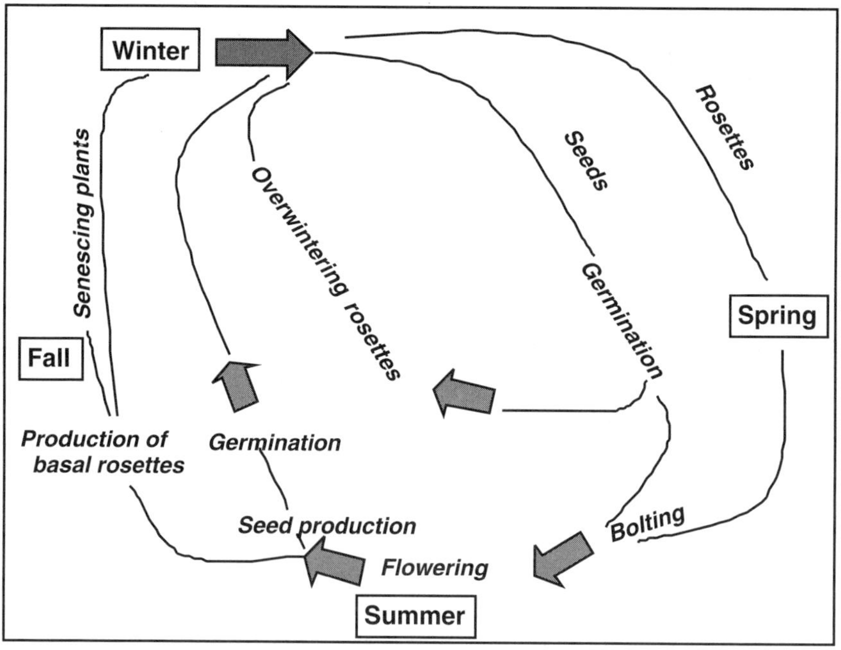

Figure 4-4. *Model of life cycle for* Boltonia decurrens.

dominates the vegetation; however, field observations indicate that in areas without a subsequent disturbance, the species is eliminated by competition within three to five years (U.S. Fish and Wildlife Service, 1990). Populations are revitalized or new ones are established when a disturbance clears a site and seedlings are generated from the soil seed bank or from seeds that have been deposited by receding floodwaters (Smith and Keevin, 1998).

The role of flooding in seed dispersal, germination, and seedling recruitment has been firmly established for *B. decurrens* in the past ten years. Field observations of the pattern of seedling establishment in clearly discernable rows on the shores of backwater lakes (T. Keevin, (U.S. Army Corps of Engineers), and M. Smith, personal observation), and laboratory studies of seed flotation and germination (Smith and Keevin, 1998), suggest that dispersal occurs by floodwater; as the water level recedes, seeds are deposited on the shores and seedlings become established in the bare mudflats (Figure 4-5). In 1994, after the Midwest flood of 1993 had cleared existing vegetation from vast areas of the Illinois floodplain, populations of *B. decurrens* that had been near extirpation exploded in size. At

Figure 4-5. *Spring 1994. Seedlings that became established on the shoreline of Meredosia Lake, Morgan County, Illinois, following the recession of floodwaters in summer 1990. (Photograph by Nancy Parker; used by permission.)*

Gilbert Lake in Jersey County, Illinois (2.6 km from the river's confluence with the Mississippi River), a population that had consisted of 50 flowering individuals in 1992 increased to more than 20,000 individuals in 1994 (Smith, 1995; Smith et al., 1998). Because of the long duration of the flood, no individuals were present at the site in 1993; therefore, the increase in population size must be attributed to seedling establishment. Whether seedlings were generated from the soil seed bank or from seeds brought in by floodwaters is uncertain; however, given the severe scouring that occurred at Gilbert Lake during the flood that removed much of the topsoil that might have held a seed bank (Smith, 1995), it is likely that seeds washed in from some of the northern populations. Because of a flood gradient along the Illinois River (increasing in severity as the river approached the area of confluence with the Mississippi) (Koellner, 1996; Smith et al., 1998), populations of *B. decurrens* north of Peoria, Illinois, produced flowering individuals that set seed in the fall of 1993 (Smith, 1995; Smith et al., 1998). Seed dispersal would have been unhampered by navigation dams, because during periods of extremely high flow the Illinois is essentially a free-flowing river due to the removal of dam wickets and gates from the channel during flooding (Sparks, 1996).

Following a major disturbance that completely eliminates flowering individuals, recolonization depends on a large influx of seeds from a nearby source (Egler, 1954) or a substantial soil seed bank (Archibold, 1978). When seed production from native species is reduced or absent, riparian habitats may be invaded by exotics (Decamps et al., 1988). Disruptions in the natural hydrologic cycle have the potential to disrupt seed production and dispersal of native herbaceous species, thereby reducing the contribution of native species to the seed bank (Decamps et al., 1988). In the absence of seed production or a seed bank, recolonization of a site by native plants is entirely dependent on an influx of seeds from other populations. In the case of *B. decurrens,* the disruption of seed production has become particularly problematic: Populations that were once contiguous are now isolated (Schwegman and Nyboer, 1985; U.S. Fish and Wildlife Service, 1990; Smith and Keevin, 1998), thus reducing the potential for recolonization of a destroyed site by seeds from neighboring populations. A significant influx of seeds is limited to years with extreme hydrologic events, such as occurred in 1993 when levees were overtopped or destroyed and the continuity of the river-floodplain system was restored (Sparks, 1996). As seeds of *B. decurrens* float for long periods of time (Stoecker et al., 1995;

Smith and Keevin, 1998), the widespread flooding in 1993 may have provided for the dispersal of seeds to new and former population sites. Field observations support this assumption, as seven new populations were reported in 1994, and several former sites experienced population booms created entirely by seedlings.

Flooding provides a regime that creates the high light environment required by *B. decurrens* for germination (Baskin and Baskin, 1988; Smith et al., 1995; Smith and Keevin, 1998) and growth (Smith et al., 1993). Seeds germinate readily on the surface of either water or moist soil if they are exposed to light; however, when covered by as little as 1 mm of silt, germination does not occur (Smith and Keevin, 1998). Even moderate shading from litter on the soil surface reduces germination by ca. 66% (Smith et al., 1995). Since 1910 the rate of sedimentation in the Illinois River has doubled, primarily due to increased row crop cultivation in the watershed (Lee and Bhomik, 1979), reducing water clarity and resulting in the deposition of silt on seasonally flooded shores of the river and backwater lakes (Lee and Stall, 1976; Lee and Bhomik, 1979). Areas formerly ideal for the establishment of seedlings of *B. decurrens* are now unsuitable for seed germination and population establishment. This situation was dramatically illustrated late in the summer of 1994, when heavy layers of silt, deposited by the 1993 flood, cracked and exposed the preflood soil surface. Seeds of *B. decurrens* were exposed and germinated, with rosettes emerging from the crevices in August (Figure 4-6). Large areas between the cracks were free of seedlings, but by September a small population of *B. decurrens* rosettes had begun to bolt. Unfortunately, germination occurred so late in the season that few individuals flowered and produced seeds (M. Smith, personal observation).

An unexpected benefit of the 1993 flood was the "flushing" of sediment from Illinois floodplain lakes and shores, resulting in a clarification of river water by 1994 (Soong and Ettinger, 2000). The observed increase in numbers of populations of *B. decurrens* along the Illinois River since 1993 (Smith, 1995) may be partially explained by this temporary improvement in water quality; however, as the system of navigation dams and agricultural practices have changed little since the flood, water quality can be expected to decline in the future (Mills et al., 1966; Bellrose et al., 1983)

Because of the alteration of the hydrology of the Illinois River, the floodplain area exposed during periods of low flow has been reduced (Bellrose et al., 1979). Although germination of *B. decurrens* readily occurs while seeds

Figure 4-6. *Seedling of* Boltonia decurrens *that germinated in a gap produced by the cracking of the silt layer that was deposited during the flood of 1993. (Photograph by Nancy Parker; used by permission.)*

are floating on floodwaters, eventually the water must recede for seedling establishment to occur (Schwegman and Nyboer, 1985; Smith and Keevin, 1998). The diversion of water into the Illinois River following the opening of the Chicago Sanitary & Ship Canal in 1900 (Miller, 1996) and the completion of navigation dams in 1933 (Waller, 1972) resulted in a doubling of total water areas (from 22,526 ha to 45,052 ha) (Bellrose et al., 1983) and a reduction of habitat suitable for moist-plant seedling establishment (Bellrose et al., 1979). The development of an extensive levee system between 1909 and 1922 (Mulvihill and Cornish, 1929) to promote the drainage of areas for agriculture served to reverse the trend somewhat; however, total water areas remained 22 percent greater than before 1900 (Bellrose et al., 1983). Many areas along the river that once had regular annual flooding, with flood waters receding in late spring, are now either under water for prolonged periods of time, are no longer flooded, or experience flood peaks during the growing season (Bellrose et al., 1979; Bellrose et al., 1983; Sparks, 1995; Sparks et al., 1998). The amount of bare, saturated shoreline created by the normal flood pulse has been reduced, limiting the areas suitable for *B. decurrens* and other species requiring similar conditions. Ironically, the survival of the flood-dependent *B. decurrens,* since the modification of the Illinois River during the twentieth century, may have been assisted by occasional severe droughts that lowered water levels and uncovered seldom exposed mudflats (Schwegman and Nyboer, 1985). During 1989, a year of extreme drought in Illinois and Missouri, six new populations were reported on these newly available areas that had been flooded for the previous eight years (U.S. Fish and Wildlife Service, 1990).

All experimental evidence indicates that the removal of competing vegetation by flood pulsing or other suitable disturbance is an integral requirement for maintaining the reproductive potential of *B. decurrens.* In a greenhouse study of the effects of light level on photosynthesis and biomass production in vegetatively produced rosettes, Smith et al. (1993) reported a 90 percent reduction in both processes when plants were grown under light levels similar to those occurring in sites following three years of succession (ca. 10 percent full sunlight), as compared with light levels of ca. 50 percent full sunlight. In addition, inflorescence numbers declined by 80 percent under the lower light level, indicating that seed production would be reduced and would negatively affect the fecundity of *B. decurrens.* In field studies, Redmond (1993) and Baker (1997) reported that reduced biomass resulted in fewer ramets produced per mature individual; therefore, both sexual and asexual reproduction are severely affected by any factor that reduces the availability of light.

The Midwest flood of 1993 affected the diversity, richness, and composition of various components of the vegetative community along the Illinois River Valley (Yin et al., 1997; Smith et al., 1998). Although there is little consensus concerning the interpretation of diversity, species diversity generally increases during the early stages of secondary succession (Odum, 1969) and is often positively correlated with an increase in richness (the number of species in a community). There are many exceptions to these generalities; however, these relationships have been reported for several riparian communities in large river systems (Klein et al., 1975; Kalliola and Puhakka, 1988; Puhakka et al., 1992). In a study comparing pre- and post-1993 flood data, Smith et al. (1998) reported a negative correlation between the size of the *B. decurrens* population and species richness and diversity (as defined by Shannon and Weaver, 1949) between years and among sites. Sites that experienced more moderate flooding had smaller reductions in diversity and richness and smaller increases in the size of the *B. decurrens* population than those that experienced severe flooding.

ADAPTATIONS TO CYCLICAL FLOODING

Historically, annual spring flooding created the open habitat required for optimal growth of *B. decurrens* by eliminating less-flood-tolerant early-successional species and clearing away the litter cover produced by dead vegetation. Field observations and laboratory studies have indicated that *B. decurrens* has physiological and morphological adaptations to flooding that provide a significant advantage over potentially competitive species. In areas where floodwaters have recently receded, *B. decurrens* is often the only surviving species (Figure 4-7). In a laboratory study comparing tolerance to root-zone saturation in *B. decurrens* and *Conyza canadensis* (an annual species that often invades *Boltonia* population sites), Stoecker et al. (1995) reported that after 28 days of treatment, biomass and survival were significantly greater in *B. decurrens* as compared with *C. canadensis*. Roots and stems of *B. decurrens* exposed to flooded conditions produced significantly more aerenchyma tissue and suffered less disintegration of

Figure 4-7. *April 1995. Flooding removed competing vegetation, leaving rosette of* Boltonia decurrens *intact. (Photograph by Nancy Parker; used by permission.)*

the primary roots than those of *C. canadensis* (Table 4-1). The authors concluded that the ability to establish an aerenchyma network from aboveground structures through the root system may have been responsible for the superior performance of *B. decurrens.*

In subsequent studies, the importance of aerenchyma formation to flood tolerance in *B. decurrens* was strengthened. Stems and roots of *B. decurrens* were examined after exposure to flooding, and it was determined that there was a significant increase in aerenchyma tissue when plants were placed under flooded conditions as compared with those grown in aerated soil (Table 4-2) (Stoecker, 1996). Stoecker et al. (1995) determined that *B. decurrens* is capable of maintaining low rates of growth and photosynthesis while completely submersed, and that plants form more iron hydroxide (FeOOH) plaque on root surfaces in flood treatments, as compared with controls (Figure 4-8). The presence of iron plaque indicates that the plants are able to detoxify the immediate rhizosphere by using photosynthetically derived oxygen (Bowes, 1987; Laan et al., 1990). Under conditions of reduced light, as would occur in heavily sedimented floodwaters, pho-

TABLE 4-1. The effect of flood treatment on root tissues in *Boltonia decurrens* and *Conyza canadensis*

Composition of stele, cortex, and aerenchyma as a percentage of total root cross-sectional area after growth for 56 days in control or flooded conditions. Means ± SE ($n = 4$); values in each column followed by the same letter form statistically indistinguishable groups by the Newman-Keuls test (*not part of comparison, only one plant measured).

		Tissue as % of Root Cross-Sectional Area		
Group	Root	Stele ($p < 0.01$)	Cortex ($p < 0.05$)	Aerenchyma ($p < 0.01$)
		Boltonia decurrens		
Control	Primary	14.8 ± 2.4[bc]	58.9 ± 6.2[b]	26.3 ± 4.5[bc]
	Lateral	7.5 ± 2.9[c]	66.3 ± 4.3[ab]	26.2 ± 4.7[bc]
Flooded	Primary	6.2 ± 0.3[c]	44.7 ± 1.8[c]	49.1 ± 2.1[a]
	Lateral	5.0 ± 1.1[c]	58.1 ± 1.2[b]	36.9 ± 2.3[ab]
		Conyza canadensis		
Control	Primary	34.7 ± 3.8[a]	60.1 ± 4.0[b]	5.2 ± 2.5[d]
	Lateral	20.4 ± 2.9[b]	71.5 + 1.5[ab]	8.1 ± 3.1[d]
Flooded	Primary	50.1 ± *	49.9 ± *	0*
	Lateral	8.6 ± 1.3[c]	77.4 ± 4.3[a]	14.0 ± 3.9[cd]

*Only one root with epidermal cells remaining; cortex was disintegrated in all roots.

Source: Stoecker et al., 1995.

tosynthesis and iron plaque formation were reduced (Figure 4-8); therefore, the clarity of floodwaters is of critical importance to the survival of submersed individuals of *B. decurrens.*

The study of the effects of flooding was extended to include growth, stomatal conductance, and fluorescence measures of flood tolerance in *B. decurrens,* as compared wtih *Aster ontarionis* (a rhizomatous perennial), *Aster pilosus* (a nonrhizomatous perennial), and *Conyza canadensis* (an annual), species with which *B. decurrens* often competes (Smith and Moss, 1998). In all comparisons, *B. decurrens* demonstrated greater survival, biomass production, and stomatal conductance than the other species (Figure 4-9). Apparently, the evolutionary development of *B. decurrens,* which is restricted to the Illinois River floodplain, resulted in morphological and physiological adaptations to the historical flood/drought cycle of the system. Anthropogenic alterations in the hydrologic regime of the Illinois River within the last 100 years, however, have reduced the competitive advantage conferred upon *B. decurrens* by its high tolerance for inundation and hypoxic soils.

TABLE 4-2. The effect of flood and light treatment of stem and root tissues in *Boltonia decurrens*

Aerenchyma as a percentage of cortex cross-sectional area in stem and root-shoot junction and as a percentage of root cross-sectional area in primary and secondary roots after 12 weeks of growth under experimental conditions (means ± SE, n = 15, for each treatment).

	Stem	Root-Shoot Junction	1° Root	2° Root
Control	2.98(0.99)	3.07(1.67)	6.27(5.21)	13.87(5.76)
Flood-Sun	31.48(4.11)	21.84(5.92)	34.78(3.89)	42.66(0.64)
Flood-Shade	40.43(5.90)	19.22(9.65)	31.89(2.33)	39.06(3.52)

Source: Stoecker, 1996.

Although *B. decurrens* is extremely flood tolerant, the quality and timing of floodwaters are critical. Flood peaks during the summer growing season (Sparks, 1995; Sparks et al., 1998) (Figure 4-10) kill seedlings and vegetative rosettes, reducing or eliminating the number of plants that flower and produce seed in the fall (Schwegman and Nyboer, 1985; Smith et al., 1998). Sediment-loaded water prevents germination and reduces the necessary light levels reaching immersed plants, eliminating photosynthetic production of oxygen for maintenance of root respiration. In addition, floods that occur after seed germination invariably scour the soil surface, uprooting seedlings. The alteration of the historical flood regime has isolated populations of *B. decurrens,* resulting in its restriction to human-disturbed alluvial soil habitats, old fields and roadsides (U.S. Fish and Wildlife Service, 1990; Smith and Keevin, 1998).

ALTERATION OF THE FLOOD PULSE

Although proposed changes to the Illinois River are hardly new, it is post-1900 anthropogenic modification that has drastically altered the flood pulse and reduced available habitat for *B. decurrens*—most notably, the construction of navigation dams to maintain a navigation channel and agricultural levees to create arable land. The Illinois River retains some characteristics of a flood pulse in unleveed portions of the river (Sparks, 1995; Sparks et al., 1998); however, significant changes in the flood pattern in these areas have developed over the past 100 years. The navigation dams and levees have contributed to the production of a series of irregular

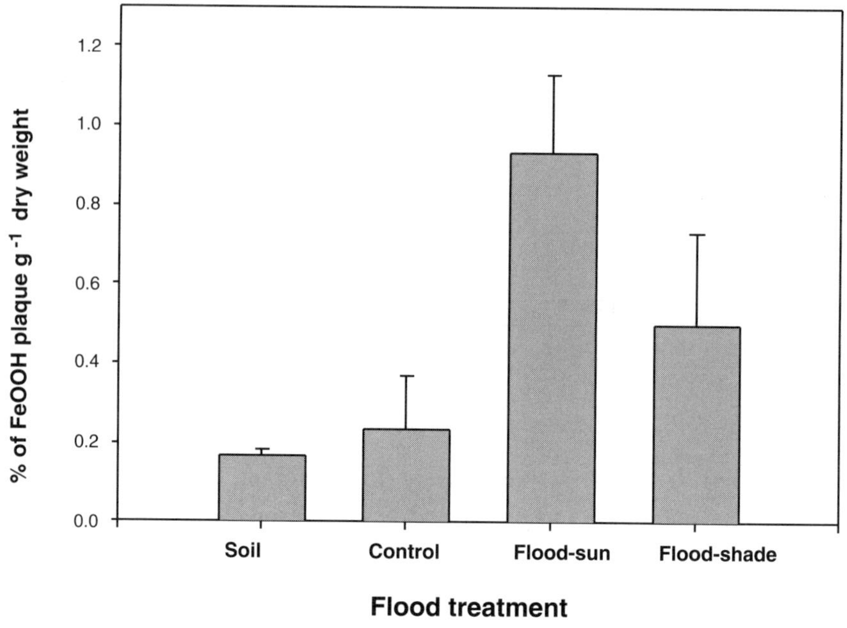

Figure 4-8. Percentage of iron hydroxide (FeOOH) plaque g^{-1} dry weight accumulated on roots of individuals of Boltonia decurrens *grown in three flood treatments, plus background amount of iron in soil (mean and SE of three replicates). (Reprinted from Stoecker, 1996, copyright © 1996 by M. Stoecker, by permission.)*

flood peaks during the summer (Figure 4-10), interrupting or shortening the growing season for many moist-soil plants (Sparks, 1995; Sparks et al., 1998, 2000). Navigation dams do not impede water during floods, as they are not engaged when water flow is high; however, when flow is low, as would normally occur during the summer, gates or wickets prevent the water from dropping below the level necessary for safe navigation (a minimum of 2.7 m). By preventing the normal drawdown of water levels during the summer, navigation dams reduce the range of natural water-level variation and eliminate much of the exposed mudflat habitats on which floodplain plant species depend. After the navigation dams were built, inundation of much of the former floodplain killed riparian vegetation, including floodplain forest, marsh, and wet-soil prairie species (Turner, 1936; Yaeger, 1949; Nelson et al., 1999).

All navigation dams reduce the range of water-level fluctuations throughout their navigation pools, thereby altering the normal flood pulse, but the effects vary, depending on the location within the pool and the type

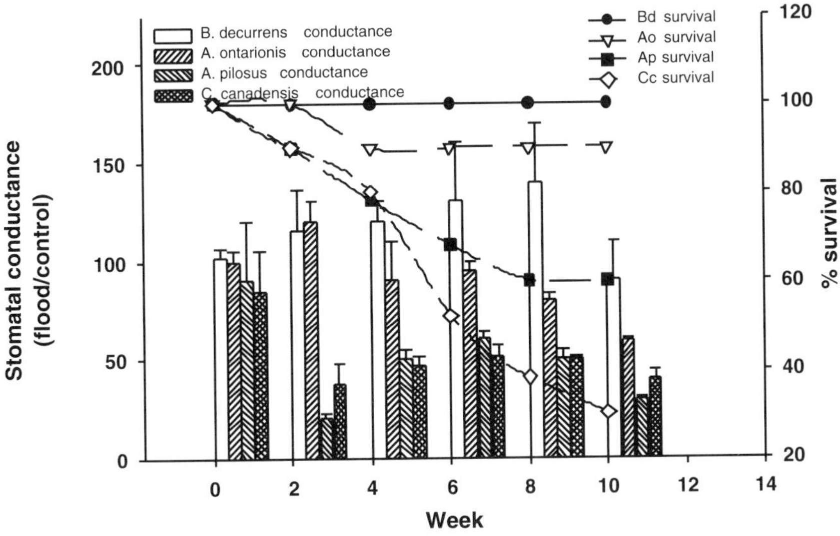

Figure 4-9. *Survival and stomatal conductance in flood plants relative to control plants of* Boltonia decurrens, Aster ontarionis, Aster pilosus, *and* Conyza canadensis *at two-week intervals, from week 0 to week 10 (for stomatal conductance, n = 3 for each species, treatment and measurement period; for survival,* n = *18 for each species and each treatment). (Reprinted from Smith and Moss, 1998, copyright © 1998 by* Journal of Applied Ecology, *by permission of Blackwell Science Ltd., Osney Mead, Oxford, U.K.)*

of dam (Bellrose et al., 1979). The wicket dams on the Illinois River at Peoria and La Grange consist of panels, hinged to a sill at the bottom of the river, that swing up and are propped in place to hold water during low flows. Because wicket dams cannot control water levels as accurately as gates, the Peoria and La Grange navigation pools exhibit frequent erratic short-term fluctuations. Water levels in the upstream portion of the pool must be at least 61 cm higher than the water at the downstream dam, so water levels near the dams drop initially and then rise suddenly as the wickets are engaged. During periods of low to moderate flooding, the areas near the dam drain, but water levels rise in the upper half of the pool. During the summer growing season, unnatural spikes in the water level occur in the upper portion of the pool, while areas near the dam have either constant water levels or unnatural drops (Sparks, 1995; Sparks et al., 1998). The erratic changes in water levels are exacerbated when malfunction of either the upstream dam or the downstream dam causes a lack of coordination between management of the wickets. The result is that not only is the magnitude of the flood pulse altered, but the natural variation in

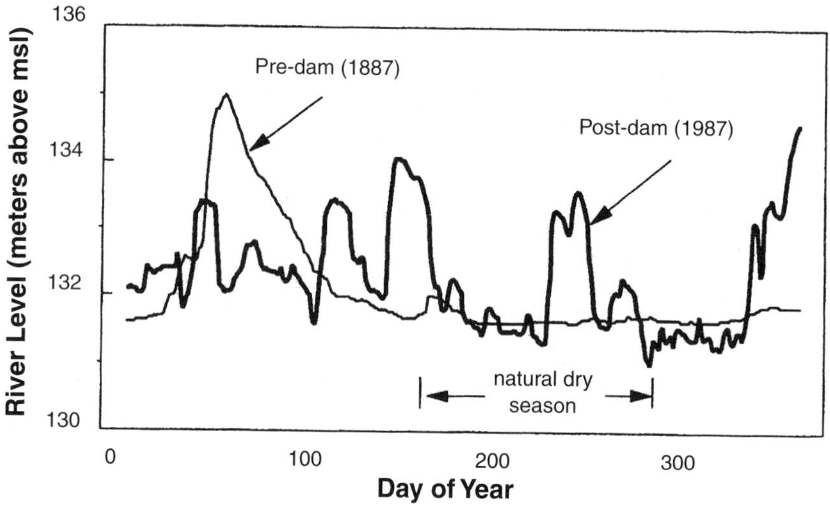

Figure 4-10. *Since the navigation dams have been in place on the Illinois River, the natural pattern of spring flood pulse followed by a summer season of low flow has been replaced by an unpredictable, chaotic pattern year-round (msl=mean surface level). (Reprinted from Sparks et al., 2000, copyright © 2000 Backhuys Publishers, by permission.)*

water level in the pool is often inverted, with the level falling unnaturally low during periods of moderate flooding and rising during the summer (Sparks et al., 1998). Inversions are especially damaging for floodplain vegetation, because abrupt, ill-timed rises in water levels during the growing season prevent the establishment of moist-soil plants. Even if seeds germinate in the spring on the briefly exposed mudflats, they are drowned before they become established. Despite its high degree of tolerance to flooding, *B. decurrens* has been eliminated from many of these sites because of the depth of the water and the timing of the altered flood pulse (Schwegman and Nyboer, 1985; Smith et al., 1998).

In areas of the floodplain not separated from the river by high levees, a relatively inexpensive modification of dam operations could improve conditions for some floodplain plants (Sparks et al., 1998, 2000). The goal in navigation pool management is to provide safe navigation during low flow periods. Traditionally, this has been achieved by keeping water levels at the midpoint of the range of water levels necessary to maintain the 2.7 m level at the upper extreme of the pool. This "hinge-point" method of managing water levels allows for a range of water levels for various flow rates.

As water flow increases, the need to artificially maintain maximum pool decreases, allowing water control managers to lower the maximum level by manipulation of the dam. This approach reduces the amount of floodplain land necessary to manage the pools, but because water control managers in the past had one value per day to use in making water management decisions (the depth at the hinge point), and because flow rates are dynamic, the standard practice was to be prudent and avoid approaching the lowest levels calculated for safe navigation (Flowers et al., 1998). This method of manipulation produced an average drawdown in the pool that was too little (0.1 to 0.5 m) and for too short a time (< 20 days) to provide viable habitat, even for flood-tolerant species.

In 1994, the Missouri Department of Conservation proposed that the St. Louis District of the U.S. Army Corps of Engineers conduct an experiment in Pool 25, on the Mississippi River, that would achieve, on a regular basis, drawdowns that might increase suitable habitat for moist-soil vegetation. The experimental procedure would provide a pool drawdown of 0.2 to 0.7 m for at least 30 days, time the drawdown for the period from May 1 to July 30, with May-June deemed the period most desirable for vegetative growth and seed production, and after the initial drawdown, allow the pool to rise at a rate of not more than 2.5 cm per day. The U.S. Army Corps of Engineers agreed to extend the study to include Pools 24, 25, and 26 (Melvin Price Lock and Dam, which controls the navigation pool from the mouth of the Illinois River to La Grange, a distance of 127.8 km). Preliminary estimates revealed that more than 1200 ha of vegetation were created in the first year of the study, at least quadruple the area of vegetation upstream from Pool 22, which was managed in the standard manner. This method, termed Environmental Pool Management (EPM) by the U.S. Army Corps of Engineers, is made possible by a computerized data collection platform that enables water management operators to view hundreds of gauges at 15-minute intervals so that river levels can be monitored more accurately and more frequently (Busse, 1998). The successful 1994 experiment was repeated in 1995–1997, leading to the continuation of the EPM program in the St. Louis District and consideration of the approach in other U.S. Army Corps of Engineers districts (Flowers et al., 1998). Although EPM does not restore the pre-1900 flood pulse, nor has it been proven to aid *B. decurrens* or other species that may require longer periods of drawdown, preliminary results indicate that in pools where it can be implemented, it improves the floodplain habitat of selected segments of the river without affecting navigation or costing additional taxpayer dollars (Flowers et al., 1998).

The EPM studies have provided valuable empirical evidence of ecological benefits gained by restoring lower summer water levels, and a modicum of a flood pulse, in three navigation pools on the Mississippi River. To consider the application of the principle on an ecosystem scale, however, it is necessary to develop a mathematical model capable of predicting biotic responses to dam operations under a variety of environmental conditions. A one-dimensional hydraulic model (UNET) (Barkau, 1995) and a floodplain forest model (SWAMP) (Phipps, 1979; Phipps and Applegate, 1983) were applied by Sparks et al. (2000) to predict the effect of changes in the summer drawdown on forest species composition in the 129 km La Grange pool on the Illinois River. When initialized using historical data, the model correctly predicted that the presettlement forest, dominated by pin oak *(Quercus palustrus)* and containing a substantial number of other species, would be converted into a contemporary forest dominated by willow *(Salix nigra)* when subjected to the pattern of changing water levels of the past 130 years. The model is to be extended to herbaceous data in the future. If it continues to accurately reflect the effect of changes in pool water levels on floodplain vegetation, it may be of use in more widely implementing dam management to restore a flood pulse to portions of the river.

Except in extreme years when levees are overtopped or undermined, a natural flood pulse does not occur when levees confine water flow to the main channel of the river. More than 50 percent of the Illinois River is restricted by federal levees (as distinguished from low natural or constructed levees that are overtopped by moderate flooding) (Mulvihill and Cornish, 1929; Starrett, 1971; Thompson, 1989); therefore, in half of the floodplain, management for moist-soil plants is possible only if water levels in areas isolated behind the levees are manipulated on a schedule planned to coincide with plant germination and growth. Typically, this is done either by installing pumps where complete drainage of backwater areas is a management goal (Fredrickson and Taylor, 1982; Theiling, 1995) or by allowing water to flow in from the river, during periods of high flow, through gates and culverts in the levees. Both types of management require construction and repair of levees, installation and maintenance of water control structures, and periodic dredging. As the goal of management is most often to attract and feed selected game species for hunters (e.g., geese and mallards), water levels may be manipulated to permit row-crop cultivation of high-energy foods for large concentrations of waterfowl during the winter, to the detriment of nontargeted organisms (Fredrickson and Taylor, 1982; Sparks, 1995). A number of these projects have been initiated along the Illinois River (Meeks, 1969; Theiling, 1995), but the benefits to fugitive plant species, even when their life history requirements are known and given consideration, are limited. The isolation of plant populations from the open river severely curtails their normal means of seed dispersal and gene flow, increasing the danger of extinction (Pickett and Thompson, 1978). Even short-distance seed dispersal is disrupted in many cases, because a group of small impoundments is more easily managed than a single large one (Fredrickson and Taylor, 1982), resulting in further habitat fragmentation and a greater risk of extinction (Simberloff and Abele, 1976).

The barrier to lateral flow created by high, man-made levees has resulted in the conversion of 82 percent of the drainage basin of the Illinois River to agricultural use (Warner, 1998; Arnold et al., 1999). The Illinois Environmental Protection Agency has identified row-crop agriculture as the primary contributor to sediment load (ca. 13.8 million tons of sediment annually) (Critical Trends Assessment Program, 1994), and the increased sedimentation in the backwater lakes and on the floodplains of the Illinois River is of serious concern to biologists and resource managers (Bellrose et al., 1979, 1983; Steffeck et al., 1980; Sparks, 1984). Navigation dams further slow water flow in the naturally sluggish Illinois River (Mills et al., 1966), and levees create increased sedimentation in unleveed areas where the sediment that has been suspended in the faster flowing channel precipitates out of the more slowly moving currents of backwater areas. Wave action created by barge traffic (Bhomik, 1981) resuspends sediments before they settle out of river currents. Turbidity is exacerbated further by the loss of rooted aquatics, which once stabilized the bottoms of shallow lakes, and by the disappearance of many of the floating aquatics that served to dampen wave action. The loss of floodplain forests early in this century, which once furnished windbreaks, is another contributing factor to wave action damage, and the loss of wet marsh and slough vegetation removed the natural filter strip that slowed runoff and absorbed sediment (Bellrose et al., 1983). Although the filling in of backwater lakes might be expected to render more habitats, such as bare mudflats, suitable for colonization by fugitive plant species, it has not happened. As lakes disappear and marshy areas are drained, the newly created land becomes a target of development projects. The former floodplain area that existed along the shores of Lake Peoria, for instance, once accommodated populations of *B. decurrens* (U.S. Fish and Wildlife Service, 1990; Smith, 1991). Silt deposits along the shoreline and the concomitant reduction in the occurrence of floods have resulted in the occupation of the area by a marina, a condominium development, and a casino-hotel complex (M. Smith, personal observation). Clearly, even plants protected by federal listing are at risk in such areas.

Until agricultural runoff from the entire basin is reduced, no long-term, systemwide reduction of sedimentation is likely to occur (Bellrose et al., 1983). It has been suggested, however, that modifications of flood control structures and navigation dam operation can result in an improvement of channel and floodplain habitat by redirecting or stabilizing sediment deposits (Theiling, 1995; Sparks et al., 1998, 2000). The manipulation of navigation dams to extend low-flow drawdowns would allow the growth of vegetation on the currently mobile sediment base and the stabilization and compaction of sediments during the period when the areas are exposed. Introduced flow, island projects, and modification of channel maintenance structures could redirect sediment loads to create or protect some areas of floodplain habitat (Theiling, 1995); however, these remedies are site-specific and cannot effect the restoration of health, or restore the flood pulse, to the entire river system.

RESTORATION OF THE FLOOD PULSE TO THE ILLINOIS RIVER VALLEY

Various terms have been used to describe improvements made to river environments, all of them suggesting some level of restoration of ecosystem function (Brookes and Shields, 1996). True restoration, sensu Cairns (1991), requires the recreation of the structure and function of an ecosystem to pristine conditions. In contrast to the paucity of useful historical data existing for the majority of large, temperate river-floodplain systems (Gore and Shields, 1995), sufficient historical hydrological and ecological data are available concerning the Illinois River system to form a reasonable comparison with the predisturbance regime (Kofoid, 1903; Turner, 1934; Bellrose et al., 1983; Sparks et al., 1998). Because of social and economic restraints, however, the complete reversal of anthropogenic disturbance is not likely. The current and proposed changes in the river system are a blend of naturalization, which recognizes that human utilization of natural resources is a component of the current environment (Rhoads and Herricks, 1996); enhancement, which is regarded as any improvement of a structural or functional attribute (National Research Council, 1992, 1995; Middleton, 1999); and rehabilitation, which refers to the attainment of a limited number of predetermined structural and functional aspects by providing new chemical and physical structure (Gore and Shields, 1995). To be politically and economically feasible, any systemwide changes in the Illinois River must find a balance between navigation interests, recreational users, and the rehabilitation and protection of the natural biota (Theiling, 1995; Sparks et al., 2000). The reconnection of the Illinois River to its floodplain and the restoration of the historical flood pulse are not likely to occur (Sparks et al., 2000).

PROTECTION FOR *B. DECURRENS* UNDER THE ENDANGERED SPECIES ACT

The Endangered Species Act of 1973, as amended in 1978 (§1533f), states that recovery is "the process by which the decline of a threatened or endangered species is arrested or reversed, and threats to its survival are neutralized, so that its long-term survival in nature can be ensured." The Endangered Species Act requires federal protection of habitat, stating that critical habitat must be designated "to the maximum extent prudent and determinable." Section 9 of the Act extends habitat protection by prohibiting the "taking of an endangered species of fish or wildlife," a portion of the Act that has raised concerns among private landowners because of its application to habitat modification (Mann and Plummer, 1995). Recognition of the relationship between disappearing habitat and the extinction of species is inherent in the Endangered Species Act and embraces the current scientific understanding of the crucial role of habitat in the maintenance of organisms (Morrison et al., 1992). Although other federal laws have been designed to protect wildlife, the Endangered Species Act is the most powerful conservation tool created to protect endangered species and their habitats. Plants, however, are less adequately protected by the Endangered Species Act than animals. Endangered animals are protected whether they occur on federal or private land; however, Section 9 of the Act offers protection for plants only if they occur on federal land or if their destruction is in violation of a law or regulation of a state. No limitations are placed on the modification of habitat that occurs under private ownership. A committee of the National Research Council commissioned to study the scientific basis of the Endangered Species Act concluded that there is no scientific reason to establish differences between the protections offered to plants and animals (National Research Council, 1995) and that the Act fails to protect endangered plants from habitat destruction. Plants that are endemic to dynamic habitats are particularly vulnerable. Fugitive plant species require large contiguous areas so that populations may move when succession has rendered a site unsuitable for plant growth (Hutchinson, 1951). The restriction of protection to areas controlled by the federal government may cause fragmentation, resulting in the disruption of dispersal routes (Nilsson et al., 1993; Jansson et al., 2000), loss of genetic variability (Pease et al., 1989), and, ultimately, extinction of the species (Quinn and Hastings, 1987; Gilpin and Hanski, 1991).

In 1990 a team composed of representatives of the United States Army Corps of Engineers, the Illinois Department of Conservation, the United States Fish and Wildlife Service, and the Missouri Department of Conservation formulated a Federal Recovery Plan for *B. decurrens* (U.S. Fish and Wildlife Service, 1990). Recovery and delisting, as called for in the plan, are based on establishing 12 self-sustaining populations (defined as populations found to be stable or expanding during a five-year monitoring period). A personal communication from John Schwegman, a member of the recovery team, included the observation that *B. decurrens* "did not reproduce sexually in dense weedy areas and that vegetative regeneration ceased after four or five years"; however, the committee did not fully understand the implications of this information. Research conducted subsequent to the development of the Recovery Plan provides ample evidence that the species does not form stable populations, and that the size of any discrete population at one point in time is not predictive of future population size—even for a single annual cycle (Smith et al., 1993, 1995, 1998). Stochastic fluctuations in population size and number confound attempts to make decisions based on populations that might appear to be stable or increasing within a five-year period. For a fugitive species that does not persist at a site without a suitable disturbance, it is difficult to protect established populations. The usual procedure for endangered species protection, as practiced by state and federal conservation agencies, is to restrict access to—and disturbance of—population sites. Barring severe flooding that would reverse succession in the protected populations, this greatly increases the probability of extinction for *B. decurrens* or any other species dependent on a regular flood pulse. Although the Federal Recovery Plan called for management studies to determine the efficacy of burning, mowing or disking to simulate natural disturbance, our subsequent studies have indicated that without the selective disturbance provided by regular flooding, these practices are not sufficient to maintain seedling establishment or vegetative rosette production in *B. decurrens* (Smith, 1995). Records that we and the Illinois Department of Natural Resources have kept over the past 20 years provide ample evidence of extreme fluctuations in population size, number, and location (Schwegman and Nyboer, 1985; Smith, 1993, 1995; Smith et al., 1998) and the relationship between population dynamics and the flood-drought cycle (Smith et al., 1997, 1998, 1999).

The narrative in the Recovery Plan indicates that the threat to *B. decurrens* is most likely due to disappearing habitat, but no criteria or actions in the plan address questions concerning how much of its range should be protected or the need for dispersal corridors among populations. Research conducted in the ten years since the recovery plan was written has confirmed that to protect *B. decurrens,* large, contiguous stretches of the floodplain must be reconnected to the river to provide suitable habitat—an approach more effective than trying to protect specific populations or population sites. The role of floodwaters in seed dispersal and colonization and renewal of population sites is now established (Smith and Keevin, 1998; Smith et al., 1998); however, to date, no studies of the genetics of the species have been called for by any state or federal agency.

The record-breaking floods of 1993 on the lower Illinois River (a 500-year record) and 1995 on the upper Illinois River (a 100-year record) opened vast areas of the floodplain for colonization by the species, and the reestablishment of agricultural levees has taken several years; thus, the species appears to be in the process of recovery. An extremely dry winter (1999) and spring (2000) may have provided more areas for population establishment because of the reduction of total water area, enhancing the illusion that recovery actions have rescued *B. decurrens* from danger of extinction. Progress based on the Recovery Plan has been restricted, for the most part, to the collection of some useful information about the species, little of which has been utilized in management plans. A comprehensive, informed strategy has yet to be formulated, and most of the recovery of *B. decurrens* is due to the fortuitous combination of climatological factors.

POLICIES AND PROSPECTS FOR THE FUTURE

The United States has never viewed flooding from a systemwide perspective. Government agencies and individual communities have spent vast sums of money on a fragmented system of structural flood control measures. The design of the district levee system in Illinois, whereby each district is given the authority to impose taxes to build and repair flood control structures, results in a patchwork of levees, with each district giving top priority to its own immediate problems (Mulvihill and Cornish, 1929; Bellrose et al., 1979; Thompson, 1989). Downstream effects are seldom considered, as there is little incentive for the coordination of flood control efforts (Galloway, 1995). What may prevent a flood in one district may ensure flooding immediately downstream, thereby exacerbating flood losses on a large segment of the river. This is a recurring pattern throughout flood-prone areas in the United States. In 1999, the National Wildlife Federation conducted a study of the cost to U.S. taxpayers of maintaining the National Flood Insurance Program and revealed that the program, which has paid more than $40 billion in flood damages over the past 5 years, has incurred debts of almost $2 billion dollars since its inception 18 years ago that cannot be covered by insurance premiums. Properties sustaining repetitive losses amount to only 2 percent of all loss claims and 40 percent of total flood insurance payments. Many of these losses are incurred by the continued occupation of high-risk floodplains. The Flood Insurance Reform Act of 1994 addressed some of the issues involved, but nothing was done to relieve the cost of repetitive-loss claims. The government continues to offer financial incentives that encourage repetitive rebuilding in areas not suitable for construction and habitation (Galloway, 1995; National Wildlife Federation, 1999). As long as this practice continues, reconnection of the Illinois River to its floodplain is not possible.

Although current practices are inconsistent with the restoration of a systemwide flood pulse, the time may be right to plan strategies for restoring the natural hydrology to a portion of the Illinois River. The Midwest flood of 1993 focused government and public attention on the dangers inherent in depending on structural control of flooding (Galloway, 1995). The costs, including taxpayer dollars, anguish, and displacement, were so high that regulatory agencies and the general public have begun to seek alternatives for floodplain management. Federal and state funds have been made available for voluntary buyouts and the acquisition of easements of particularly flood-prone properties (National Wildlife Federation, 1999; Sparks et al., 2000); however, private conservation organizations have taken the most immediate, direct action that may preserve enough of the Illinois River to allow time for public agencies to develop effective cooperative programs that would reverse the trend of the past 150 years. The Nature Conservancy (TNC), in cooperation with the Wetlands Initiative of Chicago, the United States Fish and Wildlife Service, and the Illinois Department of Natural Resources, has developed a strategic plan for the Illinois River that includes the entire river system. In 1997, TNC purchased 1157 acres of former floodplain land in Brown County, Illinois, on which it plans to restore the natural hydrology and vegetation of the area. This year, TNC spent $18.3 million to purchase a 7527-acre commercial farm that was once the site of Thompson and Flag Lakes, formerly among the most productive backwater lakes in the Illinois River Valley. The acquisition is TNC's Illinois chapter's largest purchase and the biggest private conservation deal in Illinois history (Doug Blodgett, Great Rivers Area Director, TNC, personal communication). Because private ownership carries with it a significant measure of control over floodplain issues, these areas may serve as models to government agencies for the restoration of the flood pulse on public lands.

In the areas of the Illinois River Valley where the river-floodplain connection is to be reestablished, valuable information will be gained concerning the effects of the restoration of natural hydrology on native species. Sufficient historical data exist about the biota of the Illinois River Valley to provide useful comparisons with information collected from these limited efforts, which could be valuable in formulating more extensive restoration plans. *Boltonia decurrens* is a target species in both of the TNC projects, and a study of the dynamics of populations in the restored areas will provide the first direct evidence of the response of a fugitive floodplain native plant to a more natural flood regime.

ACKNOWLEDGEMENTS

This work was funded by grants to M. Smith from NSF (DEB 9509763; DED 9321517).

REFERENCES

Archibold, O. W. 1978. Buried viable propagules as a factor in postfire regeneration in northern Saskatchewan. *Canadian Journal of Botany* **57**:54–58.

Arnold, T. L., D. J. Sullivan, M. A. Harris, F. A. Fitzpatrick, B. C. Scudder, P. M. Ruhl, D. W. Hanchar, and J. S. Stewart. 1999. *Environmental Setting of the Upper Illinois River Basin and Implications for Water Quality.* Report #98-4268. U.S. Geological Survey, Urbana, IL.

Baker, M. A. 1997. Effects of intra- and interspecific rosette competition on *Boltonia decurrens.* Master's thesis, Southern Illinois University at Edwardsville.

Barkau, R. L. 1995. One Dimensional Unsteady Flow Through a Full Network of open Channels. (Revised by CEWRC-HEC U.S.A.C.E.) Hydrologic Engineering Center, Davis, CA.

Baskin, C. C., and J. M. Baskin. 1988. Germination ecophysiology of herbaceous plant species in a temperate region. *American Journal of Botany* **75**:286–305.

Bayley, P. B. 1991. The flood pulse advantage and the restoration of river-floodplain systems. *Regulated Rivers: Research and Management* **6**:75–86.

Bayley, P. B. 1995. Understanding large river-floodplain ecosystems. *BioScience* **45**:153–158.

Bellrose, F. C., F. L. Paveglio, and D. W. Steffeck. 1979. Waterfowl populations and the changing environment of the Illinois River Valley. *Illinois Natural History Survey Bulletin* **32**:1–54.

Bellrose, F. C., S. P. Havera, F. L. Paveglio, and D. W. Steffeck. 1983. The fate of lakes in the Illinois River Valley. *Illinois Natural History Survey Biological Notes* **32**:9–12.

Bhomik, N. G. 1981. *Waves and Drawdown Generated by River Traffic on the Illinois and Mississippi Rivers.* Upper Mississippi River Basin Commission, Minneapolis, MN.

Bowes, G. 1987. Aquatic plant photosynthesis: Strategies that enhance carbon gain. In *Plant Life in Aquatic and Amphibious Habitats.* ed. by R. M. M. Crawford, pp. 79–98. Blackwell Scientific Publishing, Boston.

Brookes, A., and F. D. Shields. 1996. Perspectives on river channel restoration. In *River Channel Restoration: Guiding Principles for Sustainable Projects,* ed. by A. Brookes and F. D. Shields, pp. 1–19. John Wiley & Sons, Chichester, U.K.

Busse, D. 1998. Environmental pool management. In *Infrastructure,* ed. by H. N. Cook and R. G. Goodwin, pp. 35–39. John Wiley & Sons, New York.

Cairns, J. J. 1991. The status of the theoretical and applied science of restoration ecology. *The Environmental Professional* **13**:186–194.

Critical Trends Assessment Program. 1994. *The Changing Illinois Environment: Critical Trends.* ILENR/RE-EA-94/05 (SR) 20M/1994. Illinois Department of

Energy and Natural Resources and The Nature of Illinois Foundation, Springfield, IL.

Decamps, H., M. Fortune, F. Gazelle, and G. Pautou. 1988. Historical influences of man on the riparian dynamics of a fluvial landscape. *Landscape Ecology* **1:**163–173.

Egler, F. E. 1954. Vegetation science concepts I. Initial floristic composition, a factor in old field vegetation development. *Vegetatio* **4:**412–417.

Flowers, R. B., C. Strauser, and D. Busse. 1998. Environmental management of navigation pools on the Upper Mississippi River. Section I, Subject 5. In *29th PIANC Permanent International Navigation Congress,* pp. 2–9. Permanent International Navigation Congress, The Hague, the Netherlands.

Fredrickson, L. H., and T. S. Taylor. 1982. *Management of Seasonally Flooded Impoundments for Wildlife.* Report #148. U.S. Fish and Wildlife Service, Washington, DC.

Galloway, G. E. J. 1995. *Sharing the Challenge: Floodplain Management into the 21st century.* Report #22. Water Resources Center, University of Illinois, Urbana-Champaign.

Gilpin, M. E., and I. Hanski. 1991. *Metapopulation Dynamics: Empirical and Theoretical Investigations.* Academic Press, New York.

Gore, J. A., and F. D. Shields. 1995. Can large rivers be restored? *BioScience* **45:**142–152.

Harper, J. L. 1977. *Population Biology of Plants.* Academic Press, London.

Herkert, J. R. 1991. *Endangered and Threatened Species of Illinois: Status and Distribution.* Vols. 1 and 13. Illinois Endangered Species Protection Board, Springfield, IL.

Hus, H. 1908. An ecological cross section of the Mississippi River in the region of St. Louis, Missouri. Ph.D. diss., Washington University, St. Louis, MO.

Hutchinson, G. E. 1951. Copepodology for the ornithologist. *Ecology* **32:** 571–577.

Jansson, R., C. Nilsson, and B. Renofalt. 2000. Fragmentation of riparian floras in rivers with multiple dams. *Ecology* **81:**899–903.

Junk, W. J., P. B. Bayley, and R. E. Sparks. 1989. The flood pulse concept in river-floodplain systems. In *Proceedings of the International Large River Symposium (LARS),* ed. by D. P. Dodge, pp. 110–127. Canadian Special Publications in Fisheries and Aquatic Sciences 106. Department of Fisheries and Oceans, Ottawa.

Kalliola, R., and M. Puhakka. 1988. River dynamics and vegetation mosaicism: A study of the River Kamajohka, northernmost Finland. *Journal of Biogeography* **15:**703–719.

Klein, W. M., R. H. Daley, and J. Wedum. 1975. *Environmental Inventory and Assessment of Navigation Pools 24, 25, and 26, Upper Mississippi and Lower*

Illinois Rivers: A Vegetational Study. Report # Y-75-1. U.S. Army Corps of Engineers, St. Louis, MO.

Koellner, W. H. 1996. The flood's hydrology. In *The Great Flood of 1993,* ed. by S. A. Changnon, pp. 68–100. Westview Press, Boulder, CO.

Kofoid, C. A. 1903. The plankton of the Illinois River, 1894–1899, with introductory notes upon the hydrography of the Illinois River and its basin. Part 1: Quantitative investigations and general results. *Illinois State Natural History Survey Bulletin* **6:**95–252.

Kurz, D. R. 1981. *Status Report on* Boltonia asteroides *var.* decurrens *in Illinois.* Illinois Department of Conservation, Springfield, IL.

Laan, P. M., M. Tosserams, C. W. P. M. Blom, and B. W. Veen. 1990. Internal oxygen transport in Rumex species and its significance for respiration under hypoxic conditions. *Plant and Soil* **122:**39–46.

Lee, M. T., and N. G. Bhomik. 1979. *Sediment Transport in the Illinois River.* Report # 176A. Illinois State Water Survey Contract Report, Springfield, IL.

Lee, M. T., and J. B. Stall. 1976. *Sediment Conditions in Backwater Lakes Along the Illinois River—Phase 2.* Report # 176B. Illinois State Water Survey Contract Report, Springfield, IL.

Mann, C. C., and M. L. Plummer. 1995. *Noah's Choice: The Future of Endangered Species.* Alfred A. Knopf, New York.

Meeks, R. L. 1969. The effect of drawdown date on wetland plant succession. *Journal of Wildlife Management* **33:**817–821.

Middleton, B. 1999. *Wetland Restoration: Flood Pulsing and Disturbance Dynamics.* John Wiley & Sons, New York.

Miller, D. L. 1996. *City of the Century: The Epic of Chicago and the Making of America.* Simon and Schuster, New York.

Mills, H. B., W. C. Starrett, and F. C. Bellrose. 1966. *Man's Effect on the Fish and Wildlife of the Illinois River.* Report # 57. Natural History Survey, Champaign, IL.

Missouri Department of Conservation. 1999. *Missouri Species of Conservation Concern Checklist.* Missouri Department of Conservation, Jefferson City, MO.

Morgan, S. 1980. *Status Report on* Boltonia asteroides *var.* decurrens *in Illinois.* Report to the U.S. Fish and Wildlife Service by the Missouri Department of Conservation, Jefferson City, MO.

Morrison, M. L., B. G. Marcot, and R. W. Mannan. 1992. *Wildlife-Habitat Relationships, Concepts and Applications.* University of Wisconsin Press, Madison.

Mulvihill, W. F., and L. D. Cornish. 1929. *An Engineering Study of the Flood Situation in the State of Illinois.* Illinois Division of Waterways, Springfield, IL.

National Research Council. 1992. *Restoration of Aquatic Ecosystems: Science, Technology and Public Policy.* National Academy Press, Washington, DC.

National Research Council. 1995. *Science and the Endangered Species Act.* National Academy Press, Washington, DC.

National Wildlife Federation. 1999. *Higher Ground: A Report Analyzing 18 years of National Flood Insurance Program Statistics.* National Wildlife Federation, http://www.nwf.org/ pubs/higherground/recs.html, Washington, DC.

Nelson, J. C., R. E. Sparks, L. DeHaan, and L. Robinson. 1999. *Presettlement and Contemporary Vegetation Patterns Along Two Navigation Reaches of the Upper Mississippi River.* Report # USGS/BRD/BSR-1998-0003. United States Geological Survey, Biological Resources Division, Washington, DC.

Nilsson, C., E. Nilsson, M. E. Johansson, M. Dynesius, G. Grelsson, S. Xiong, R. Jansson, and M. Danvind. 1993. *Process Structuring Riparian Vegetation,* pp. 419–431. Council of Scientific Research Integration, Trivandrum, India.

Odum, E. P. 1969. The strategy of ecosystem development. *Science* **164:**262–270.

Pease, C. M., R. Lande, and J. J. Bull. 1989. A model of population growth, dispersal and evolution in a changing environment. *Ecology* **70:**1657–1664.

Phipps, R. L. 1979. Simulation of wetlands forest vegetation dynamics. *Ecological Modeling* **7:**257–288.

Phipps, R. L., and L. H. Applegate. 1983. Simulation of management alternatives in wetland forests. In *Application of Ecological Modeling in Environmental Management,* ed. by S. E. Jørgensen and M. J. Mitsch, pp. 311–339. Elsevier Science Publishers, Amsterdam, The Netherlands.

Pickett, S. T. A., and J. N. Thompson. 1978. Patch dynamics and the design of nature reserves. *Biological Conservation* **13:**27–37.

Puhakka, M., R. Kalliola, M. Rajasilta, and J. Salo. 1992. River types, site evolution and successional vegetation patterns in Peruvian Amazonia. *Journal of Biogeography* **19:**651–665.

Quinn, J. F., and A. Hastings. 1987. Extinction in subdivided habitats. *Conservation Biology* **1:**198–208.

Redmond, A. 1993. Population study of *Boltonia decurrens,* a federally threatened species. Master's thesis, Southern Illinois University at Edwardsville, Edwardsville.

Rhoads, B. L., and E. E. Herricks. 1996. Naturalization of headwater streams in Illinois: Challenges and possibilities. In *River Channel Restoration: Guiding Principles for Sustainable Projects,* ed. by B. L. Rhoads and E. E. Herricks, pp. 332–367. John Wiley & Sons, Chichester, U.K.

Schwegman, J. E., and R. W. Nyboer. 1985. The taxonomic and population status of *Boltonia decurrens* (Torr. and Gray) Wood. *Castanea* **50:**112–115.

Shannon, C. E., and W. Weaver. 1949. *The Mathematical Theory of Communication.* University of Illinois Press, Urbana.

Simberloff, D. S., and L. G. Abele. 1976. Island biogeography theory and conservation practices. *Science* **191:**285–286.

Smith, M. 1990. *Report on Basic Life History Characteristics of* Boltonia decurrens (Decurrent False Aster). U.S. Army Corps of Engineers, St. Louis, MO.

Smith, M. 1991. *Life History Research for Decurrent False Aster.* Illinois Department of Conservation, Springfield, IL.

Smith, M. 1993. *Effects of the Flood of 1993 on the Decurrent False Aster* (Boltonia decurrens). U.S. Army Corps of Engineers, St. Louis, MO.

Smith, M. 1995. *Post-Flood Assessment of the Decurrent False Aster.* U.S. Fish and Wildlife Service, Twin Cities, MN.

Smith, M., and T. Keevin. 1998. Achene morphology, production and germination, and potential for water dispersal in *Boltonia decurrens* (decurrent false aster), a threatened floodplain species. *Rhodora* **100:**69–81.

Smith, M., and S. Moss. 1998. Root-zone flooding as a strategy to reduce competition in populations of the threatened floodplain species, *Boltonia decurrens:* Stomatal conductance and Fv/Fm as indicators of stress. *Journal of Applied Ecology* **35:**553–561.

Smith, M., T. Brandt, and J. Stone. 1995. Effect of soil texture and microtopography on germination and seedling growth in *Boltonia decurrens* (Asteraceae), a threatened species. *American Journal of Botany* **80:**854–864.

Smith, M., T. Keevin, and P. Mettler-McClure. 1997. Relating site characteristics of a threatened floodplain plant species to flood frequency and duration. In *Research on Agricultural Chemicals in Illinois Groundwater: Status and Future Directions,* ed. by M. Davis, pp. 109–125. Illinois Groundwater Consortium, Southern Illinois University Carbondale, Carbondale, IL.

Smith, M., Y. Wu, and O. Green. 1993. Effect of light and water-stress on photosynthesis and biomass production in *Boltonia decurrens* (Asteraceae), a threatened species. *American Journal of Botany* **80:**859–864.

Smith, M., T. Keevin, P. Mettler-McClure, and R. Barkau. 1998. Effect of the flood of 1993 on *Boltonia decurrens,* a rare floodplain plant. *Regulated Rivers: Research and Management* **14:**191–202.

Smith, M., R. Pearson, E. Esselman, K. Victory, J. Thompson, and B. Denother. 1999. Developing a GIS database to monitor plant population dynamics in the floodplain: A model based on *Boltonia decurrens,* a threatened species. In *Research on Agricultural Chemicals in Illinois Groundwater: Status and Future Directions,* ed. by M. Davis, pp. 135–152. Illinois Groundwater Consortium, Southern Illinois University Carbondale, Carbondale, IL.

Soong, T. W., and W. H. Ettinger. 2000. After the 1993 flood: A water and surficial bed sediment quality scenario on the Illinois and upper Mississippi Rivers. *Journal of the American Water Resources Association* **36:**105–121.

Sousa, W. P. 1984. The role of disturbance in natural communities. *Annual Review of Ecology and Systematics* **15:**353–391.

Sparks, R. E. 1984. The role of contaminants in the decline of the Illinois River: Implications for the upper Mississippi. In *Proceedings of the 15th Annual*

Meeting of the Mississippi River Research Consortium, ed. by J. G. Wiener, R. V. Anderson, and D. R. McConville, pp. 25–28. Butterworth Publishers, Stoneham, MA.

Sparks, R. E. 1995. Need for ecosystem management of large rivers and their floodplains. *BioScience* **45:**168–182.

Sparks, R. E. 1996. Ecosystem effects: Positive and negative outcomes. In *The Great Flood of 1993,* ed. by S. A. Chagnon, pp. 132–162. Westview Press, Boulder, CO.

Sparks, R. E., and A. Spink. 1998. Disturbance, succession and ecosystem processes in rivers and estuaries: Effects of extreme hydrologic events. *Regulated Rivers: Research and Management* **14:**155–159.

Sparks, R. E., J. C. Nelson, and Y. Yin. 1998. Naturalization of the flood regime in regulated rivers. *BioScience* **48:**706–720.

Sparks, R. E., J. B. Braden, M. Demissie, P. Mitra, D. W. Schneider, D. C. White, and R. Xia, eds. 2000. *Technical Support of Public Decisions to Restore Floodplain Ecosystems: A Status Report on the Illinois River Project, USA.* Backhuys Publishers, Leiden, The Netherlands.

Sparks, R. E., P. B. Bayley, S. L. Kohler, and L. L. Osborne. 1990. Disturbance and recovery of large floodplain rivers. *Environmental Management* **14:** 699–709.

Starrett, W. C. 1971. Man and the Illinois River. In *International Symposium on River Ecology and the Impact of Man,* ed. by R. T. Oglesby, C. A. Carlson, and J. A. McCann, pp. 131–167. Academic Press, Amherst, MA.

Steffeck, D. W., F. L. Paveglio, F. C. Bellrose, and R. E. Sparks. 1980. Effects of decreasing water depths on the sedimentation rate of Illinois River bottomland lakes. *Water Resources Bulletin* **16:**553–555.

Stoecker, M. 1996. Flood tolerance in the federally threatened species, *Boltonia decurrens.* Master's thesis, Southern Illinois University, Edwardsville, IL.

Stoecker, M. A., M. Smith, and E. D. Melton. 1995. Survival and aerenchyma development under flooded conditions of *Boltonia decurrens,* a threatened floodplain species, and *Conyza canadensis,* a widely distributed competitor. *American Midland Naturalist* **134:**117–126.

Theiling, C. H. 1995. Habitat rehabilitation on the Upper Mississippi River. *Regulated Rivers: Research and Management* **11:**227–238.

Thompson, J. 1989. *Case Studies in Drainage and Levee District Formation and Development on the Floodplain of the Lower Illinois River, 1890s to 1930s.* Report # 16. Illinois Water Resources Center, Champaign, IL.

Torrey, J., and A. Gray. 1841. *Flora of North America.* Vol. 2. Wiley and Putnam, New York.

Turner, L. M. 1934. Grassland in the floodplain of Illinois Rivers. *The American Midland Naturalist* **15:**770–780.

Turner, L. M. 1936. Ecological studies in the Lower Illinois River Valley. *Botanical Gazette* **97:** 689–727.

U.S. Fish and Wildlife Service. 1990. *Decurrent False Aster Recovery Plan.* U.S. Fish and Wildlife Service, Twin Cities, MN.

Vogl, R. J. 1980. The ecological factors that produce perturbation-dependent ecosystems. In *The Recovery Process in Damaged Ecosystems,* ed. by J. Cairns, pp. 63–94. Ann Arbor Science, Ann Arbor, MI.

Waller, R. 1972. The Illinois Waterway from conception to completion, 1908–1933. *Journal of the Illinois State Historical Society* **65:**125–141.

Warner, K. L. 1998. *Water-Quality Assessment of the Lower Illinois River Basin: Environmental Setting.* Report # 97-4165. U.S. Geological Survey, Urbana, IL.

White, P. S., and S. T. A. Pickett. 1985. Natural disturbance and patch dynamics: An introduction, In *The Ecology of Natural Disturbance and Patch Dynamics,* ed. by S. T. A. Pickett and P. S. White, pp. 3–13. Academic Press, San Diego, CA.

Yaeger, L. E. 1949. Effect of permanent flooding in a river timber area. *Illinois Natural History Survey Bulletin* **25:**33–65.

Yin, Y., J. C. Nelson, and K. S. Lubinski. 1997. Bottomland hardwood forest along the Upper Mississippi River. *Natural Areas Journal* **17:**164–173.

5

Conservation and Restoration of Semiarid Riparian Forests: A Case Study from the Upper Missouri River, Montana

Michael L. Scott and Gregor T. Auble

U.S. Geological Survey, Midcontinent Ecological Science Center, Fort Collins, Colorado

For large flow-controlled rivers, full restoration of natural, regionally specific stream flow patterns is not likely. In these situations, scientific understanding of the physical and biotic processes responsible for creating and sustaining riverine and riparian ecosystem values is critical to making wise decisions about partial restoration efforts in the context of competing social and economic values. A case study of riparian cottonwood forest dynamics from the Wild and Scenic Reach of the Missouri River in Montana, USA, illustrates the difficulties of such efforts. This reach of the Missouri River has become the nexus of debate over how future land and water man-

Flood Pulsing in Wetlands: Restoring the Natural Hydrological Balance, edited by Beth A. Middleton.
ISBN 0 471-41807-2

agement can best protect this resource. Of particular concern is how flow management and livestock grazing are influencing the distribution and abundance of riparian cottonwoods and associated wildlife species.

Historical accounts are consistent with contemporary observations that cottonwood forests are sparse, and dendrogeomorphic analyses indicate that a majority of existing trees were established following infrequent, large flood pulses. Multiyear monitoring of seedling demographics in exclosures and paired control sites document the relative importance of mortality factors such as ice scour and livestock grazing. Livestock grazing greatly reduces seedling densities, but long-term survival of stems in constrained channel reaches is correlated with flood-related establishment on alluvial surfaces above the zone of frequent ice scour. Reconstruction of unregulated peak flows indicates that although the frequency of large flood pulses has not been influenced by upstream dams, the magnitude of these events has been reduced from 40 to 50 percent. Retrospective analyses of cottonwood patches demonstrate that postdam recruitment is limited to unconstrained channel reaches, typically associated with channel islands and tributary junctions, where stems are most likely to survive subsequent ice disturbance. Riparian bird surveys demonstrate that avian diversity along the upper Missouri is significantly related to the flow-related geomorphic processes responsible for the establishment and survival of new cottonwood and willow patches, combined with a reduction in or elimination of grazing, following establishment of these patches.

Flow models and flood damage curves for the upper Missouri River indicate that at present there is enough operational flexibility in the system to produce large flood pulses without greatly compromising other values. However, although flood pulses are necessary, they are not sufficient. In addition to grazing, future human development on the floodplain of the upper Missouri and its tributaries could increasingly limit the potential to produce flood pulses. Thus, riparian forest restoration on the upper Missouri River will require integrated land and water management activities throughout the entire basin.

INTRODUCTION

Human demands for water in arid and semiarid regions of North America have led to intensive modification of the hydrologic landscape. In the American Southwest and Great Plains, dams and reservoirs are capable of storing 2.5 to 3.8 times the mean annual runoff (Graf, 1999). These dams were typically designed and operated to optimize water storage and release for flood control, irrigation, and power generation. The unanticipated consequences of these water management measures include fragmentation of river corridors (Dynesis and Nilsson, 1994) and modification of the natural patterns of sediment and water flow (Williams and Wolman, 1984), which create the disturbance patches and topographic diversity that maintain riverine and riparian ecosystems (Brinson 1990; Ligon et al., 1995; Poff et al., 1997). Riparian forests dominated by cottonwood (*Populus* spp.) have experienced widespread decline and structural simplification along flow-controlled rivers (Rood and Mahoney, 1990). Flow depletions downstream of dams induce drought stress and mortality in existing trees, and modification of spring flood pulses reduces or eliminates conditions required for the establishment of new seedlings (Johnson et al., 1976; Rood and Mahoney, 1990; Johnson, 1992). The importance of pulsed flood disturbances in maintaining and restoring the ecological integrity of wetland and riparian ecosystems (Middleton, 1999) is related to the frequency, magnitude, duration, timing, and recession rates of these hydrologic events (Stromberg et al., 1993; Hughes, 1994; Mahoney and Rood, 1993; Cordes et al., 1997; Scott et al., 1997; Shafroth et al., 1998).

Prescribing more natural flood pulses to flow-controlled rivers, for the purpose of restoring riparian forest ecosystems, requires a general understanding of the ecology of riparian species, as well as a region- or even basin-specific knowledge of the long-term interplay of physical processes that drive patterns of recruitment and mortality. Full restoration of natural stream flow patterns may not be possible along large flow-controlled rivers, and flow prescriptions for natural resource values must be made in the context of competing social and economic values (Schmidt et al., 1998; Stromberg, in press). In this chapter we review the ecology of cottonwood species and the physical processes that govern the pattern and extent of riparian forests in semiarid landscapes. We conclude with a discussion on the process of developing flow prescriptions for the upper Missouri River in Montana, a flow-controlled reach where the role of large flood pulses and ice-related disturbance in structuring riparian forests is particularly evident.

RIPARIAN FORESTS IN DRY REGIONS

Establishment and Persistence of Woody Riparian Species

Riparian sites in arid and semiarid regions are typically the only portions of the landscape moist enough to support tree growth. Species of cottonwood are typically the dominant native trees of riparian ecosystems throughout arid and semiarid regions of North America (Rood and Mahoney, 1990). Cottonwoods require continually moist substrates for establishment (Read, 1958; Friedman et al., 1995) and are generally restricted to alluvial soils with shallow groundwater (Meinzer, 1927; Robinson, 1958). Cottonwoods are susceptible to drought-induced cavitation of the xylem vessels (Tyree et al., 1994) and are less drought-tolerant than several eastern deciduous forest species (Kaylor et al., 1935; Albertson and Weaver, 1945). The ruderal traits of cottonwood contribute to the persistence of this drought-sensitive species in drought-prone regions (Friedman et al., 1997); these traits include the production of small wind- and water-dispersed seeds, rapid germination and growth, and the ability to tolerate inundation and intense physical disturbance during floods. Cottonwood forest cover has declined where severe drought, channel incision, or land and water management activities have decreased water availability by reducing surface flows or depleting alluvial groundwater aquifers (Albertson and Weaver, 1945; Cooke and Reeves, 1976; Rood and Heinze-Milne, 1989; Rood et al., 1995; Stromberg et al., 1996; Scott et al., 1999, 2000; Shafroth et al., 2000).

The environmental conditions associated with the successful establishment of cottonwoods have been well described. An abundant crop of seeds is released in early summer, soon after peak discharge in unregulated streams (Fenner et al., 1985). The seeds can germinate immediately, but lose germinability under field conditions within a few weeks (Moss, 1938; Ware and Penfound, 1949; Kapustka, 1972). Freshly deposited alluvium typically provides ideal substrate for germination and establishment. Because they are small-seeded and intolerant of shade, cottonwoods are unlikely to establish from seed under an existing stand of trees (Johnson et al., 1976) or herbs (Friedman et al., 1995). This trait often leads to even-aged stands. Young seedlings require a continuously moist substrate during at least the first week of growth (Moss, 1938; Read, 1958). Root growth during the first month is slow (van Haverbeke, 1990), but later in the first growing season seedlings of some cottonwood species are able to extend a taproot deep enough to survive declines in the water table of as much as 1 m (Fenner et al., 1985; Mahoney and Rood, 1991; Segelquist et al., 1993). Bottomland cottonwoods are tolerant of burial and are able to sprout from stems or roots (Nanson and Beach, 1977; Krasny et al., 1988). However, extensive mortality of young plants has been reported as a result of floods (Zimmerman, 1969) or ice scour (Johnson, 1994; Smith and Pearce, 2000). On coarse substrates in dry regions, early establishment and growth of cottonwood seedlings may require water tables within 1 to 2 m of the establishment surface (McBride and Strahan, 1984; Mahoney and Rood, 1992; Segelquist et al., 1993; Stromberg et al., 1996), although, in some settings, seedlings may not establish contact with the alluvial water table for the first few years following establishment (Cooper et al., 1999). Depth to the water table may increase as a result of subsequent floodplain accretion or channel incision (Everitt, 1968; Hereford, 1986), and cottonwood species have been observed at sites where depth to the water table is 7 to 9 m (Robinson, 1958). However, mature riparian cottonwoods are typically found in riparian settings where depth to the water table is ≤ 3.5 m (Busch et al., 1992; Stromberg et al., 1996; Smith et al., 1998). The architecture of cottonwood root systems reflects specific groundwater environments (Yeager, 1935), and established trees on sites with shallow groundwater have correspondingly shallow, laterally extensive root systems (Sprakling and Read, 1979; Scott et al., 2000; Shafroth et al., 2000). Although species of cottonwood may utilize moisture from unsaturated soil layers (Snyder and Williams, 2000), they are relatively drought-sensitive (Rood et al., 2000). Cottonwood stands on coarse alluvial substrates with shallow water tables may be particularly vulnerable to drought stress and mortality following sustained water table declines as small as 1 m (Scott et al., 1999, 2000; Shafroth et al., 2000).

Fluvial Geomorphic Processes

The conditions required for the initial establishment of early successional woody species such as cottonwood include bare, moist, mineral substrates that remain relatively free of subsequent disturbance. These are restrictive requirements in arid and semiarid riverine landscapes and are met most frequently and extensively by flow-induced channel change. Fluvial geomorphic processes like channel narrowing and meandering, as well as vertical accretion of alluvial surfaces by flood deposition, produce bare, moist alluvial substrates suitable for the establishment of early successional riparian species such as cottonwood and willow. These processes produce distinctive spatial and temporal patterns of cottonwood/willow forest establishment (Bradley and Smith, 1986; Friedman et al., 1996; Scott et al., 1996; Auble and Scott, 1998, Dykaar and Wigington, 2000).

In riparian systems the associated watercourse is an important determinant of site conditions. The river provides moisture that can serve as either a subsidy or stress to woody vegetation. Throughout much of the riparian zone in arid and semiarid regions, supplemental moisture from the river is a subsidy that provides sufficiently mesic conditions for trees in a landscape where the surrounding uplands are too zeric to support them. Brinson et al. (1981) illustrate this subsidy effect by comparing upland to riparian sites. On upland sites, there is a clear pattern of increasing basal area with higher precipitation. However, the basal area of a riparian site is largely unrelated to precipitation because of the supplemental moisture provided by the river. In wetter regions, excessive moisture limits the distribution of trees through the stresses associated with waterlogged and anoxic soils. Species differences in tolerance to anoxia and moisture stress produce much of the distinctive patterns of zonation in bottomland hardwood riparian forests (Wharton et al. 1982). Similar limitations resulting from excessive moisture occur in the riparian zones of arid and semiarid regions, although the effects on distribution may be evident only on the streamward portion of the riparian zone (Friedman and Auble 2000).

The river may also serve as an important transport vector for dispersing seeds (Nilsson et al., 1991) and carrying organic and inorganic materials to and from the site. Transport and flushing of salts is especially important where disconnection from the river may produce concentrations of salts from evapotranspiration. The river episodically disturbs riparian sites by killing plants or removing biomass. Geomorphic change such as channel narrowing and widening, lateral migration and avulsion, and aggradation and degradation can alter the way stream flow affects conditions at any fixed site. Herbivory, including sheep and cattle grazing, may strongly influence riparian vegetation at some sites (Chaney et al., 1990; Green and Kauffman, 1995). Finally, the introduction of exotic species has fundamentally altered the mix of plant species at most southwestern riparian sites and may have defined new trajectories of response to some combinations of physical site conditions (Ohmart et al., 1988).

Locally degraded riparian sites occur along rivers that have been fundamentally altered and no longer provide appropriate physical site conditions for the original vegetation. The upper Missouri River in Montana serves as an example of how a scientific understanding of the physical processes that shape riparian vegetation can lead to specific restoration prescriptions, and the social and political implications of implementing such measures.

THE UPPER MISSOURI RIVER, MONTANA: A CASE STUDY

Study Area

Within its upper basin (Figure 5-1) the Missouri River traverses the highly dissected topography of the Missouri Breaks and, in places, occupies a relatively narrow, postglacial valley incised 150 to 560 m below the surrounding landscape (Alden, 1932). The landscape surrounding the river consists of relatively undisturbed prairie vegetation dominated by black greasewood, (*Sarcobatus vermiculatus* (Hook.) Torr.), big sagebrush (*Artemisia tridentata* Nutt.), and mixed grasses, including western wheat grass (*Agropyron smithii* Rydb.), blue grama (*Bouteloua gracilis* (H.B.K.)), downy brome (*Bromus tectorum* L.), quackgrass (*Agropyron repens* (L.) Beauv.), green needlegrass (*Stipa viridula* Trin.), smooth brome (*Bromus inermis* Leyss.), and needle-and-thread (*Stipa comata* Trin. and Rupr.), which occur on exposed ridges and flats. North-facing slopes contain isolated stands of Douglas fir (*Pseudotsuga menziesii* (Mirbel) Franco), ponderosa pine (*Pinus ponderosa* Laws.), and juniper (*Juniperus scopulorum* Sarg.), whereas south-facing slopes contain scattered grasses, ponderosa pine, or bare exposures of shale. The principal tree species along the Missouri River is plains cottonwood (*Populus deltoides* Marsh. Subsp. *monilifera* (Ait.) Eckenw.). Box elder (*Acer negundo* L.), green ash (*Fraxinus pennsylvanica* Marsh.), and peach-leaf willow (*Salix amygdaloides* Anderss.) occur as less common associates, particularly on islands and in former back channels that have been filled by alluvial sediments. Understory shrubs on alluvial surfaces include yellow willow (*Salix lutea* Nutt.), sandbar willow (*Salix exigua* Nutt.), western snowberry (*Symphoricarpos occidentalis* Hook.), Wood's rose (*Rosa woodsii* Lindl.), silver sagebrush

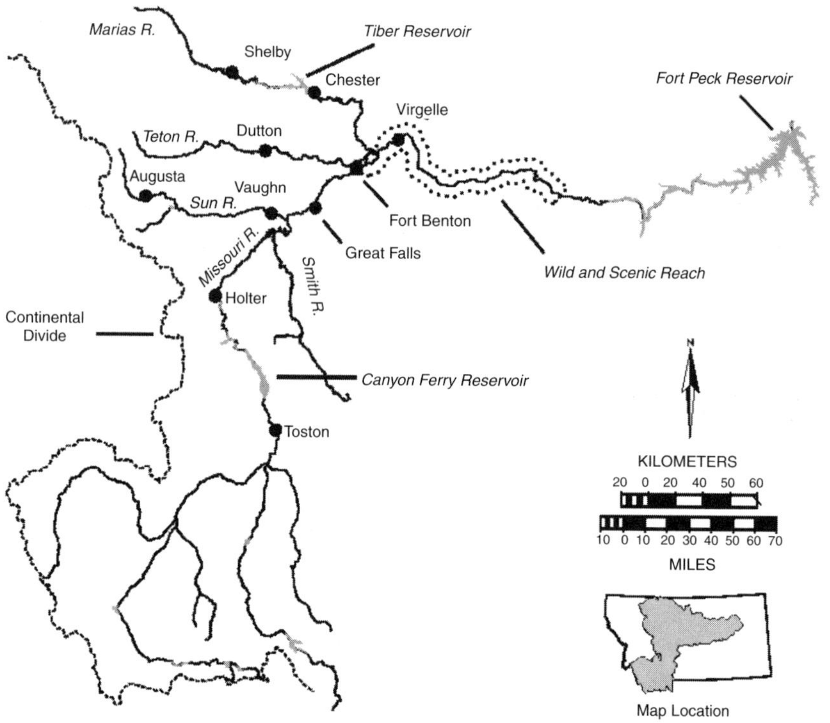

Figure 5-1. *Map of the upper Missouri River basin, showing locations of major reservoirs and hydrologic nodes used in the hydrograph synthesis. The study area within the Wild and Scenic Reach is highlighted by the line of dots adjacent to the channel. (After Bovee and Scott, in press; copyright © by* Regulated Rivers: Research and Management; *used by permission.)*

(*Artemisia cana* Pursh.), common chokecherry (*Prunus virginiana* L.), and, rarely, red osier dogwood (*Cornus stolonifera* Midix.) (Ross and Hunter, 1976).

Modern agricultural development within the river corridor is limited and primarily confined to old Missouri River terraces. Both the river corridor and the surrounding uplands have been subject to livestock grazing since settlement in the late 1800s; these locations currently include all or parts of 55 permitted grazing areas, totaling 92,848 ha of public land (U.S. Department of the Interior, 1993). However, because of a shortage of upland water sources and the steep, rugged terrain between the uplands and the bottomland, livestock within the Upper Missouri River Wild and Scenic River corridor are poorly distributed in most allotments and preferentially concentrate on the river floodplain where water and shade are available. Bureau of Land Management (BLM) monitoring data indicate heavy to severe utilization of the floodplain in most allotments (J. Frazier, personal communication). In contrast, where pastures are permitted for fall and winter grazing only or at sites inaccessible to cattle, portions of the floodplain are subject to less or no livestock grazing.

This reach of the Missouri River has a snowmelt hydrograph with annual flow peaks typically occurring in May or June. The natural flow regime has been influenced by irrigation withdrawals and upstream dams. Two large upstream dams, Canyon Ferry on the Missouri River and Tiber on the Marias River, a major tributary, were completed in 1954 and 1956, respectively. Although the seasonal timing of flows has not been altered, the magnitude of peak flows has been reduced and portions of the low flow hydrograph have increased (Scott et al., 1997; Bovee and Scott, in press).

Physiographic and Climatic Setting

The effects of fluvial processes on riparian vegetation dynamics can vary within a region because of physiographic and climatic factors (Swetnam and Bentancourt, 1998; Friedman and Auble, 2000). Within the upper Missouri River basin, extreme rainfall and abrupt, mechanical ice breakups contribute to the development of episodic spring flood pulses and winter ice floes or drives (Dightman, 1950; Smith and Pearce, 2000). The interplay of these two disturbance factors exert primary control on the pattern and extent of riparian forests in the Wild and Scenic study reach.

The Climatology of Flood Pulses

Widespread and intense rainstorms along the eastern front of the Rocky Mountains in northern Montana and southern Alberta, Canada, are part of the climatic history of the upper Missouri River basin. Records indicate the period from May 20 through June as the time when atmospheric conditions are most likely to produce heavy rainfall in central Montana (Cordell, 1974). Heavy rainstorms produced historical floods on the upper Missouri in 1894, 1906, 1908, 1916, 1927, 1938, 1948, 1953, and 1964. Although antecedent conditions of snowmelt runoff and rain-soaked soils also contributed to these floods, they have been primarily rain-induced (Bonner and Stermitz, 1967).

The importance of upslope orographic precipitation in many of the historical floods, particularly in the Marias and Teton basins, is implied by the distribution of rainfall maxima, which are often observed on the north and east slopes of mountain ranges east of and along the Continental Divide (Dightman, 1950; U.S. Geological Survey, 1957; Bonner and Stermitz, 1967). Flood-producing rainstorms resulted from similar large- and medium-scale meteorological conditions. The elements necessary to produce heavy rain in the upper Missouri basin include (1) a source of warm, moist air, (2) vertical lifting of this air mass by a combination of large and medium-scale atmospheric processes, which are (3) sustained over a period of hours or days. All previous rainstorms received a supply of moist, tropical air from the Gulf of Mexico, carried northward and then westward, upslope, over northern Montana ahead of low-pressure centers that moved slowly eastward into the western Great Plains. This moisture-laden air fed vertical air motion at the storm centers, producing rainfall totals of up to 40 cm over periods of two to three days. The sustained vertical lifting of air necessary to produce the intensity of precipitation observed in these storms resulted from a combination of (1) orographic lifting, as warm, moist upslope winds encountered the steep slopes of local mountain ranges east of the Continental Divide, (2) vorticity and horizontal convergence of air at the surface and aloft near the storm centers, and (3) frontal lifting as the upslope flow overrode colder air at the surface.

The Climatology of Ice Drives

In winter, warm, downslope winds sometimes referred to as "chinooks," may develop along the eastern slopes of the Continental Divide from New Mexico to Alaska. Chinook winds are especially pronounced along the eastern foothills of the Rocky Mountain cordillera in northern Montana and southern Alberta. These conditions typically result when the east side of the Continental Divide is under the influence of Arctic high pressure and the eastward flow of air from the Pacific intensifies, typically in association with a low-pressure trough in the Gulf of Alaska. As this flow of relatively warm Pacific air pushes up and spills over the Continental Divide, it produces standing waves. The troughs of these standing waves displace cold air at the surface. Compression of the air mass as it descends the east slope results in the warm, dry winds typical of a chinook. During a chinook, temperatures at the surface can rise 30° C in a few hours (Figure

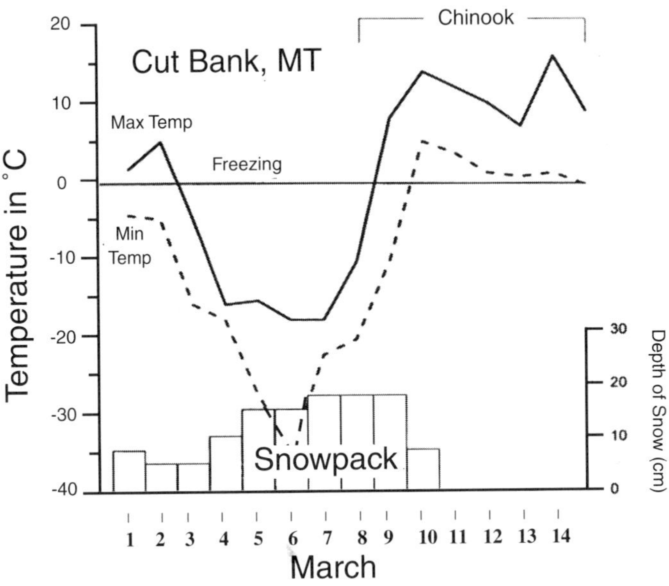

Figure 5-2. *Weather data from Cut Bank, Montana, illustrating the abrupt rise in temperatures and rapid loss of snowpack resulting from a chinook event, March 8–15 (Reprinted from Smith and Pearce, 2000, copyright © by Society of Wetland Scientists, by permission.)*

5-2), resulting in rapid melting of the snowpack. Subsequent tributary runoff causes mechanical ice breakups and ice drives in trunk streams, as described by Smith (1980) for large, northern plains rivers in Canada. The channel geometry of such rivers is typically adjusted to passage of ice drives rather than the bankfull or two-year return flows more typical of rivers in warmer regions. The passage of an ice drive, over the course of several hours, enlarges the channel and subjects vegetation between the mean 2- and 9-year flood stage to intense physical disturbance (Figure 5-3).

Geologic History and Channel Type

The influence of floods and ice drives on riparian forest dynamics of the upper Missouri are conditioned by geologic history, which influences the pattern and extent of channel movement. Geologic constraints on lateral channel movement can limit the creation of suitable sites for the establishment of cottonwood. Gregory et al. (1991) operationally defined constrained channels as reaches where valley width was less than twice the width of the active channel. The Wild and Scenic study reach, situated be-

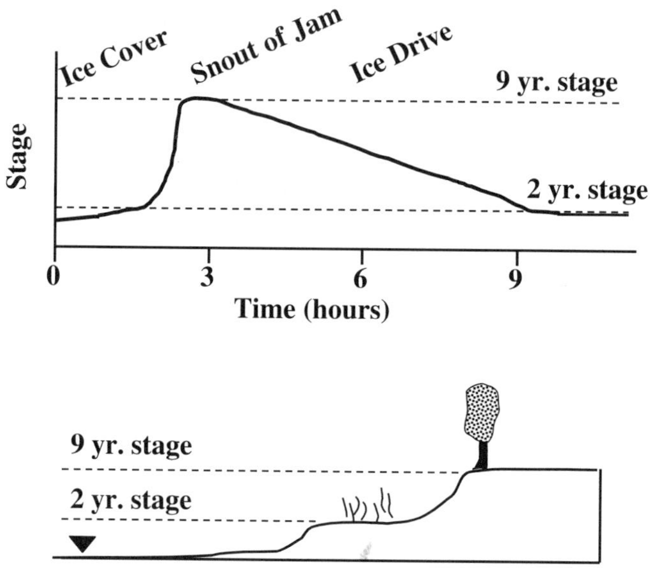

Figure 5-3. The stage range of ice with the passage of an ice drive over the course of several hours. Repeated drives enlarge the channel and subject vegetation between the two- and nine-year average return flood stage to intense physical disturbance. (After Smith, 1980, copyright © by Allen & Unwin; used by permission.)

tween the Fort Benton and Landusky stream gauges, includes constrained and unconstrained channel reaches that reflect a complex geologic history (Figure 5-4). For example, the preglacial Missouri River flowed northeast from Fort Benton in a broad valley. Another preglacial valley carried flows northwest from the Judith River and Dog Creek drainages, which joined the Missouri north of Virgelle (Figure 5-1). Portions of these preglacial valleys were filled by drift during continental glaciation (Calhoon, 1906). The modern Missouri River intersects these preglacial valleys upstream of the Virgelle gauge and at the Judith River confluence. During the last glacial advance, the flow of the Missouri was diverted southward by the ice sheet, near Virgelle, into a relatively young, narrow postglacial valley that traverses the steep sandstone and shale badlands of the Missouri Breaks. Just above the Landusky gauge, the Missouri River encounters broad exposures of easily erodable bear paw shale near river level, and the valley widens again (Porter and Wilde, 1984).

Where the course of the modern Missouri within the Wild and Scenic

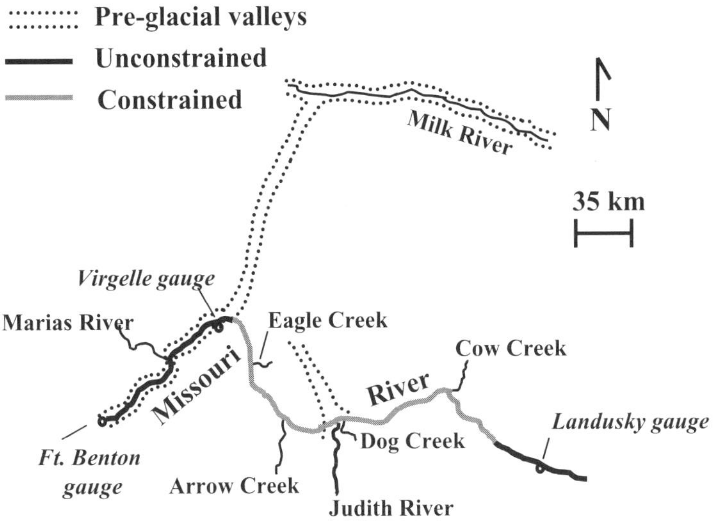

Figure 5-4. Geologic setting of the Missouri River Wild and Scenic Study Reach, situated between the Ft. Benton and Landusky (Montana) stream gauges. The study reach includes unconstrained channel reaches in relatively wide preglacial valleys and constrained reaches in a narrow postglacial valley. Upstream of Landusky, broad exposures of bear paw shale near river level are associated with increased valley width and a less constrained channel.

Reach coincides with drift-filled preglacial valleys, or river-level exposures of bear paw shale, the channel is relatively unconstrained (Figure 5-4); that is, the active channel width is several times less than the valley width. In contrast, the postglacial Missouri River channel between Virgelle and Landusky consists of a series of entrenched meanders, constrained by exposures of sandstone and shale badlands. Here the channel width is only slightly less than the valley width (Figure 5-5*A* and *B*). Where channels are relatively unconstrained, cottonwood recruitment sites are created more frequently and extensively by lateral channel movement and vertical sediment accretion, over a range of flows. However, where lateral channel movement is constrained, recruitment sites are generally small, scattered, and confined to elevated flood deposits following large, infrequent flood pulses (Scott et al., 1996). Within larger constrained reaches, unconstrained zones typically occur near tributary junctions, channel islands, and overflow channels (Figure 5-6). Riparian vegetation along the Animas River in Colorado has greater patch diversity at tributary

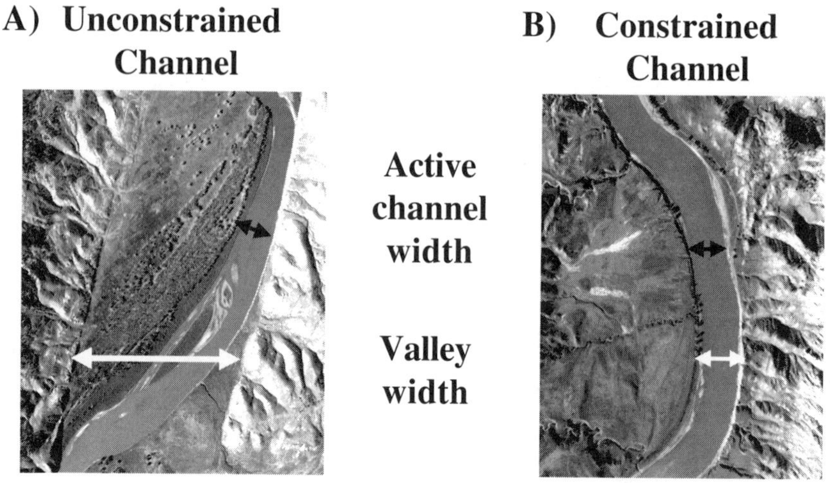

Figure 5-5. Plan view aerial photos from the study reach depicting unconstrained (A) and constrained (B) channel reaches. (Source of aerial photographs: U.S. Bureau of Land Management, Denver Federal Center, Denver, Colorado.)

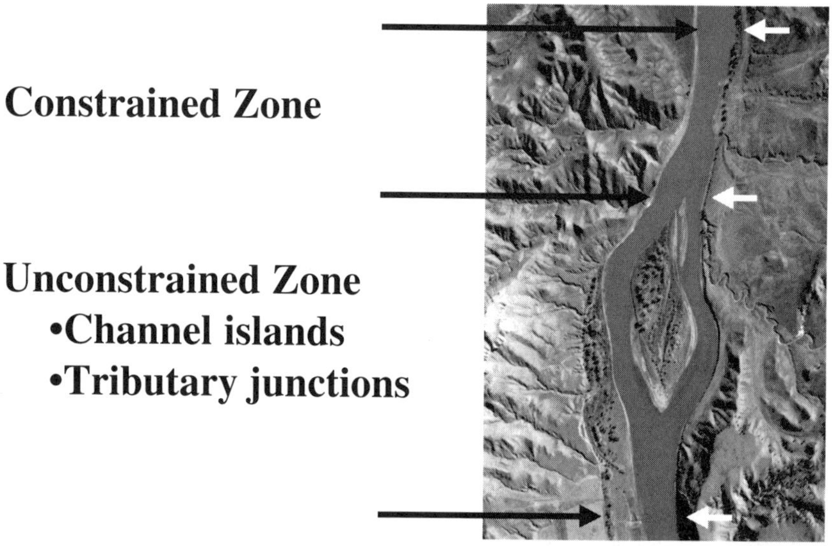

Figure 5-6. Plan view aerial photo illustrating localized variation in channel constraint within the larger constrained reach. These unconstrained reach segments are primarily associated with tributary junctions and channel islands. (Source of aerial photographs: U.S. Bureau of Land Management, Denver Federal Center, Denver, Colorado.)

junctions and less constrained channel reaches, where flood disturbances were most likely to create new geomorphic surfaces (Walford and Baker, 1995).

Presettlement descriptions of a generally sparse bottomland forest in constrained reaches, and more abundant forest cover in unconstrained reaches, are qualitatively similar to the observations of conditions throughout the study reach today. With the exception of the Judith River confluence, where they describe channel islands "covered with cottonwood" and a wide bottomland "enriched" by trees, the Lewis and Clark journals (Coues, 1893) make repeated reference to the limited occurrence of cottonwood and other woody riparian vegetation, both upstream and downstream of the Judith River (Figure 5-4). After covering 30 up-river kilometers to their camp at the mouth of Eagle Creek (Figure 5-4), they report, "The river today has been from about 150 to 250 yards wide, with but little timber." In contrast, above Coal Banks Landing, where the river occupies a broad drift-filled valley, they report that its "timber increases in quantity, the low ground became more level and extensive," and they "came-to for the night in a handsome, low cottonwood plain on the south."

Retrospective and Demographic Analyses of Cottonwood Establishment and Mortality

Because cottonwoods colonize bare, freshly deposited alluvial surfaces, they are important indicators of fluvial geomorphic processes. Dendrochronological measurements of the spatial and temporal patterns of cottonwood establishment have provided important insights into the patterns and processes of channel movement, rates of floodplain formation, and the timing of specific fluvial events (Everitt, 1968, 1979; Nanson and Beach, 1977; Bradley and Smith,1986; Gottesfeld and Gottesfeld, 1990; Stromberg et al., 1991; Friedman et al., 1996). A dendrogeomorphic approach was used to examine the relation between the origin of cottonwood forest patches and historic flows along the Wild and Scenic Reach of the Missouri River. Relating cottonwood establishment to specific flow events requires precise aging of stems at the original establishment surface. Because cottonwoods typically establish in depositional environments, stems may be prostrated or flood-trained and establishment surfaces buried by subsequent sediment accretion (Everitt, 1968). Thus, precise aging of cottonwood stems requires excavating each stem, determining the point of establishment using morphological and stratigraphic information, and sectioning the stem in multiple locations to verify the establishment date based on wood anatomy and ring count (Figure 5-7).

Using this approach, Scott et al. (1997) determined that 72 percent of all sampled cottonwood stems ≥ 1 m in height had established in the year of a flood pulse exceeding 1400 m^3 sec^{-1}, measured at the Fort Benton gauge, or in the two years following (Figure 5-8). Floods of this magnitude had an estimated recurrence interval of 9.3 years for the 112-year period of recorded flow. Although seedling, sapling, and pole-sized cottonwood stems establish across a range of elevations, all trees had established ≥ 1.75 m above the lower limit of perennial vegetation (Figure 5-9), or the active channel shelf of Osterkamp and Hedman (1982). Further, many saplings and poles established at lower elevations revealed evidence of repeated past disturbances (Figure 5-10). In the absence of recent channel downcutting (Scott et al., 1997), the combined processes of episodic flood pulses and winter ice-drive disturbance best explain the observed patterns of cottonwood establishment and survival at higher elevations and repeated physical disturbance and mortality at lower elevations.

To refine our understanding of cottonwood establishment and mortality factors, such as ice scour and livestock grazing, we began demographic monitoring in 1996 at a set of eight sites subject to a range of livestock grazing intensities. Three permanent transects of 1 × 5 m plots were censused annually, with empirical rating curves used to relate elevations of plots to stream discharge as measured at the upstream Virgelle gauge (Auble and Scott 1998). The year 1997 was a year of higher than average flow, producing high densities of new seedlings over a wide range of bank positions (Figure 5-11). Subsequent years of moderate flow (1998 and 1999) and the drought year of 2000 allowed good survivorship and recruitment from seedling to sapling stages of those 1997 seedlings established at higher bank positions. These higher locations were not strongly disturbed by either flows or ice scour in the years 1998 through 2000.

The strong association between new seedling establishment and the water's edge during the period of seed release and germinability (germination window) can be seen by comparing the wide distribution of new seedlings in 1997, and to a lesser extent in 1998, with the very limited and low positions of seedlings established in the dry years of 1999 and 2000. The shape of the hydrograph also influences establishment. The year 1998 had an abnormal and unsuitable hydrograph, with flows increasing throughout

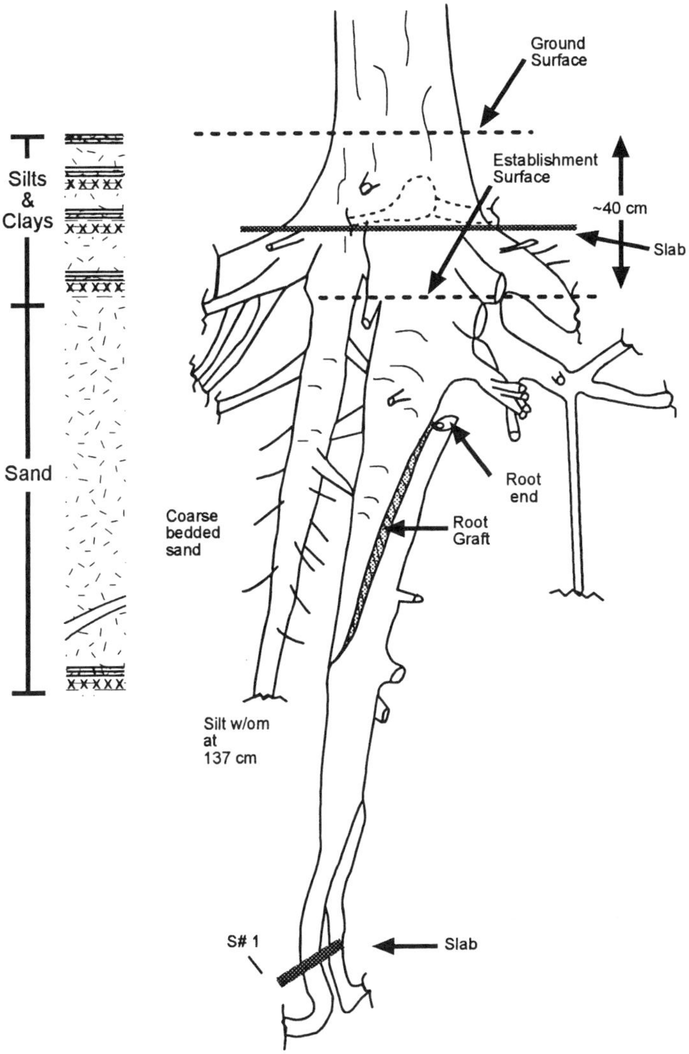

Figure 5-7. *Field sketch of excavated cottonwood stem showing the relationship between depositional stratigraphy and stem morphology. The establishment surface of this tree corresponded with the top of a thick, coarse-bedded sand, above which were distinct units of silts, fining upward to clays. Slabs were sectioned from the tree at positions indicated by the patterned arrows to determine a precise maximum tree age. (From Scott, et al.; unpublished data.)*

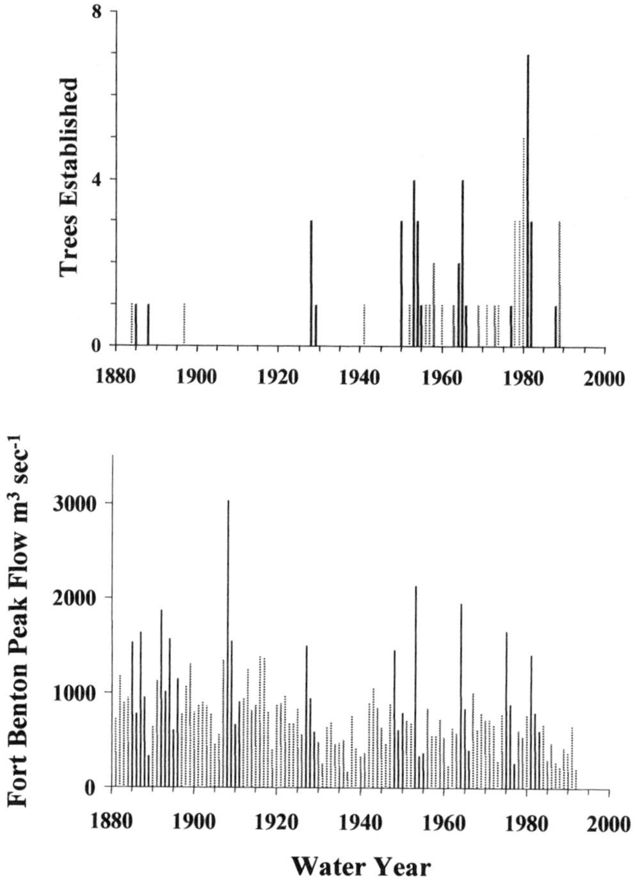

Figure 5-8. *The lower histogram plots maximum daily discharge in m^3 sec^{-1}, by water year, for the period of record at the Fort Benton (Montana) gauge. Flood years, in which discharge exceeded 1400 m^3 sec^{-1}, and the following two years are represented by solid bars. The upper histogram represents the number of trees established in each water year, with solid bars representing establishment in flood years. (After Scott et al., 1997; copyright © by the Ecological Society of America; used by permission.)*

much of the germination window (Figure 5-11). This produced relatively low densities of new seedlings, as each day's increased flow tended to remove seedlings established in previous days.

Cattle grazing is reducing densities of new seedlings, survivorship of seedlings, and net recruitment from the seedling to sapling size class. Some results of the grazing influence can be seen in the pooled densities of Figure 5-11 where the loss of saplings at higher positions between 1999 and 2000 was due to a combination of drought stress and grazing that con-

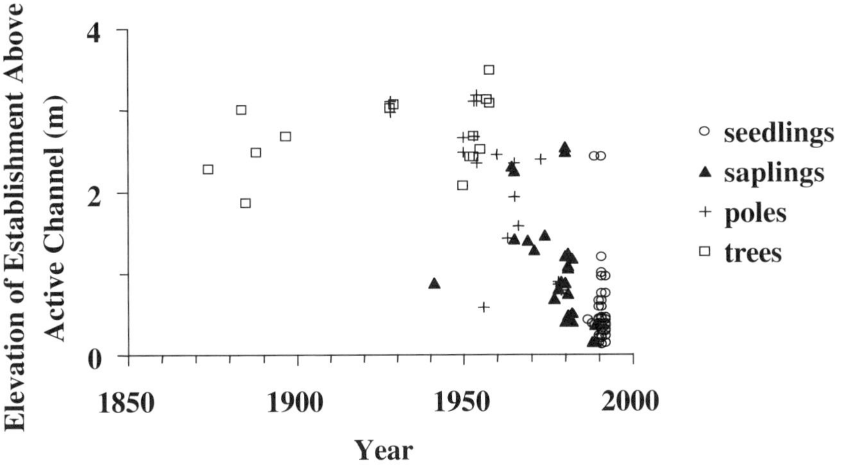

Figure 5-9. Establishment elevation in meters above the active channel shelf (Osterkamp and Hedman, 1982) is plotted by establishment year for four size classes of cottonwood stems. (After Scott et al., 1997; copyright © by the Ecological Society of America; used by permission.)

verted some short saplings back into the old seedling size class. Repeated census of seedlings, in combination with measurement of interannual variation of controlling variables, has revealed much about the interplay of factors influencing cottonwood regeneration along this laterally constrained reach. However, the most telling comparison is between the high positions of the lowest scour line resulting from the severe winter ice of 1995–1996 and the lowest existing tree at any of the eight sites, relative to the positions of seedlings established and surviving in years of more moderate flow that we have observed so far. These relative positions suggest that it is unlikely that many of the seedlings we are currently tracking will survive the next episodic ice scour or ultimately recruit to the tree size class, even in the absence of heavy grazing.

On the basis of the aforementioned findings, we propose the following conceptual models to explain observed patterns of flood-dependent cottonwood establishment and survival for constrained and unconstrained reaches of the Wild and Scenic Missouri River. In constrained reaches, successful recruitment is dependent on large flood pulses that exceed the stage of the mean one-in-nine-year flow (Figure 5-12). In the first year of such a flow, germination occurs on surfaces between the peak flow and

Figure 5-10. A flood- or ice-trained cottonwood stem from the study reach. Between the establishment point and ground surface, this stem was prostrated twice (A and B) and sheared once (C), producing multiple stems at the ground surface, which had accreted more than 3 m in 40 years. (Photograph by Michael Scott.)

base flow stages. In future years, with smaller peaks, frequent flow and ice drive disturbances repeatedly damage and ultimately remove stems below the approximate nine-year return stage. Trees that survive above this elevation are subject to ice scarring during occasional extreme ice drive events (Figure 5-13). In unconstrained reaches, successful recruitment is less dependent on large flood pulses (Figure 5-14). In the first year following a moderate flood pulse, germination occurs below the nine-year return stage, between the peak and base flow stages, and stems are repeatedly damaged by flow and ice drives (see Figure 5-10). However, some stems survive, particularly in the absence of heavy grazing, as lateral channel movement and vertical accretion of sediments effectively remove surviving stems from the zone of frequent physical disturbance (Everitt, 1968).

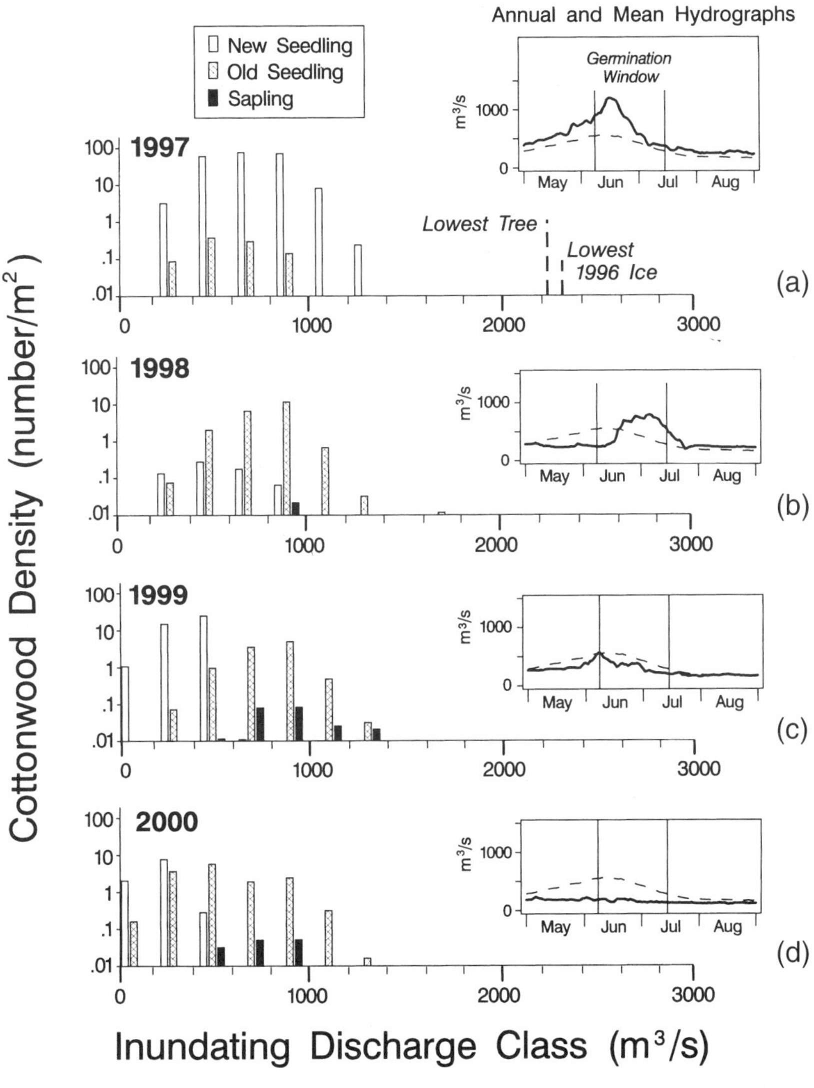

Figure 5-11. *Distribution of cottonwood seedlings and saplings along gradient of discharge required to inundate a plot. Densities are the means of all plots within a class of inundating discharges for new or current year seedlings; older, < 1 m tall seedlings that germinated in previous years; and saplings > 1 m tall but less than 2.5 cm diameter at breast height. The inundating discharge position of the lowest tree (> 2.5 cm diameter at breast height) at any of the sites and the lowest position of ice scour observed in the severe winter conditions of 1995–1996 are indicated by dashed vertical lines on the cottonwood distribution histogram for 1997. The inset hydrographs of discharge at the Virgelle gage (a-d) show the daily discharges for May–August of each year as a solid line and the long-term, 1935–2000, mean discharges for the each day as a dashed line. The germination window of primary cottonwood seed release and germinability is bounded by vertical lines on the inset hydrographs. (From Auble and Scott, unpublished data.)*

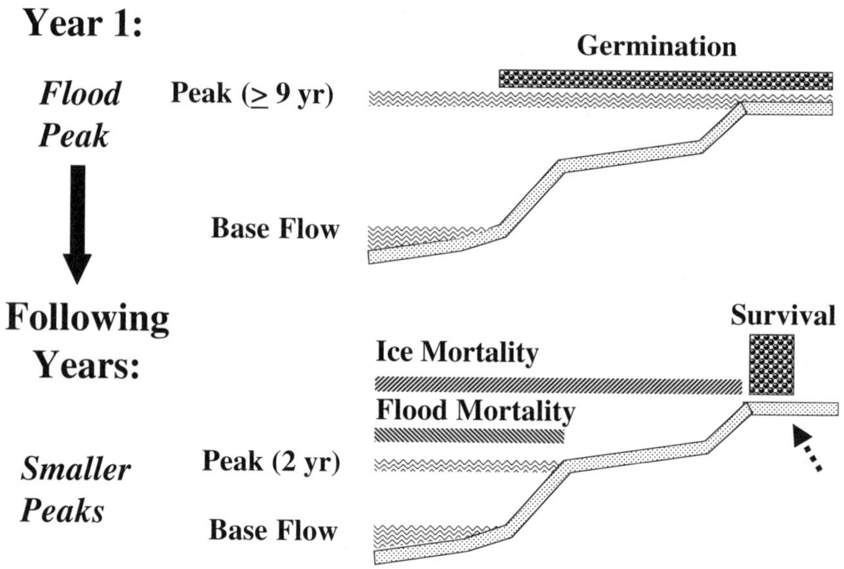

Figure 5-12. A conceptual model to explain the pattern of episodic cottonwood establishment and survival in a constrained reach of the Missouri River. In the first year of a flood peak that exceeded the stage of the average one-in-nine-year flow, germination occurred on surfaces between the peak flow and base flow stages. In subsequent years, with smaller peaks, frequent flow and ice drive disturbances repeatedly damaged and ultimately removed stems below the approximate nine-year return stage. Trees that survived above this elevation were subjected to ice scaring during occasional extreme ice drive events. (After Auble and Scott, 1998; copyright © by Society of Wetland Scientists; used by permission.)

Effects of Flow Regulation on Flood Pulses

Although in-channel dams produce variable downstream impacts, based in part on their design and operation, most reduce the frequency and magnitude of flood pulses (Williams and Wolman, 1984). Given the dependence of cottonwood recruitment on flood pulses, the nature and extent of flood pulse alteration resulting from flow regulation is key to assessing long-term impacts and recommending restoration strategies for the upper Missouri River.

Flows ≥ 1400 m^3 sec^{-1} at the Fort Benton gauge were significantly associated with successful recruitment of cottonwood stands within the Wild and Scenic Reach of the Missouri (Scott et al., 1997). However, because the Marias River joins the Missouri below Fort Benton, the gauge at Virgelle better represents flow conditions in the Wild and Scenic Reach

Figure 5-13. *A stand of mature trees in a constrained channel reach in the study area. The* dashed arrows *and* solid lines *at the left edge of the photograph define the surface above which long-term survival is likely, in spite of occasional stem scaring* (solid arrows) *by extreme ice drives. (Photograph by Michael Scott.)*

(Figure 5-1). By correlating flows between these two gauges, a discharge range between 1850 and 2000 m^3 sec^{-1}, at Virgelle, best reflects the discharge threshold at which established cottonwoods are likely to survive future ice drive disturbance (Bovee and Scott, in press). A mass balance flow model was used to reconstruct unregulated daily peak flows in the postdam period.

Comparison of reconstructed and recorded maximum daily discharges at the Virgelle gauge indicated that flow regulation in the postdam period has reduced the magnitude but not the frequency of flood pulses in the Wild and Scenic Reach of the upper Missouri River (Figure 5-15). However, flood pulse reduction resulting from flow regulation has been dramatic. In the two years in which flows exceeded the 1850 to 2000 m^3 sec^{-1} threshold, 1964 and 1975, maximum daily discharge was reduced 52 and 41 percent, respectively (Bovee and Scott, in press). Based on this model, reductions in the magnitude of large flood pulses would be predicted to lower the average elevation at which cottonwoods establish, subjecting

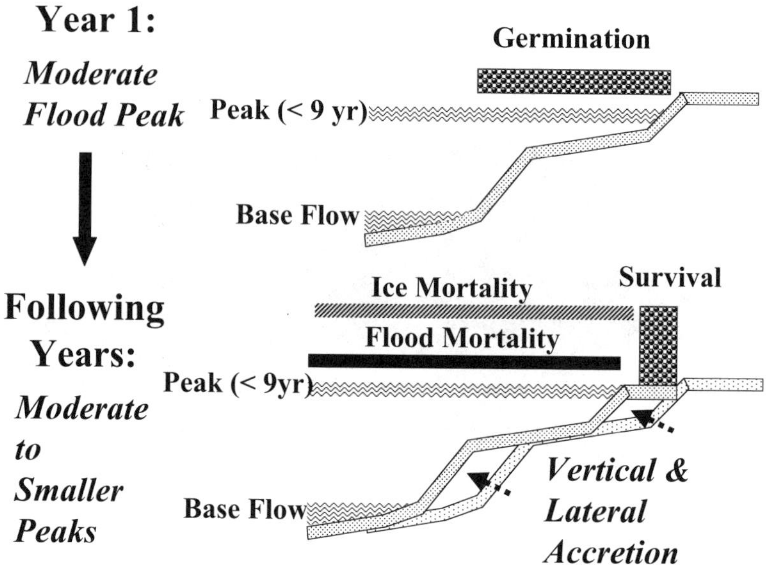

Figure 5-14. A conceptual model to explain the pattern of more frequent and extensive cottonwood establishment, physical damage, and survival in an unconstrained reach of the Missouri River. In the first year of a flood peak with a stage less than the stage of the average one-in-nine-year flow, germination occurs on surfaces between the peak flow and base flow stages. In subsequent years, of equal or smaller flood peaks, lateral channel movement and vertical sediment accretion raise new alluvial surfaces to near the nine-year return stage, reducing the frequency of physical disturbance by flow and ice drives, and thus allowing some survival. As in constrained reaches, surviving trees would be subject to ice scaring during occasional extreme ice drive events. (After Auble and Scott, 1998, copyright © by Society of Wetland Scientists; used by permission.)

stems to more frequent flow and ice drive disturbance and potentially limiting future recruitment to unconstrained reaches where stems established at lower elevations are likely to survive. The importance of large flood pulses in positioning cottonwoods above the zone of ice drive disturbance is suggested by a comparison of recorded daily peak water and ice stages at the Virgelle gauge for the postdam period (1954–1997) (Figure 5-16). Ice was recorded in most years, and in 61 percent of these years the ice stage exceeded the flow stage. The only year in which the recorded flood stage was not subsequently eclipsed by ice was 1964 (Figure 5-16). Of the seven sampled cottonwood stems that established in association with this event, all were positioned more than 1.5 m above the active channel shelf

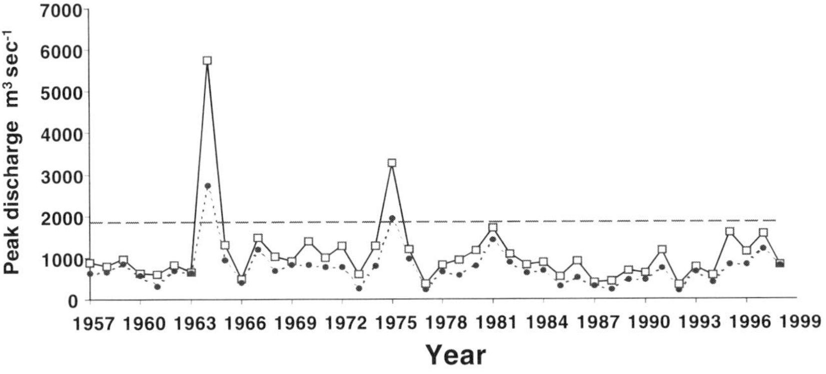

Figure 5-15. *Comparison of reconstructed unregulated* (solid line) *and recorded* (dotted line) *maximum daily discharges at Virgelle, Montana. Discharges shown in relation to a discharge of 1850 m^3 sec^{-1}* (dashed line), *the estimated minimum discharge required to trigger successful cottonwood establishment in the Wild and Scenic Reach of the Missouri River. (From Bovee and Scott, in press; copyright © by* Regulated Rivers: Research and Management; *used by permission.)*

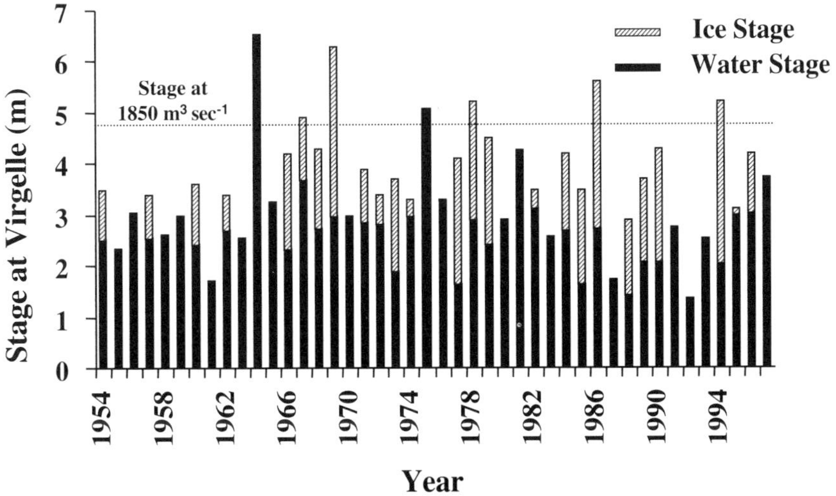

Figure 5-16. *The histogram represents the recorded daily peak water* (solid bars) *and ice* (dashed bars) *stages, in meters, at the Virgelle, Montana gauge, by water year, for the post-dam period. The* dashed line *represents the stage corresponding to a discharge of 1850 m^3 sec^{-1} at the gauge.*

and are most likely to escape subsequent ice damage and survive to become trees (Figure 5-9).

Changes in the Fluvial Disturbance Regime and Cottonwood Recruitment Patterns

We examined postdam recruitment patterns for 22 randomly selected cottonwood stands, which were identified in 1988 as uniform patches having stems in either sapling (2.5 to 12.2 cm in diameter at breast height)- or pole (12.5 to 22.3 cm)-size classes (Hansen, 1989). All stands originated in the postdam period across a range of elevations and within unconstrained channel reaches. Three of the 14 sapling stands were lost and 11 survived to 1999, at which point 89 percent of the stems remained in the sapling-size class. Six stands (55 percent) showed damage by ice and 5 (45 percent) were damaged by beaver. Intense physical disturbance was indicated at 4 of 11 stands (36 percent) where stems had been prostrated. Two of the 8 original pole stands were lost, and 62 percent of the live stems in the remaining stands were in the sapling-size class by 1999. Ice damage was evident in 4 (67 percent) and beaver damage in 2 (33 percent) of the stands. Only 1 stand had stems that had been prostrated by flow or ice. These results indicate that the reduced magnitude of flood peaks in the postdam period has subjected established cottonwoods to repeated physical disturbance and created a potential bottleneck in recruitment of stems into tree-size classes, further exaggerating the episodic nature of successful cottonwood tree recruitment.

In summary, successful forest establishment is dependent on flood pulses that position stems above the elevation of frequent ice drive disturbance. Postdam cottonwood recruitment is limited to unconstrained channel reaches, including islands, where stand development is impeded by physical disturbance including ice and beaver damage. Management efforts to maintain or restore cottonwood forests in the Wild and Scenic Missouri River must recognize the long-term implications of changes in the interplay of flood pulses and ice drives. In addition to human effects on these factors, their relative magnitudes and frequencies may be influenced by environmental change over longer time scales. Because cottonwoods are relatively long-lived organisms, their present-day distribution may not accurately reflect contemporary fluvial geomorphic processes (Hughes, 1994).

Insight into longer-term recruitment patterns of cottonwood in the Wild and Scenic Reach is provided by photographic comparisons from portions of the study reach. The first set of photographs includes constrained and unconstrained reaches near Iron City Island (Figure 5-17). In the constrained reach, there has been a loss of mature trees on high alluvial surfaces with little apparent change in channel configuration over the 97-year interval, despite eight floods, including the 1908 flood of record, with an estimated discharge >5000 m^3 sec^{-1} at Virgelle (Figure 5-1). A back channel evident in the early photograph has been partially filled by alluvial sediments and is now occupied by a stand of trees. Dendrogeomorphic examination of the stand indicates that the majority of trees established in 1928, one year following a flood pulse estimated to have exceeded the recruitment threshold of 1850 to 2000 m^3 sec^{-1} at Virgelle (Scott et al., 1997; Bovee and Scott, in press). In the reach containing Iron City Island, there has been tree loss, and subsequent establishment of a new stand has accompanied bank erosion and island expansion by lateral and vertical accretion.

The second set of photographs illustrates the attachment of a channel island near the old power plant ruins (Figure 5-18). The head of a low channel island, separated from a steep cut bank by an open back channel, is evident in the 1925 photo. In the photo of 1977, the back channel has narrowed considerably as the island, now heavily vegetated, appears to have accreted laterally and vertically. By 2000, island attachment was completed as the back channel filled with alluvial sediments. Two cottonwoods appeared near the head of the island. Sandbar willow dominated the remaining island surface and had expanded over much of the former back channel to the base of the scarp.

Although observed channel stasis in the constrained reach and channel narrowing and island accretion in unconstrained reaches may be localized, surveyed channel maps of the river from the end of the 1800s (Missouri River Commission, 1988) depict a wide channel with numerous open back channels and unvegetated channel islands, implying that the patterns observed in the historical photos may be widespread throughout the study reach. This suggests that contemporary cottonwood forest distribution reflects a continuous, long-term adjustment of the fluvial disturbance regime in response to climate as well as human-caused changes in flow (Hughes 1994). For example, open side channels, and linear stands of mature cottonwoods on high alluvial surfaces in constrained reaches, may be physical and biological legacies of a previous fluvial disturbance regime

Figure 5-17. A photographic comparison from the study reach depicting a 97-year time period (1903–2000). In the downstream constrained reach, trees have been lost from high alluvial surfaces with little accompanying channel change (A). In contrast, a stand of trees established in a formerly open back channel (B). Finally, tree loss and subsequent establishment of a new stand has accompanied bank erosion and island expansion by vertical and lateral accretion in an unconstrained reach (C). (1903 photograph by T.W. Stanton, U.S. Geological Survey Photo Archives, Denver, Colorado; 2000 photograph by Michael Merigliano, used by permission.)

(Sprugel, 1991; Naiman et al., 1995). Wetter conditions at the end of the 1880s have been associated with more extensive recruitment of cottonwood than is occurring today along some rivers in semiarid areas in North America (Baker, 1990; Cordes, 1991). Decreases in the frequency and magnitude of fluvial disturbances on the Upper Missouri River resulting

Figure 5-18. *A photographic comparison depicting a site within the study reach at three points in time—1925, 1977, and 2000. Because of bank erosion, the later two photographs were not taken from precisely the same location as the first photograph. The photographs document the process of island attachment by the filling of a back channel with alluvial sediments. A relatively open back channel in 1925 has narrowed as vegetation has established on the expanding alluvial surfaces by 1977. Finally, by 2000 the channel has filled and sand bar willow, which dominate the island, have expanded onto or across the former channel to the toe of the upland scarp. (1925 photograph by A. J. Collier, U.S. Geological Survey Photo Archives, Denver, Colorado; 1977 photograph by Mike Gilkerson, U.S. Bureau of Land Management, used by permission; 2000 photograph by Michael Scott).*

from climate change in the mid to late 1800s may have initiated a process of channel narrowing by infilling existing back channels. For this reason, combined with the effects of flow regulation in the mid-1950s, channel narrowing would be expected to continue, with confinement of cottonwood recruitment to sites in unconstrained reaches outside the primary ice drive path. Efforts to manage and restore riparian forests such as those on the upper Missouri should incorporate an understanding of forest patterns that result from processes acting over longer time scales.

Relationships Between Riparian Forest Structure and Avian Diversity

Flow-related patterns of channel change control the size and distribution of forest patches of riparian cottonwood (Everitt, 1968; Johnson et al., 1976; Bradley and Smith, 1986; Stromberg et al., 1991; Scott et al., 1996; Dykaar and Wigington, 2000), and the composition and structural complexity of these patches change with riparian succession (Johnson et al., 1976; Boggs and Weaver, 1994). Human endeavors such as flow modification and livestock grazing alter these processes and thus have direct impact on the habitat characteristics of riparian forests. Although limited in aerial extent (Knopf et al., 1988), these riparian forests provide important habitat for native vertebrates, including a diverse avian fauna (Brinson et al., 1981; Finch and Ruggiero, 1993). More than 233 bird species have been reported within the upper Missouri River corridor in Montana (U.S. Department of the Interior, 1993), including a broad diversity of migratory, nongame bird species.

Within the study reach, breeding bird species composition, richness, and abundance were examined in relation to variation in vegetation structure across 34 forested and nonforested riparian habitat patches. The patches included structurally complex cottonwood stands with a relatively well developed shrub understory (cottonwood-shrub patches) and stands of cottonwoods with comparatively little or no woody understory (cottonwood patches). Treeless areas on adjacent alluvial surfaces dominated by grasses and shrubs were identified as a third patch type (shrub-steppe patches). Differences in tree and tall shrub cover effectively distinguished these three patch types (Table 5-1, A). Further, the matrix of landscape features surrounding forested riparian habitat patches was relatively uniform, reflecting the relatively undisturbed nature of the river corridor (Table 5-1, B).

TABLE 5-1. Habitat (A) and landscape (B) variables for riparian vegetation patch types

Means (± S.E.) for measured variables among Riparian vegetation patch types (*n*) within the study reach of the Missouri River, Montana, U.S.A. Different lowercase letters for the variables denote statistically significant differences ($p < 0.05$, Bonferroni adjustment).

(A.) Vegetation Structure and Patch Dimensions

Variables	Cottonwood-Shrub (12)	Cottonwood (10)	Shrub-Steppe (12)
Tree cover (%)	23.6 (1.7) [a]	13.2 (2.7) [b]	0.0 (0.0) [c]
Tall shrub cover (%)	26.6 (3.2) [a]	3.4 (1.0) [b]	0.4 (0.2) [b]
Low shrub cover (%)	22.6 (5.6)	22.8 (3.7)	21.7 (3.0)
Herb cover (%)	68.3 (3.2) [a]	77.5 (3.8) [a,b]	83.0 (2.4) [b]
Patch width (m)	92.4 (19.4) [a]	47.3 (8.0) [b]	
Patch length (m)	627.4 105.7) [a]	408.5 (48.2) [b]	
Patch area (ha)	5.8 (1.2) [a]	1.9 (0.4) [b]	

(B.) Landscape Variables

Variables	Cottonwood-Shrub (12)	Cottonwood (10)
Distance to adjacent cottonwoods (m)	20.3 (3.8)	29.9 (6.6)
Nonforested floodplain (%)	17.7 (2.7)	22.7 (2.7)
Upland grass (%)	45.3 (2.9)	44.4 (5.0)
Upland forest (%)	8.9 (1.7)	8.5 (1.9)
Wetted channel (%)	13.6 (0.9)	12.5 (0.5)
Agriculture (%)	3.1 (2.8)	1.5 (1.4)
Cottonwood forest (%)	4.8 (1.1)	3.1 (2.8)
Other (%)	6.7 (0.9)	6.6 (1.2)

Source: Scott et al., in review.

Species richness (number of species) and bird diversity (Shannon index of diversity) increased with increasing structural complexity of vegetation ($F_{1,32} = 75.49$, $p < 0.0001$; $F_{1,32} = 79.76$, $p < 0.0001$, respectively). Bird species composition, as represented by Axis 1 of a detrended correspondence analysis (DCA), was significantly correlated with vegetation strata diversity ($r_{s,33} = 0.98$, $p < 0.0001$; Figure 5-19). Bird abundance in canopy and tall shrub foraging guilds increased significantly with increasing tree and tall shrub cover ($F_{1,22} = 34.68$, $p < 0.0001$; $F_{1,20} = 22.22$, $p < 0.0001$, respectively; Figure 5-20 *A* and *B*). Seventeen bird species were strongly associated ($p < 0.10$) with structurally complex cottonwood-shrub patches, whereas only three bird species were strongly associated with nonforested shrub-steppe patches.

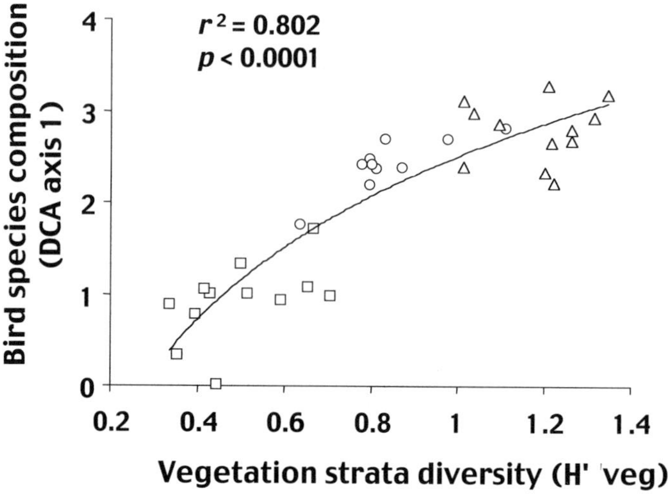

Figure 5-19. *Relation between the detrended correspondence analysis (DCA) axis 1 gradient in bird species composition and vegetation strata diversity (H' veg; $H' = -\Sigma p_i \ln p_i$, where p_i is the proportion of vegetation cover found in the ith stratum). DCA site scores are in units of average standard deviation of species turnover; four units is about one complete turnover. Cottonwood shrub patches are denoted by triangles, cottonwood patches by circles, and shrub-steppe patches by squares. For measured habitat variables associated with each patch type, see Table 5-1. (From Scott et al., in review, copyright © by the Society for Conservation Biology; used by permission.)*

Structural complexity of habitat patches was positively related to sediment accretion ($t_{33} = 3.31$, $p < 0.002$) and vegetation establishment ($t_{20.7} = -3.63$, $p < 0.002$) over a 35-year postdam period (1953 to 1988). Riparian succession in semiarid environments typically begins with the establishment of species of cottonwood and willow on freshly deposited alluvial substrates (Johnson et al., 1976; Rood and Mahoney, 1990; Scott et al., 1996; Merigliano, 1998). In portions of the Northern Great Plains, succession terminates in grassland in the absence of subsequent fluvial disturbance (Hansen et al., 1991; Boggs and Weaver, 1994; Friedman et al., 1997). Structural diversity of semiarid riparian forests reaches a maximum after approximately 90 years with development of a mature cottonwood canopy and shrub understory. With the attrition of mature cottonwoods, structure declines as the stand gives way to shrubs and, ultimately, to upland grasses (Boggs and Weaver, 1994).

Structural differences between cottonwood-shrub, cottonwood, and shrub-steppe reflect local geological controls on the frequency and extent

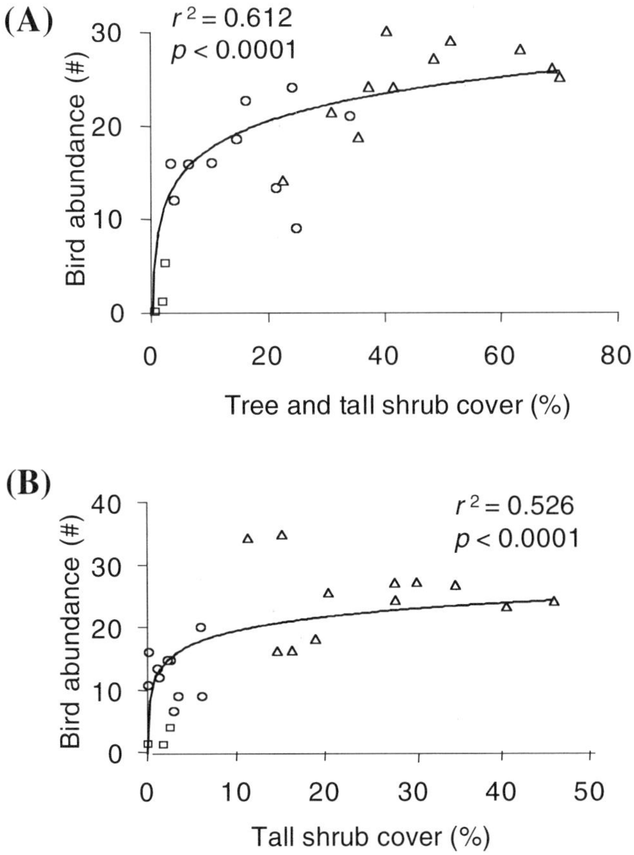

Figure 5-20. *Relation between* (A) *bird abundance (number of birds) in the canopy foraging guild versus tree and tall shrub cover and* (B) *bird abundance (number of birds) in the tall shrub foraging guild versus tall shrub cover. (From Scott et al., in review, copyright © by the Society for Conservation Biology; used by permission.)*

of cottonwood and willow establishment, subsequent riparian succession, and grazing history. Cottonwood-shrub patches, but not cottonwood nor shrub-steppe, are associated with recent geomorphic activity and establishment of woody vegetation. Loss of tree cover on shrub-steppe sites further suggests that these patches represent the expected climax vegetation on long-stable alluvial surfaces in this region (cf. Boggs and Weaver, 1994). The relative lack of geomorphic change and associated woody vegetation establishment at cottonwood sites in the postdam period suggests that many of these patches will succeed to shrub-steppe, with a predicted loss of forest structure and avian diversity.

In addition to the role of fluvial geomorphic processes in controlling the pattern and extent of riparian forests, long-term overgrazing by livestock can alter riparian vegetation succession by limiting reproduction and altering vegetation structure and composition (Knopf and Cannon, 1982; Kauffman and Krueger, 1984; Taylor, 1986; Schultz and Leininger, 1990). Measured structural differences between cottonwood-shrub, cottonwood, and shrub-steppe patches, such as extent of trees, tall shrubs, and herbaceous plants, were significantly related to grazing activity ($t_{19.6} = 3.75$, $p < 0.001$). Chronic, long-term grazing may accelerate succession of riparian forest to shrub- or grass-dominated associations by reducing or eliminating establishment of tree and shrub species like cottonwood, willow, green ash, box elder, and other associated woody species (Szaro and Pase, 1983; Hansen et al., 1991; Boggs and Weaver, 1994).

Conserving avian diversity along the Upper Missouri River will require protection of the flow-related geomorphic processes responsible for establishment of new cottonwood and willow patches, in combination with reduction or elimination of livestock grazing following the establishment of these patches. Geomorphic activity contributes directly to the structural complexity of forest habitats, resulting in greater avian abundance and diversity. Long-term grazing by livestock simplifies the structure of riparian vegetation, thereby significantly affecting the avian community.

Flood Pulse Restoration

Restoring the frequency, magnitude, duration, and timing of flood pulses would benefit important natural resource values, including riparian forests, avian diversity, and native fish species in the Upper Missouri River basin (Hesse et al., 1993; Rood and Mahoney, 1995; Scott et al., 1997; Scott et al., in review). A mass balance flow model (see the earlier section "Effects of Flow Regulation on Flood Pulses") and flood damage curves developed by the U.S. Army Corps of Engineers (Betty Peake, personal communication) were used to examine several alternative flow scenarios to deliver flows ≥ 1850 m^3 sec^{-1} to the Wild and Scenic Reach (Figure 5-21; Table 5-2). Scenario 0 was the no-regulation alternative, represented by the unregulated peak flow reconstruction. This alternative represents removal of Canyon Ferry and Tiber Dams (Figure 5-1) or their operation as run-of-river (outflow = inflow) facilities. Scenario 1 examined the delivery of additional water from Canyon Ferry Dam, required to achieve a flow of 1850

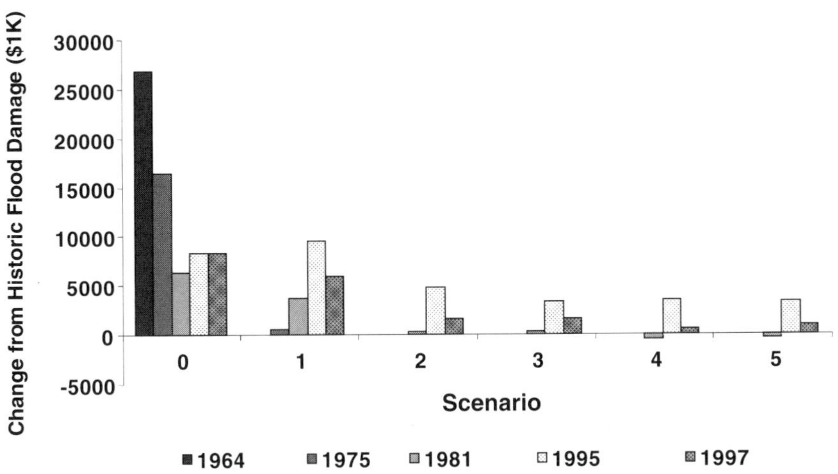

Figure 5-21. *Comparison of flood damage indices for six alternative upper Missouri River reservoir release scenarios. The flood damage index is the difference between damage estimates for the scenario and damage estimates for floods that occurred historically in the years listed. Damage estimates are in thousands of 1998 dollars. For details of release scenarios, see Table 5-2. (Reprinted from Bovee and Scott, in press, copyright © by* Regulated Rivers: Research and Management, *by permission.)*

m^3 sec^{-1} at Virgelle, given recorded releases and flow inputs from all other sources. Under Scenario 2, the delivery of additional water was distributed among Canyon Ferry Dam, Gibson Dam (Sun River), and Tiber Dam. Tiber Dam was the sole source of makeup water under Scenario 3. The final scenarios examined reductions in water released from Canyon Ferry Dam, with increased releases from Tiber Dam (Scenario 4) and Tiber and Gibson Dams combined (Scenario 5). Table 5-2 summarizes operational conditions for each facility and scenario.

Restoring or enhancing the magnitude of flood pulses by delivery of additional water from Tiber Reservoir is physically and economically feasible. The most effective opportunities for flood pulse restoration seem to occur during years when global and continental weather patterns deliver warm, moist air to central Montana from the Gulf of Mexico. These weather patterns are routinely monitored and recognizable, providing reservoir operators with sufficient flexibility to anticipate when conditions might be favorable for flood pulse restoration.

The geography of the basin adds further flexibility for restoring flood

TABLE 5-2. Reservoir release scenarios

To achieve minimum threshold discharge of 1850 m^3 sec^{-1} to the Wild and Scenic Reach of the upper Missouri River.

Scenario	Year	Canyon Ferry	Gibson	Tiber
0	All	Outflow = inflow	Outflow = inflow	Outflow = inflow
1	1975[a]	Add 62 m^3 sec^{-1}	Same as historic	Same as historic
1	1981	Add 400 m^3 sec^{-1}	Same as historic	Same as historic
1	1995	Add 1096 m^3 sec^{-1}	Same as historic	Same as historic
1	1997	Add 640 m^3 sec^{-1}	Same as historic	Same as historic
2	1975[a]	Same as historic	Same as historic	Add 62 m^3 sec^{-1}
2	1981	Same as historic	Same as historic	Add 400 m^3 sec^{-1}
2	1995	Add 283 m^3 sec^{-1}	Add 142 m^3 sec^{-1}	Add 578 m^3 sec^{-1}
2	1997	Same as historic	Add 71 m^3 sec^{-1}	Add 569 m^3 sec^{-1}
3	1975[a]	Same as historic	Same as historic	Add 62 m^3 sec^{-1}
3	1981	Same as historic	Same as historic	Add 400 m^3 sec^{-1}
3	1995	Same as historic	Same as historic	Add 1096 m^3 sec^{-1}
3	1997	Same as historic	Same as historic	Add 640 m^3 sec^{-1}
4	1975[a]	Same as historic	Same as historic	Add 62 m^3 sec^{-1}
4	1981	Reduce 142 m^3 sec^{-1}	Same as historic	Add 547 m^3 sec^{-1}
4	1995	Reduce 142 m^3 sec^{-1}	Same as historic	Add 1144 m^3 sec^{-1}
4	1997	Reduce 283 m^3 sec^{-1}	Same as historic	Add 923 m^3 sec^{-1}
5	1975[a]	Same as historic	Same as historic	Add 62 m^3 sec^{-1}
5	1981	Reduce 142 m^3 sec^{-1}	Add 85 m^3 sec^{-1}	Add 462 m^3 sec^{-1}
5	1995	Reduce 142 m^3 sec^{-1}	Add 142 m^3 sec^{-1}	Add 1096 m^3 sec^{-1}
5	1997	Reduce 142 m^3 sec^{-1}	Add 142 m^3 sec^{-1}	Add 640 m^3 sec^{-1}

[a] Delivers discharge ≥ 2000 m^3 sec^{-1} to the Virgelle gauge.

Source: Bovee and Scott, in press.

pulses. The best opportunity to provide a large flood pulse to the Wild and Scenic Reach occurs when tributary flow accretions are high. The lag time for flow peaks from major tributaries and Tiber Dam is about the same, allowing flow releases from Tiber to arrive at Virgelle coincident with runoff spikes from other tributaries. Installation of telemetering equipment could automatically alert dam operators of large inflows at any or all important upstream gauges.

Although flood pulse restoration using releases from Tiber Dam is feasible from a water supply and routing perspective, none of the alternatives we examined is risk-free. Each calls for very large releases (up to 1100 m^3 sec^{-1}) from Tiber Dam. Based on flood control considerations, maximum releases from Tiber are currently restricted to 283 m^3 sec^{-1}, the estimated bank-full discharge. Larger releases should be evaluated in light of costs and benefits. Periodic large flood pulses below Tiber Dam would help restore some degree of natural channel movement and cottonwood recruitment along the Marias River (Rood and Mahoney, 1995). However, flows exceeding bank-full capacity could result in downstream channel erosion and flooding. Increased channel and bank erosion would influence wildlife resources, such as fish and shorebird habitat, and could pose problems for infrastructure such as bridges and roads. For flood pulse restoration to be feasible over the long term, certain conditions must be in place. First, it is imperative to restrict floodplain development along the Marias below Tiber Dam. Second, because Tiber Dam was authorized as a flood control facility when it was approved by Congress, the operational policies of the U.S. Bureau of Reclamation would have to be modified. Third, there must be a mechanism to compensate existing riparian landowners for property loss and flood damage resulting from large releases from Tiber Dam.

CONCLUSIONS

For large flow-controlled rivers, full restoration of natural, regionally specific stream flow patterns may not be possible, and flow prescriptions for natural resource values must be balanced against competing economic and social values (Schmidt et al., 1998; Stromberg, in press). A scientific understanding of the physical and biotic processes responsible for creating and sustaining riverine and riparian ecosystem values must be clearly articulated.

Because of its remote character, scenic qualities, historical significance, and wildlife resources, portions of a 240 km reach of the Upper Missouri River were designated as Wild and Scenic. The river and adjoining public lands, administered by the U.S. Bureau of Land Management (BLM), have received increasing use by canoeists, sightseers, motorized boaters, hunters, and fishermen in recent years. As a result of conflicts between traditional grazing uses and emerging recreational uses, the Wild and Scenic Missouri River has become the nexus of a debate over how future land and water management can best protect this resource. Of particular concern is how management of flows from upstream dams and livestock grazing is influencing the distribution and abundance of riparian cottonwoods and associated wildlife species.

Historical accounts are consistent with contemporary observations that cottonwood forests are sparse and dominated by stands of mature trees (Behan, 1981). Dendrogeomorphic analyses indicate that a majority of existing trees were established following infrequent large flood pulses. Livestock grazing greatly reduces seedling densities, but long-term survival of stems in constrained channel reaches is correlated with establishment on alluvial surfaces above the zone of frequent ice scour.

Reconstruction of unregulated flow in the upper Missouri River basin indicates that although the frequency of large flood pulses has not been influenced by the operation of upstream dams, the magnitude of these events has been reduced from 40 to 50 percent. Retrospective analyses of cottonwood patches demonstrate that postdam recruitment is limited to unconstrained channel reaches, typically associated with channel islands and tributary junctions, where stems are most likely to survive subsequent ice disturbance.

Avian diversity is dependent on the fluvial geomorphic processes responsible for the establishment and survival of new cottonwood and willow patches, as well as land-use practices that reduce or eliminate livestock grazing following establishment of these patches. However, photographic comparisons suggest that because of possible climate-related changes in the frequency and magnitude of fluvial disturbance events, contemporary patterns of flood-related forest establishment should not necessarily be expected to match historical patterns.

Flow models and flood damage curves for the upper Missouri River indicate that there is enough operational flexibility in the system to produce flood pulses without greatly compromising other values. In addition to grazing, future human development on the floodplain of the upper Missouri and tributaries could increasingly limit the potential to produce flood pulses. Thus, riparian forest restoration on the upper Missouri River will require integrated land and water management activities throughout the entire basin.

REFERENCES

Albertson, F. W., and J. E. Weaver. 1945. Injury and death or recovery of trees in prairie climate. *Ecological Monographs* **15:**395–433.

Alden, W. C. 1932. *Physiography and Glacial Geology of Eastern Montana and Adjacent Areas.* U.S. Geological Survey Professional Paper 174. U.S. Government Printing Office, Washington, DC.

Auble, G. T., and M. L. Scott. 1998. Fluvial disturbance patches and cottonwood recruitment along the upper Missouri River, Montana. *Wetlands* **18:**546–556.

Baker, W. L. 1990. Climatic and hydrologic effects on the regeneration of *Populus angustifolia* James, along the Animas River, Colorado. *Journal of Biogeography* **17:** 59–73.

Behan, M. 1981. The Missouri's stately cottonwoods: How can we save them? *Montana Magazine.* (September): 76–77.

Boggs, K., and T. Weaver. 1994. Changes in vegetation and nutrient pools during riparian succession. *Wetlands* **14:**98–109.

Bonner, F. C., and F. Stermitz. 1967. *Floods of June 1964 in Northwestern Montana.* Water-Supply Paper 1840-B:1-242. U.S. Geological Survey, Washington, DC.

Bovee, K. D., and M. L. Scott. In press. Effects of flow regulation on the upper Missouri River: Implications for flood pulse restoration. *Regulated Rivers: Research and Management.*

Bradley, C.E., and D.G. Smith. 1986. Plains cottonwood recruitment and survival on a prairie meandering river floodplain, Milk River, southern Alberta and northern Montana. *Canadian Journal of Botany* **64:**1433–1442.

Brinson, M.M., B.L. Swift, R.C. Plantico, and J.S. Barclay. 1981. *Riparian Ecosystems: Their Ecology and Status.* U.S. Fish and Wildlife Service Biological Report 81. U.S. Government Printing Office, Washington, DC.

Brinson, M. M. 1990. Riverine Forests. In *Forested Wetlands,* ed. by A. E. Lugo, M. M. Brinson, and S. Brown, 87–141. Elsevier, Amsterdam.

Busch, D. E., N. L. Ingraham, and S. D. Smith. 1992. Water uptake in woody riparian phreatophytes of the southeastern United States: A stable isotope study. *Ecological Applications* **2:**450–459.

Calhoon, F. H. H. 1906. *The Montana Lobe of the Keewatin Ice Sheet.* Professional Paper 50, Series-B. U.S. Geological Survey, Washington, DC.

Chaney, E., W. Elmore, and W. S. Platts. 1990. *Livestock Grazing on Western Riparian Areas.* Northwestern Resource Information Center, Eagle, ID.

Cooke, R. U., and R. W. Reeves. 1976. *Arroyos and Environmental Change in the American Southwest.* Clarendon Press, Oxford, U.K.

Cooper, D. J., D. M. Merritt, D. C. Anderson, and R. A. Chimner. 1999, Factors controlling the establishment of Fremont cottonwood seedlings on the upper Green River, USA. *Regulated Rivers: Research and Management* **15:**419–440.

Cordell, G. V. 1974. The climate of Montana. In *Climates of the States,* Vol. 2, Western States, by The National Oceanic and Atmospheric Administration, U.S. Department of Commerce, pp. 743–762. Water Information Center, Inc., Port Washington, NY.

Cordes, L. D. 1991. The distribution and age structure of cottonwood stands along the lower Bow River. In *The Biology and Management of Southern Alberta's Cottonwoods,* ed. by S. B. Rood and J. M. Mahoney, pp. 13–23. University of Lethbridge, Alberta, Canada.

Cordes, L. D., F. M. R. Hughes, and M. Getty. 1997. Factors affecting the regeneration and distribution of riparian woodlands along a northern prairie river: The Red Deer River, Alberta, Canada. *Journal of Biogeography* **24:**675–695.

Coues, E. 1893. *The History of the Lewis and Clark Expedition.* Vol. 1. E. Coues, ed. Reprint. Dover Publications, New York.

Dightman, R. A. 1950. Montana Marias basin rainstorm, June 16–17, 1948. *Monthly Weather Review* **78:**6–12.

Dykaar, B. B., and P. J. Wigington Jr. 2000. Floodplain formation and cottonwood colonization patterns on the Willamette River, Oregon, USA. *Environmental Management* **25:**87–104.

Dynesis, M., and C. Nilsson. 1994. Fragmentation and flow regulation of river systems in the northern third of the world. *Science* **266:**753–762.

Everitt, B.L. 1968. Use of the cottonwood in an investigation of the recent history of a floodplain. *American Journal of Science* **266:**417–439.

Everitt, B.L. 1979. The cutting of Bull Creek arroyo. *Utah Geology* **6:**39–44.

Fenner, P., W.W. Brady, and D.R. Patton. 1985. Effects of regulated water flows on regeneration of Fremont cottonwood. *Journal of Range Management* **38:** 135–138.

Finch, D. M., and L. F. Ruggiero. 1993. Wildlife habitats and biological diversity in the Rocky Mountains and Northern Great Plains. *Natural Areas Journal* **13:**191–203.

Friedman, J. M., and G. T. Auble. 2000. Floods, flood control, and bottomland vegetation. In *Inland Flood Hazards: Human, Riparian and Aquatic Communities,* ed. by E. E. Wohl, pp. 219–237. Cambridge University Press, Cambridge, U.K.

Friedman, J. M., W. R. Osterkamp, and W.M. Lewis Jr. 1996. Channel narrowing and vegetation development following a Great-Plains flood. *Ecology* **77:** 2167–2181.

Friedman, J. M., M. L. Scott, and G. T. Auble. 1997. Water management and cottonwood forest dynamics along prairie streams. In *Ecology and Conservation of Great Plains Vertebrates,* ed. by F. L. Knopf and F. Samson, pp. 49–71. Springer-Verlag, New York.

Friedman J. M., M. L. Scott, and W. M. Lewis Jr. 1995. Restoration of riparian forest using irrigation, artificial disturbance, and natural seedfall. *Environmental Management* **19:**547–557.

Gottesfeld, A. S., and L. M. J. Gottesfeld. 1990. Floodplain dynamics of a wandering river, dendrochronology of the Morice River, British Columbia, Canada. *Geomorphology* **3:**159–179.

Graf, W. L. 1999. Dam nation: A geographic census of American dams and their large-scale hydrologic impacts. *Water Resources Research* **35:**1305–1311.

Green, D. M., and J. B. Kauffman. 1995. Succession and livestock grazing in a northeastern Oregon riparian ecosystem. *Journal of Range Management* **48:** 307–313.

Gregory, S. V., F. J. Swanson, W. A. McKee, and K. W. Cummins. 1991. An ecosystem perspective of riparian zones. *BioScience* **41:**540–551.

Hansen, P. L. 1989. *Inventory, Classification, and Management of Riparian Sites Along the Upper Missouri National Wild and Scenic River.* Final Report. Montana Riparian Association, Montana Forest and Conservation Experiment Station, School of Forestry, University of Montana, Missoula, MT.

Hansen, P. L., K. Boggs, R. Pfister, and J. Joy. 1991. *Classification and Management of Riparian-Wetland Sites in Montana.* Montana Riparian Association, University of Montana, Missoula, MT.

Hereford, R. 1986. Modern alluvial history of the Paria River drainage basin, southern Utah. *Quaternary Research* **25:**293–311.

Hesse, L. W., G. E. Mestl, and J. W. Robinson. 1993. Status of selected fishes in the Missouri River in Nebraska with recommendations for their recovery. In *Restoration Planning for the Rivers of the Mississippi River Ecosystem,* ed. by L. W. Hesse, C. B. Stalnaker, N. G. Benson, and J. R. Zuboy, pp. 327–340. Biological Report 19, October 1993. U.S. Department of the Interior, National Biological Survey, Washington, DC.

Hughes, F. M. R. 1994. Environmental change, disturbance, and regeneration in semi-arid floodplain forests. In *Environmental Change in Drylands: Biogeographical and Geomorphical Perspectives,* ed. by A. C. Millington, K. Pye, pp. 321–345. John Wiley & Sons, New York.

Johnson, W. C. 1992. Dams and riparian forests: Case study from the upper Missouri River. *Rivers* **3:**229–242.

Johnson, W. C. 1994. Woodland expansion in the Platte River, Nebraska: Patterns and causes. *Ecological Monographs* **64:**45–84.

Johnson, W. C., R. Burgess, and W. Keammerer. 1976. Forest overstory vegetation and environment on the Missouri River floodplain in North Dakota. *Ecological Monographs* **46:**59–84.

Kapustka, L. A. 1972. Germination and establishment of *Populus deltoides* in eastern Nebraska. Master's thesis, Department of Botany, University of Nebraska, Lincoln.

Kauffman, J. B., and W. C. Krueger. 1984. Livestock impacts on riparian ecosystems and streamside management implications: A review. *Journal of Range Management* **37:**430–438.

Kaylor, J. F., C. C. Starring, and C. P. Ditman. 1935. A survey of past plantings. In *Possibilities of Shelterbelt Planting in the Plains Region,* pp. 39–47. U.S. Department of Agriculture, U.S. Forest Service, Washington, DC.

Knopf, F. L., and R. W. Cannon. 1982. Structural resilience of a willow riparian community to changes in grazing practices. In *Wildlife-Livestock Relationships Symposium,* ed. by J. M. Peek, and P. D. Dalke, pp. 198–207. University of Idaho, Forest, Wildlife and Range Experiment Station, Moscow, ID.

Knopf, F. L., R. R. Johnson, T. Rich, F. B. Samson, and R. C. Szaro. 1988. Conservation of riparian ecosystems in the United States. *Wilson Bulletin* **100:** 272–284.

Krasny, M. E., K. A. Vogt, and, J. C. Zasada. 1988. Establishment of four Salicaceae species on river bars in interior Alaska. *Holarctic Ecology* **11:**210–219.

Ligon, F. K., W. E. Dietrich, and W. J. Trush. 1995. Downstream ecological effects of dams. *BioScience* **45:**183–192.

Mahoney, J. B., and S. B. Rood. 1991. A device for studying the influence of declining water table on poplar growth and survival. *Tree Physiology* **8:** 305–314.

Mahoney, J. B., and S. B. Rood. 1992. Response of hybrid poplar to water table decline in different substrates. *Forest Ecology and Management* **54:**141–156.

Mahoney, J. M., and S. B. Rood. 1993. A model for assessing the effects of altered river flows on the recruitment of riparian cottonwoods. In *Riparian Management: Common Threads and Shared Interests,* ed. by B. Tellman, H. J. Cortner, M. G. Wallace, L. F. DeBano, and R. H. Hamre, pp. 228–232. U.S. Forest Service General Technical Report USDA Forest Service, RM-226. Albuquerque, NM.

McBride, J. B., and J. Strahan. 1984. Establishment and survival of woody riparian species on gravel bars of an intermittent stream. *American Midland Naturalist* **112:**235–245.

Meinzer, O. E. 1927. *Plants as indicators of ground water.* U.S. Geological Survey Water-Supply Paper 577. Washington, DC. U.S. Geological Survey.

Merigliano, M. F. 1998. Cottonwood and willow demography on a young island, Salmon River, Idaho. *Wetlands* **18:**571–576.

Middleton, B. 1999. *Wetland Restoration: Flood Pulsing and Disturbance Dynamics.* John Wiley & Sons, New York.

Missouri River Commission. 1988. *Map of the Missouri River.* Topographical and hydrological surveys, published in 84 sheets, 1892–1895. Third printing. U.S. Army Corps of Engineers, Omaha, NE.

Moss, E.H. 1938. Longevity of seed and establishment of seedlings in species of *Populus. Botanical Gazette* **99:**529–542.

Naiman, R. J., J .J. Magnuson, D. M. McKnight, and J. A. Stanford. 1995. *The Freshwater Imperative.* Island Press, Washington, DC.

Nanson, G. C., and H. F. Beach. 1977. Forest succession and sedimentation on a meandering-river floodplain, northeast British Columbia, Canada. *Journal of Biogeography* **4:**229–251.

Nilsson, C., M. Gardfjell, and G. Grelsson. 1991. Importance of hydrochory in structuring plant communities along rivers. *Canadian Journal of Botany* **69:** 2631–2633.

Ohmart, R. D., B. W. Anderson, and W. C. Hunter. 1988. *The Ecology of the Lower Colorado River from Davis Dam to the Mexico-United States International Boundary: A Community Profile.* U.S. Fish and Wildlife Service Biological Report 85 (7.19). U.S. Fish and Wildlife Service, Fort Collins, CO.

Osterkamp, W. R., and E. R. Hedman. 1982. *Perennial-Streamflow Characteristics Related to Channel Geometry and Sediment in Missouri River Basin.* U.S. Geological Survey Professional Paper 1242. U.S. Geological Survey, Washington, DC.

Poff, N. L., J. D. Allen, M. B. Bain, J. R. Karr, K. L. Prestegaard, B. D. Richter, R. E. Sparks, and J. C. Stromberg. 1997. The natural flow regime. *BioScience* **47:**769–784.

Porter, K. W., and E. M. Wilde. 1984. *Preliminary Geologic Map of the Zortman 30 x 60-Minute Quadrangle.* U.S. Geological Survey and Montana Bureau of Mines and Geology, Butte, MT.

Read, R. A. 1958. *Silvicultural Characteristics of Plains Cottonwood.* Rocky Mountain Forest and Range Experiment Station Paper No. 33. USDA Forest Service, Lincoln, NE.

Robinson, T .W. 1958. *Phreatophytes.* Geological Survey Water-Supply Paper 1423. U.S. Geological Survey, Washington, D.C.

Rood, S. B., and S. Heinze-Milne. 1989. Abrupt downstream forest decline following river damming in southern Alberta. *Canadian Journal of Botany* **67:**1744–1749.

Rood, S. B., and J. M. Mahoney. 1990. Collapse of riparian poplar forests downstream from dams in western prairies: Probable causes and prospects for mitigation. *Environmental Management* **14:**151–164.

Rood, S. B., and J. M. Mahoney. 1995. River damming and riparian cottonwoods along the Marias River, Montana. *Rivers* **5:**195–207.

Rood, S. B., J. B. Mahoney, D. E. Reid, and L. Zilm. 1995. Instream flows and the decline of riparian cottonwoods along the St. Mary River, Alberta. *Canadian Journal of Botany* **73:**1250–1260.

Rood, S. B., S. Patino, K. Coombs, and M. T. Tyree. 2000. Branch sacrifice: Cavitation-associated drought adaptation of riparian cottonwoods. *Trees* **14:** 248–257.

Ross, R. L., and H. E. Hunter. 1976. *Climax Vegetation of Montana Based on Soil and Climate.* U.S. Department of Agriculture, Soil Conservation Service, Bozeman, MT.

Schmidt, J. C., R. H. Webb, R. A. Valdex, G. R. Marzolf, and L. E. Stevens. 1998. Science and values in river restoration in the Grand Canyon. *BioScience* **48:** 735–747.

Schultz, T. T., and W. C. Leininger. 1990. Differences in riparian vegetation structure between grazed areas and exclosures. *Journal of Range Management* **43:**295–299.

Scott M. L., G. T. Auble, and J. M. Friedman. 1997. Flood dependency of cottonwood establishment along the Missouri River, Montana, USA. *Ecological Applications* **7:**677–690.

Scott M. L., J. M. Friedman, and G. T. Auble. 1996. Fluvial process and the establishment of bottomland trees. *Geomorphology* **14:**327–339.

Scott, M. L., S. K. Skagen, and M. F. Merigliano. In review. Relating breeding bird diversity to geomorphic change and grazing in riparian forests. *Conservation Biology.*

Scott, M. L., G. C. Lines, and G. T. Auble. 2000. Channel incision and patterns of cottonwood stress and mortality along the Mojave River, California. *Journal of Arid Environments* **44:**399–414.

Scott, M. L., P. B. Shafroth, and G. T. Auble. 1999. Response of riparian cottonwoods to alluvial water table declines. *Environmental Management* **23:** 347–358.

Segelquist, C. A., M. L. Scott, and G. T. Auble. 1993. Establishment of *Populus deltoides* under simulated alluvial groundwater declines. *American Midland Naturalist* **130:**274–285.

Shafroth, P. B., J. C. Stromberg, and D. T. Patten. 2000. Woody riparian vegetation response to different alluvial water table regimes. *Western North American Naturalist* **60:**66–76.

Shafroth, P. B., G. T. Auble, J. C. Stromberg, and D. T. Patten. 1998. Establishment of woody riparian vegetation in relation to annual patterns of streamflow, Bill Williams River, Arizona. *Wetlands* **18:**577–590.

Smith, D. G. 1980. River ice processes: Thresholds and geomorphologic effects in northern and mountain rivers. In *Thresholds in Geomorphology,* ed. by D. R. Coates and J. D. Vitek, pp. 323–343. Allen & Unwin, London, U.K.

Smith, D. G., and C. M. Pearce. 2000. River ice and its role in limiting woodland development on a sandy braid-plain, Milk River, Montana. *Wetlands* **20:** 232–250.

Smith, S. D., D. A. Devitt, A. Sala, J. R. Cleverly, and D. E. Busch. 1998. Water relations of riparian plants from warm desert regions. *Wetlands* **18:**687–695.

Snyder, K. A., and D. G. Williams. 2000. Water sources used by riparian trees varies among stream types on the San Pedro River. Arizona. *Journal of Agricultural and Forest Meteorology.* **105:**227–240.

Sprackling, J. A., and R. A. Read. 1979. *Tree Root Systems in Eastern Nebraska.* Nebraska Conservation Bulletin, Number 37. University of Nebraska, Lincoln, NE.

Sprugel, D.G. 1991. Disturbance, equilibrium, and environmental variability: What is "natural" vegetation in a changing environment? *Biological Conservation* **58:**1–18.

Stromberg, J. C. In press. Restoration of desert riparian vegetation in the U.S. Southwest: Importance of flow regimes and fluvial dynamism. *Journal of Arid Environments.*

Stromberg, J. C., D. T. Patten, and B. D. Richter. 1991. Flood flows and dynamics of Sonoran riparian forests. *Rivers* **2:**221–235.

Stromberg, J. C., R. Tiller, and B. Richter. 1996. Effects of groundwater decline on riparian vegetation of semiarid regions: The San Pedro, Arizona. *Ecological Applications* **6:**113–131.

Stromberg, J. C., B. D. Richter, D. T. Patten, and W. G. Wolden. 1993. Response of a Sonoran riparian forest to a 10-year return flood. *Great Basin Naturalist* **53:**18–130.

Swetnam, T. W., and J. L. Bentancourt. 1998. Mesoscale disturbance and ecological response to decadal climatic variability in the American Southwest. *Journal of Climate* **11:** 3128–3147.

Szaro, R. C., and C. P. Pase. 1983. Short-term changes in a cottonwood-ash-willow association on a grazed and an ungrazed portion of Little Ash Creek in central Arizona. *Journal of Range Management* **36:**382–384.

Taylor, D. M. 1986. Effects of cattle grazing on passerine birds nesting in riparian habitat. *Journal of Range Management* **39:**254–258.

Tyree, M. T., K. J. Kolb, S. B. Rood, and S. Patiño. 1994. Vulnerability to drought-induced cavitation of riparian cottonwoods in Alberta: A possible factor in the decline of the ecosystem? *Tree Physiology* **14:**455–466.

U.S. Department of the Interior. 1993. *Upper Missouri National Wild and Scenic River Management Plan Update.* Bureau of Land Management, Lewistown, MT.

U.S. Geological Survey. 1957. *Floods of May-June 1953 in Missouri River Basin in Montana.* Geological Survey Water-Supply Paper 1320-B:69-153. U.S. Geological Survey Washington, DC.

van Haverbeke, D. F. 1990. Plains cottonwood. In *Silvics of North America.* Vol. 2, *Hardwoods.* R. M. Burns and B. H. Honkala, tech. coords., pp. 536–543. U.S. Forest Service, Washington, DC.

Walford, G. M., and W. L. Baker. 1995. Classification of the riparian vegetation along a 6-km reach of the Animas River, southwestern Colorado. *Great Basin Naturalist* **55:**287–303.

Ware, G. H., and W. T. Penfound. 1949. The vegetation of the lower levels of the floodplain of the South Canadian River in central Oklahoma. *Ecology* **30:** 478–484.

Wharton, C. H., W. M. Kitchens, E. C. Pendleton, and T. W. Sipe. 1982. *The Ecology of Bottomland Hardwood Swamps of the Southeast: A Community Profile.* FWS/OBS-81-37. U.S. Fish and Wildlife Service, Biological Services Program, Washington, DC.

Williams, G. P., and M. G. Wolman. 1984. *Downstream Effects of Dams on Alluvial Rivers.* United States Geological Survey Professional Paper 1286. U.S. Geological Survey Washington, DC.

Yeager, A. F. 1935. Root systems of certain trees and shrubs grown on prairie soils. *Journal of Agricultural Research* **51:**1085–1092.

Zimmerman, R. C. 1969. *Plant Ecology of an Arid Basin, Tres Alamos-Reding Area, Southeastern Arizona.* United States Geological Survey Professional Paper 485-D. U.S. Geological Survey Washington, DC.

6

Implications of Reestablishing Prolonged Flood Pulse Characteristics of the Kissimmee River and Floodplain Ecosystem

Louis A. Toth and Joseph W. Koebel Jr.

Kissimmee Division, Watershed Management Department, South Florida Water Management District, West Palm Beach, Florida

Andrew G. Warne

U. S. Geological Survey, Water Resources Division, Caribbean District, GSA Center, Guaynabo, Puerto Rico

Joanne Chamberlain

Kissimmee Division, Watershed Management Department, South Florida Water Management District, West Palm Beach, Florida

Flood Pulsing in Wetlands: Restoring the Natural Hydrological Balance, edited by Beth A. Middleton.
ISBN 0 471-41807-2

The Kissimmee River basin lies in the central portion of the southern peninsula of Florida, where, before alteration in the 1960s, it formed the headwaters of a contiguous aquatic and wetland hydroscape that extended from central Florida to Florida Bay. The Kissimmee basin empties into Lake Okeechobee, the second largest freshwater lake (1730 km²) in the conterminous United States. Lake Okeechobee provided a hydrologic link between the Kissimmee basin and the historic Everglades, which originated as overflow from the southern shore of the lake and formed a 100 km wide, 10,000 km² wetland corridor to the southern tip of Florida. The Kissimmee basin has a 4229 km² headwater catchment with 26 interconnected lakes and a 1963 km² lower drainage basin that extends from the largest and southernmost headwater lake, Lake Kissimmee, and includes the Kissimmee River channel, a 2 to 5 km wide floodplain, and lateral tributary watersheds.

Between 1962 and 1971 the Kissimmee River and floodplain were channelized to provide flood protection for the developing upper basin watershed. A 9 m deep, 100 m wide drainage canal was excavated through the river-floodplain corridor and divided by levees into a series of five impoundments. As part of this flood protection project, lakes in the upper basin were connected by canals and partitioned into flood storage reservoirs. A network of water control structures was constructed to regulate water levels and flows through both the upper and lower basins.

The impacts of the flood control project, including loss of wetlands and associated fish and wildlife resources and water quality degradation, stimulated a grassroots restoration movement that provided the impetus for 20 years of restoration-related studies. These studies led to the development of a plan to restore more than 100 km² of river-floodplain ecosystem, including 70 km of river channel and 11,000 ha of floodplain wetlands. The restoration project, which will dechannelize the central portion of the lower basin, requires the reestablishment of historical flood pulse characteristics, including continuous discharge, seasonal high and low flow regimes, frequent overbank flows, slow stage recession rates, and prolonged floodplain inundation. The Kissimmee River restoration plan was authorized for implementation by the 1992 Water Resources Development Act (PL 102-580) and is currently under way.

The ongoing restoration of the Kissimmee River is of unprecedented scale and represents a prototypal application of use of the "natural flood pulse" for restoration of a river-floodplain ecosystem. In this chapter we describe how geomorphic features of the Kissimmee River basin influenced historical flood pulse characteristics and how hydrogeomorphic linkages will be re-created in the restored system. We also predict how restored flood pulse attributes will drive expected ecological responses.

HYDROGEOMORPHOLOGY OF THE KISSIMMEE RIVER BASIN

The historic Kissimmee River drained a depressional watershed (Huber et al., 1976) dominated by marsh and lake storage, with a poorly developed dendritic drainage pattern, an extremely low slope, and prolonged seasonal rainfall and flooding. The drainage basin is essentially a low-gradient swale formed by late tertiary marine coastal sedimentation and erosion processes, subterranean carbonate solution, and subsidence (White, 1970). A surficial silica-rich sand mantle was deposited during the late Miocene to Pleistocene sea level highs, when longshore currents and marine waves generated a series of sandy beach ridges with adjacent lagoons and shallow marine shelves that now compose the highland ridges, interridge lows, and inland plains of central Florida. The basin is underlain by a carbonate platform in which dissolution has induced karst landforms, particularly in the upper basin, where extensive sinkhole development occurred as groundwater levels dropped during Pleistocene sea level lows. The carbonate basement also holds the major regional (Floridan) aquifer, which contributes to baseflow during dry periods.

The lower Kissimmee basin follows a south-sloping fault block (Bond et al., 1981) that occupies a Pleistocene lagoonal valley. The former marine landform has been subsequently modified only slightly by fluvial processes, and the current landscape is a heterogeneous mixture of relict beach plain, karst, and fluvial features. Channel slope (0.06-0.09 m km^{-1}) was low and typical of much higher order rivers (Warne et al., 2000). However, the complex of abandoned channels suggests high rates of channel migration, cutoff, and avulsion (abandonment of one channel system and concomitant development of another), which were likely due to the noncohesive sandy substrates (Warne et al., 2000). Although the river did not transport large sediment loads, noncontiguous levees developed along one or both banks of the channel and typically rose 1.0 to 1.5 m above the surface of the floodplain. The more limited karst landforms in the lower basin are attributable to the thick sands that overlie the carbonate basement and mask its dissolution. However, peripheral areas of the floodplain include small shallow depressions that may be manifestations of subsurface sinkholes. Although these are not evident in the more central portions of the floodplain, they may be hidden by the peat and muck deposits that mantle these areas. Most of the floodplain is very flat, with a lateral elevation gradient of about 2 m from the river channel to the surrounding uplands. The transition between the floodplain and the uplands is distinct and typically occurs over 100 to 150 m with 2 to 3 m of relief.

The hydrodynamics of the Kissimmee River were typical of subtropical rivers, with peak flows most commonly occurring at the end of the summer-fall rainy season. Mean annual rainfall for the basin is 124 cm yr^{-1}, with 60 percent of the annual precipitation occurring between June and September. Mean annual floods had discharges of 92 m^3 sec^{-1} with a return period of 3.1 years at the gauging station (Figure 6-1) below Lake Kissimmee (river km 3) and 199 m^3 sec^{-1} with a return period of 2.7 yr at the gauging station just upstream of Lake Okeechobee (river km 127). The Kissimmee's unimodal flood pulse reflected the seasonal precipitation pattern and typically peaked in October (Figure 6-2).

The retention, distribution, and discharge of surface water were largely controlled by geomorphic features. About 60 percent of the average annual discharge was contributed by the upper basin, where the absence of a well-developed fluvial drainage network led to prolonged floodwater retention times. Although several small streams delivered runoff to a few of the headwater lakes, surface water generally was detained in the lakes and wetlands and slowly funneled into Lake Kissimmee, primarily as overland sheet flow through wetland sloughs. The delivery of surface water from Lake Kissimmee to the Kissimmee River was largely controlled by a constricted outlet at the southern end of the lake (Figure 6-1). This constriction is likely attributable to basement tectonic features and acted as a natural weir that induced high water levels in the upper basin and dampened and prolonged peak discharge to the river.

The lower basin had a more well developed fluvial network, including 166 km of meandering, somewhat anastomosing river channel and at least 22 lateral tributary watersheds that enhanced peak discharges and annual floods in lower reaches of the river. However, the tributary watersheds had very low drainage densities (about 1.0 km km^{-2}), intersected the floodplain

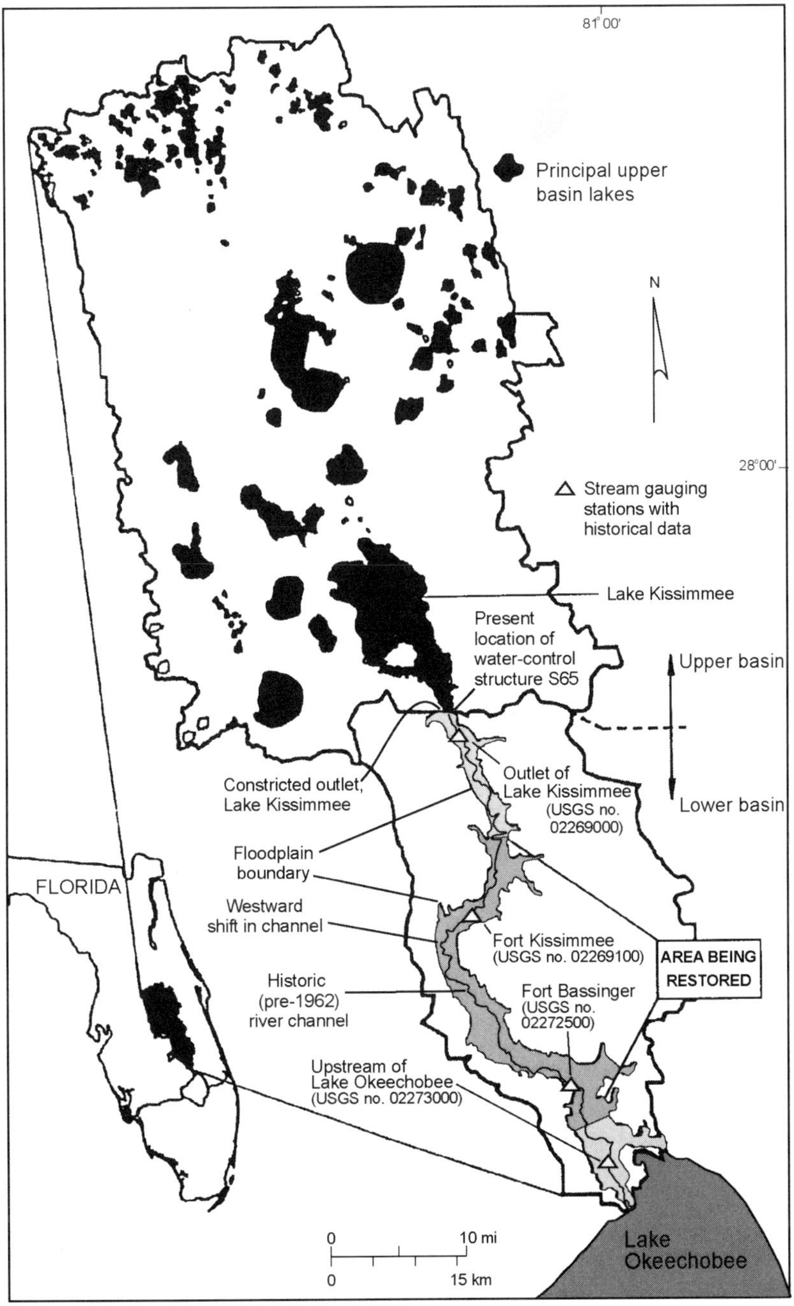

Figure 6-1. Upper and lower Kissimmee River drainage basins, locations of historic (U.S. Geologic Survey) gauging stations, and portion of the river and floodplain that will be restored.

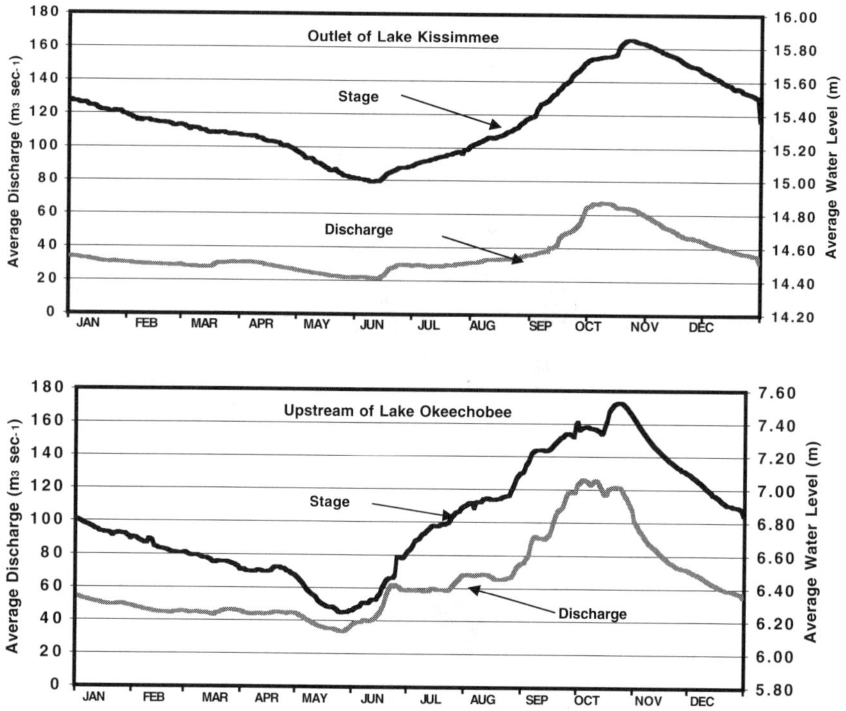

Figure 6-2. Average annual flood pulse of the Kissimmee River. Graphs show mean daily stage and discharge at gauging stations at the outlet of Lake Kissimmee (river km 3) (1933–59) and upstream of Lake Okeechobee (river km 127) (1928–59). Hydrographs are based on daily stage and discharge data from USGS gauges 02269000 and 02273000. (Toth, unpublished.)

as wetland sloughs without defined channels, and contributed predominately subsurface runoff to the river (Huber et al., 1976).

Discharges from Lake Kissimmee (Figure 6-3a) were continuous and consisted of high (>85 m^3 sec^{-1}) pulses, which occurred 10 times during the historical period of record (1933–1960), moderate pulses (28 to 56 m^3 sec^{-1}), which occurred about 28 percent of the time, and low flow regimes which, except for rare droughts (e.g., 1955–56), were dry-season transitions between the flood pulses. Discharges from Lake Kissimmee fell below 8.5 m^3 sec^{-1} during only 5 percent of the historical period of record. Most high and moderate discharge pulses lasted about a year, typically beginning in June-July, peaking at the end of the wet season, September-

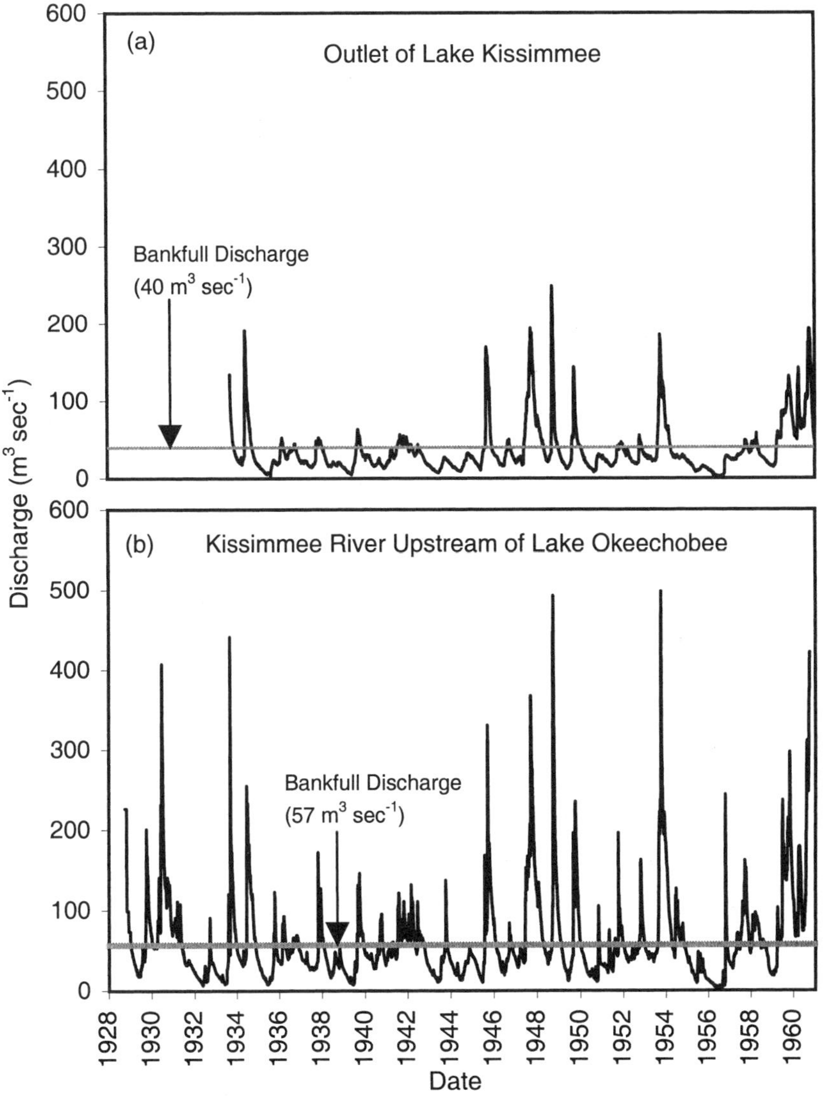

Figure 6-3. *Historic discharge hydrographs at gauging stations at the outlet of Lake Kissimmee (river km 3) and upstream of Lake Okeechobee (river km 127). Horizontal line on each hydrograph shows out-of-bank discharge threshold. (Warne et al., 2000). Hydrographs are based on daily discharge data from USGS gauges 02269000 and 02273000. (Toth, unpublished.)*

October, and ending in June. Discharge pulses took an average of 80 days to reach peak flows and a mean of 192 days to recede.

Although the lower basin constitutes only 30 percent of the total basin drainage area, peak discharges and mean annual floods were disproportionately higher at the downstream gauging station (river km 127) than at the outlet of Lake Kissimmee, particularly during periods of moderate upper basin inflows (e.g., October 1951, 1952, 1953, and 1957; Figure 6-3). High discharge pulses near the confluence with Lake Okeechobee exceeded 200 m^3 sec^{-1}, and moderate discharge pulses ranged from 80 to 200 m^3 sec^{-1}. Although lower basin tributary systems produced discharge pulses more frequently and of higher amplitude than inflow pulses from Lake Kissimmee, the average duration of discharge pulses (197 days) was 75 days shorter at the downstream gauging station.

Historical stage hydrographs reflected the influence of upper and lower basin geomorphic characteristics on discharge pulses. Moderated inflows from the upper basin watershed produced a somewhat smoother hydrograph at the gauging station immediately downstream of Lake Kissimmee than at downstream locations where more spike-like hydrographs showed the influence of lateral tributary inflows and localized rainfall events (Figure 6-4). The average stage recession rates associated with descending limbs of discharge hydrographs increased from 0.67 cm day^{-1} at the gauging stations downstream of Lake Kissimmee and at Fort Kissimmee (river km 34), to 0.78 cm day^{-1} at Fort Bassinger (river km 93) and was 1.67 cm day^{-1} at the lowermost gauging station.

Differences in surface water slopes along the river are consistent with higher stage recession rates at downstream locations. The slope of the water surface profile in the upper reach of the river (i.e., between Lake Kissimmee and Fort Kissimmee) increased from about 5 to 7 cm km^{-1} with increasing headwater stages and discharges (Figure 6-5). Thus, water levels in this upper reach were controlled primarily by upper basin inflows. The water surface slope between Fort Kissimmee and Fort Bassinger ranged between 7 and 8 cm km^{-1} (mean = 7.6 cm km^{-1}) and was not correlated with either upstream or downstream stages or discharges (Figure 6-6). Farther downstream (between the Fort Bassinger and downstream gauges), the water surface slope declined from about 8 cm km^{-1} to 5 cm km^{-1} with increasing downstream stages (Figure 6-7). This was due to downstream stages increasing more rapidly than Fort Bassinger stages during high discharges, which indicates that there was a bottleneck near the confluence with Lake

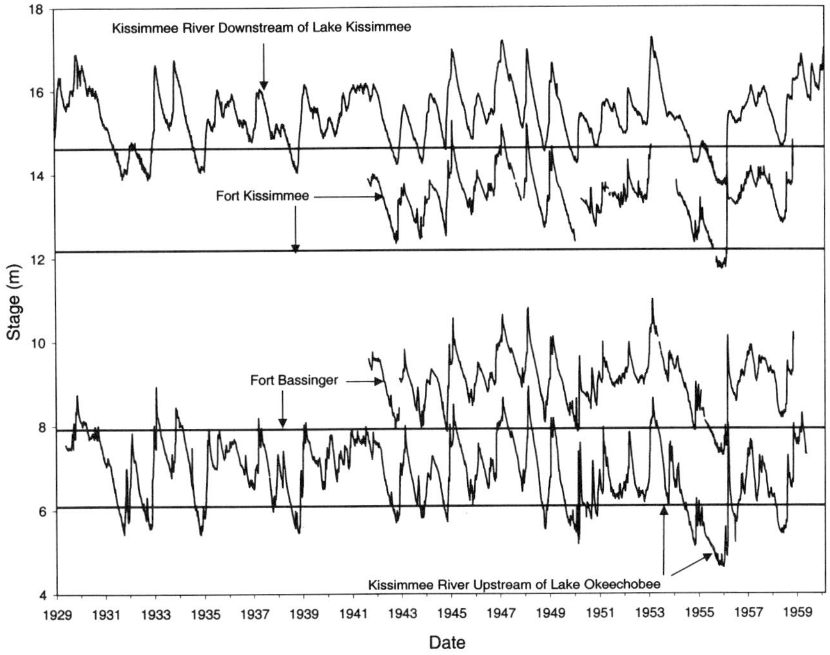

Figure 6-4. *Historic stage hydrographs at gauging stations at the outlet of Lake Kissimmee (river km 3), Fort Kissimmee (river km 34), Fort Bassinger (river km 93), and upstream of Lake Okeechobee (river km 127). Horizontal lines on each hydrograph show the floodplain inundation threshold. Hydrographs are based on daily stage data from USGS gauges 02269000, 02269100, 02272500, and 02273000. (Toth, unpublished.)*

Okeechobee that controlled rates of drainage and associated river stages along this lower reach of the river.

Based on estimated bankfull discharges (40 m^3 sec^{-1} downstream of Lake Kissimmee and 57 m^3 sec^{-1} near Lake Okeechobee), the river was completely out of bank for 50 percent of the historical period of record along its upper reach and 25 percent of the time at the downstream gauge (Figure 6-3). Bankfull discharge in the Kissimmee River was nearly equivalent to mean annual discharge and almost an order of magnitude lower than other river systems with similar drainage areas (Warne et al. 2000).

However, discharge pulses in the historic Kissimmee often supplemented rather than initiated floodplain inundation. During 6 of the 10 high discharge pulses and 12 of the 19 moderate discharge pulses from the upper basin, some of the floodplain at the gauging station downstream of

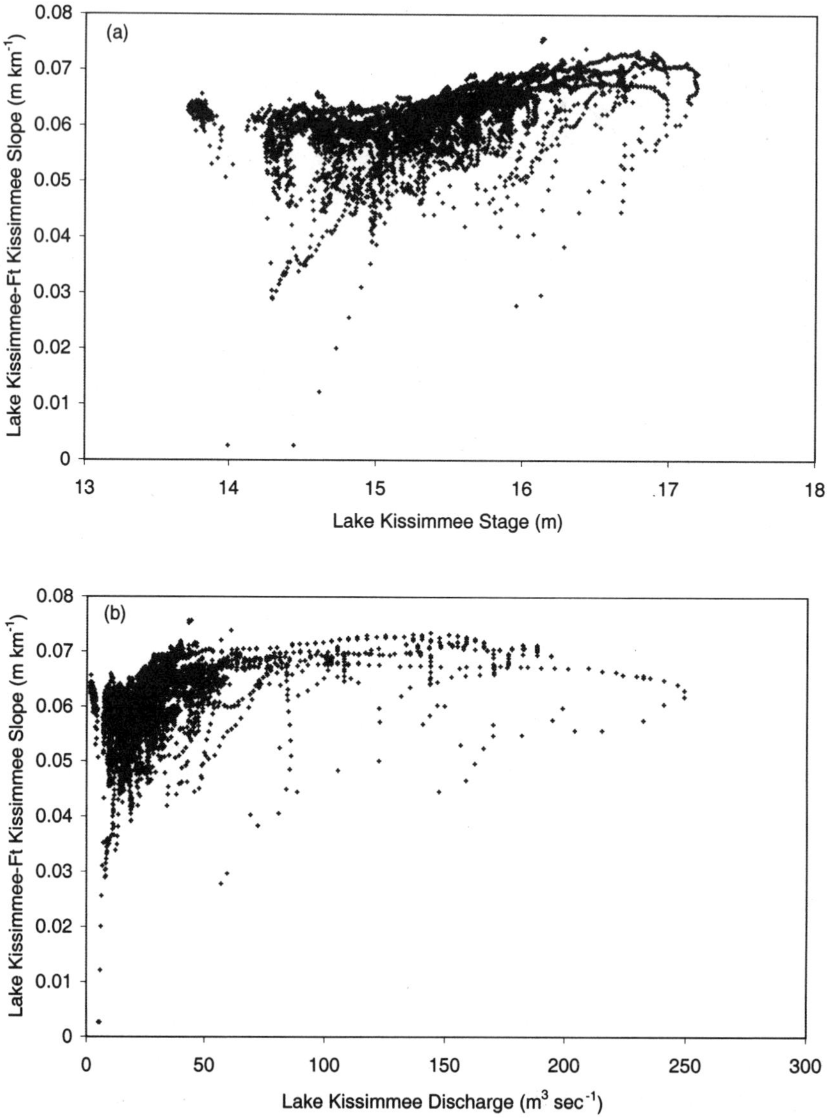

Figure 6-5. *Slopes of daily water surface profiles along the upper reach of the Kissimmee River relative to daily stage* (a) *and discharge* (b) *at the gauging station at the outlet of Lake Kissimmee. Slopes were derived from the difference in mean daily stages (1942–59) over 31 km of river channel between the Lake Kissimmee and Fort Kissimmee gauges (based on daily stage data from USGS gauges 02269000 and 02269100). (Toth, unpublished.)*

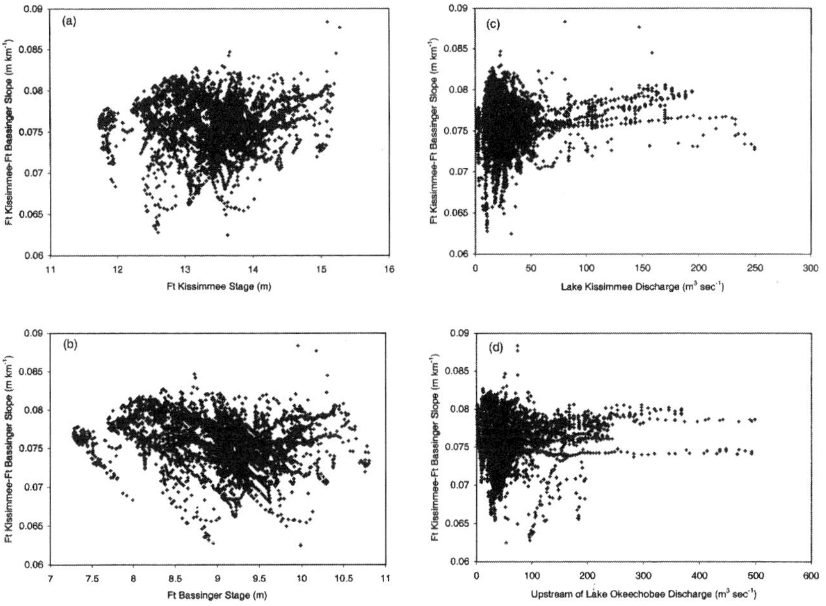

Figure 6-6. *Slopes of daily water surface profiles along the middle reach of the Kissimmee River relative to daily stages at the gauging stations at* (a) *Fort Kissimmee and* (b) *Fort Bassinger and to daily discharges at the gauging stations at the* (c) *outlet of Lake Kissimmee and* (d) *upstream of Lake Okeechobee. Slopes were derived from the difference in mean daily stages (1942–59) over 59 km of river channel between the Fort Kissimmee and Fort Bassinger gauges (based on daily stage data from USGS gauges 02269000 and 02269100). (Toth, unpublished.)*

Lake Kissimmee was already inundated (Figure 6-4). More than half (53 percent) of the floodplain at this location was inundated for 58 percent of the historic period of record and 75 percent of the floodplain was flooded 41 percent of the time. The entire floodplain at this location was dry only when discharges were < 14 m^3 sec^{-1}, which occurred during only 13 percent of the historic period of record.

Hydroperiods were even longer downstream at Fort Kissimmee, where 77 percent of the floodplain had mean annual hydroperiods of 265 days and was inundated for 76 percent of the historical period of record, including spans of at least 362 consecutive days 59 percent of the time. Peripheral areas of the floodplain had more variable hydroperiods but generally were inundated during at least a portion of the wet season (July to November). Depths exceeded 1 m 50 percent of the time along the por-

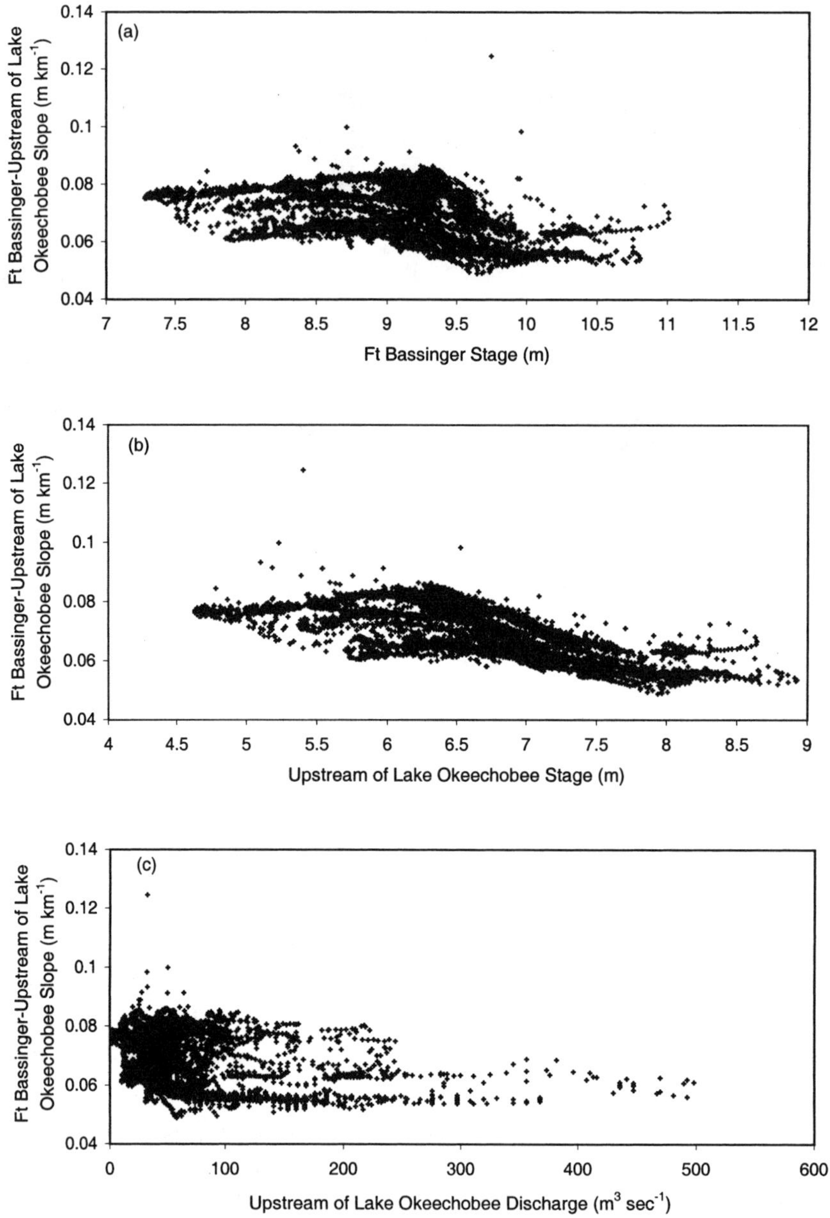

Figure 6-7. *Slopes of daily water surface profiles along the lower reach of the Kissimmee River relative to daily stages at the gauging stations* (a) *at Fort Bassinger and* (b) *upstream of Lake Okeechobee, and to* (c) *daily discharge at the gauging station upstream of Lake Okeechobee. Slopes were derived from the difference in mean daily stages (1942–59) over 34 km of river channel between the gauges at Fort Bassinger and upstream of Lake Okeechobee (based on daily stage data from USGS gauges 02272500 and 02273000). (Toth, unpublished.)*

tion (15 percent) of the floodplain that flanked the river channel. Except during the 1956 drought, stages remained above these inner floodplain elevations (Figure 6-4) during periods of low inflows from Lake Kissimmee. Prolonged hydroperiods at Fort Kissimmee may have been due to the confluence of a large (256 km^2) lateral tributary watershed. Moreover, drainage at this location likely was impeded by a major westwardly shift in the longitudinal course of the river and floodplain (Figure 6-1), which coincides with an east-west cross-peninsular lineament (White, 1970; Bond et al., 1981).

Historic hydroperiods at Fort Bassinger were similar to frequencies of floodplain inundation along the upper reach of the river (downstream of Lake Kissimmee). Fifty-eight percent of the floodplain at Fort Bassinger was inundated for 47 percent of the historical period of record, and 85 percent of the floodplain was inundated 24 percent of the time. Farther downstream, stage hydrographs and associated floodplain inundation reflected the more rapid rates of drainage along the lower reach of the river, but the entire floodplain adjacent to the lowermost gauging station was dry just 30 percent of the time (Figure 6-4).

FLOOD PULSE ECOLOGY

Prior to channelization, the river channel was flanked by 16,000 ha of floodplain wetlands that extended to an upland boundary delimited by a relatively narrow stand of live oak *(Quercus virginiana)* and cabbage palm *(Sabal palmetto)*. The river and floodplain supported populations of resident and migratory waterfowl, a diverse wading bird community, 39 fish species, and the endangered bald eagle *(Haliaeetus leucocephalus)*, snail kite *(Rostrhamus sociabilis)*, and wood stork *(Mycteria americana)* (Toth, 1993).

Because of the prolonged high amplitude of its flood pulse, the hydrology of the historic Kissimmee River and floodplain was unique among North American river-floodplain ecosystems and supported a mosaic of four dominant wetland types. Backwater lakes and ponds adjoined the river channel, and three wetland plant communities prevailed, with spatial distributions that varied according to lengths and depths of inundation along the lateral elevation gradient across the floodplain (Toth et al., 1995). The central portion of the floodplain was exposed to prolonged, deep hydroperiods and was covered by willow *(Salix caroliniana)* and buttonbush *(Cephalanthus occidentalis)* shrub communities and a predominantly herbaceous broadleaf marsh composed of pickerelweed *(Pontederia cordata),* arrowhead *(Sagittaria lancifolia),* cutgrass *(Leersia hexandra),* and maidencane *(Panicum hemitomon).* Water depths within these habitats varied spatially with topographic heterogeneity along and across the floodplain, and temporally with the dynamics of the flood pulse. Prolonged inundation provided conditions that led to dense vegetation and accumulation of plant litter that likely sustained the slow stage recession and drainage rates. Peripheral floodplain elevations received the full amplitude of the flood pulse and had shorter and shallower hydroperiods that selected for more diverse wet prairie communities composed of a mixture of forbs, grasses, rushes, and sedges.

Protracted inundation of the floodplain provided vast habitat for small-bodied fishes, particularly livebearers (Poecillidae), and for larvae and juveniles of larger riverine species (Florida Game and Fresh Water Fish Commission, 1957) that were afforded refugia from predators by dense vegetation. Use of the floodplain by larger-bodied fish species such as bass (Centrarchidae) and pike (Esocidae) may have been limited to the backwater lakes and deeper wetlands that flanked the river channel, although the habitat value of the floodplain for fish expanded during flood pulses (Trexler, 1995). Avian utilization of the historic floodplain was also linked to the dynamics of the advancing and retreating flood pulse, which provided a range of water depths for reproduction and foraging by migratory diving ducks, dabbling ducks, short- and long-legged wading birds, bald eagles, wood storks, and snail kites (Weller, 1995). However, key avian habitats were the seasonally inundated wet prairie and isolated drying pools of standing water that formed in localized topographic depressions during receding hydrographs, particularly during the dry season months. Drying pools provided concentrated sources of fish and invertebrate prey resources (Kushlan, 1986) for nesting wading birds and overwintering waterfowl. For example, 47 depressions ranging in size from 0.01 to 4.7 ha (mean = 0.4 ha) occurred in the 13.2 km^2 of floodplain adjacent to the Fort Kissimmee gauging station (Figure 6-8). Forty-two of these depressions formed isolated drying pools during the dry season (14 to 19 times during the 1942–1960 historic period of record), with an average duration of 20 to 22 days (Table 6-1), but lasting as long as 72 to 86 days.

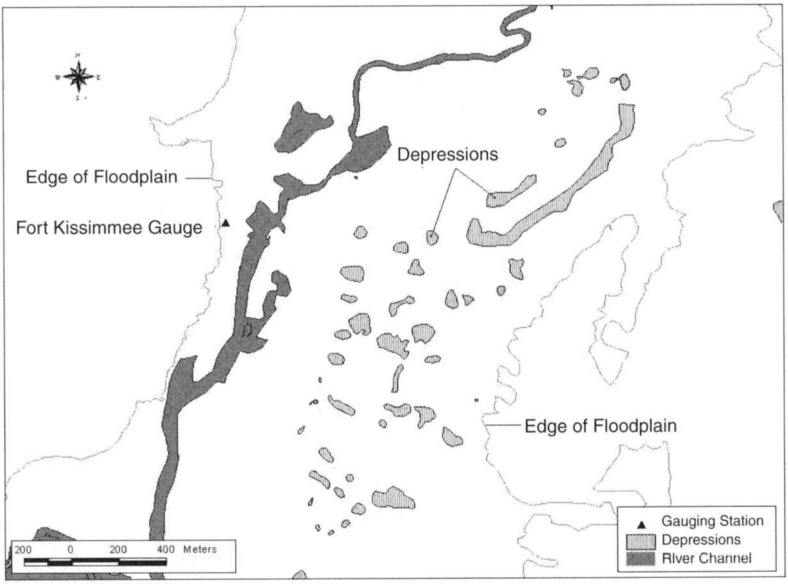

Figure 6-8. Topographic depressions on 13.2 km^2 of floodplain adjacent to the Fort Kissimmee gauging station. (Toth, unpublished.)

RESTORATION OF THE FLOOD PULSE

The ecological functionality of the river's historic flood pulse is a key element of the goal of the Kissimmee River restoration project to reestablish the ecological integrity of the river-floodplain ecosystem (Toth, 1995). The development of restoration and subsequent management plans for the river required incorporation of hydrogeomorphic controls of the flood pulse, particularly landscape features that influenced the delivery of water from contributing watersheds and the rates at which water drained within and from the river system.

Implementation of flood control regulation schedules in the upper Kissimmee basin resulted in drastic modifications to historical headwater discharge characteristics, including a reversal of seasonal high and low flow periods, and frequent and prolonged periods of no discharge to the lower basin. The headwater discharge pulse will be reestablished by modifying the flood control regulation schedule criteria for the headwater lakes and using the water control structure at the outlet of Lake Kissimmee

TABLE 6-1. Drying pools for foraging wading birds and waterfowl.

Mean duration (days) and number of times (in parentheses) that inundated topographic depressions (drying pools) adjacent the Fort Kissimmee gauging station were isolated during wet (June–September) and dry (November–April) seasons of historic (prechannelization) and simulated (restored flood pulse) periods of record. Data are based on daily stages and topography of 13.2 km^2 of floodplain adjacent the gauging station.

			Duration (and Number) of Periods Isolated			
			Historic (1942–60)		Simulated (1970–95)	
Elevation (m)	Number of Depressions	Mean Area (ha)	Wet	Dry	Wet	Dry
12.2–12.5	2	0.01	9.5(4)	9.9(7)	19.2(5)	18.5(14)
12.5–12.8	1	0.26	17.0(7)	16.2(9)	25.4(8)	22.7(17)
12.8–13.1	25	0.22	19.0(7)	20.3(14)	16.1(16)	16.0(31)
13.1–13.4	17	0.55	22.8(6)	22.1(19)	19.4(19)	14.1(12)
13.4–13.7	2	0.43	23.3(8)	37.1(21)	34.5(20)	25.9(23)

Source: Toth, unpublished data.

(S-65) to simulate the historical limited outflow capacity and associated prolonged retention and release of surface water from the upper to the lower basin (Toth et al., 1997). Operating criteria for the water control structure will be based on the historic stage–discharge relationship for the upper basin, which will reestablish continuous inflows and discharge pulses of comparable magnitude and seasonality to those of historic pulses (Figures 6-3 and 6-9). However, because of the need for some continued regulation of the upper lakes for flood control, model simulations indicate that restored discharge pulses may have a slightly shorter duration (mean duration of simulated restored discharge pulses = 220 days) than historical pulses.

The linkage between reestablished discharge pulses and restoration of the structural and functional integrity of floodplain wetlands requires the reestablishment of the associated historic slope of the water surface profile, which is presently regulated and stepped down in 2 m increments at each water control structure along the longitudinal course of the channelized river. The historic water surface profile will be reestablished primarily by the backfilling of 35 continuous kilometers of the flood control canal (Figure 6-10) and removal of two water control structures within this dechannelized reach (Koebel 1995). Disjunct remnant channels will be reconnected with new river sections (Figure 6-11), which will be carved

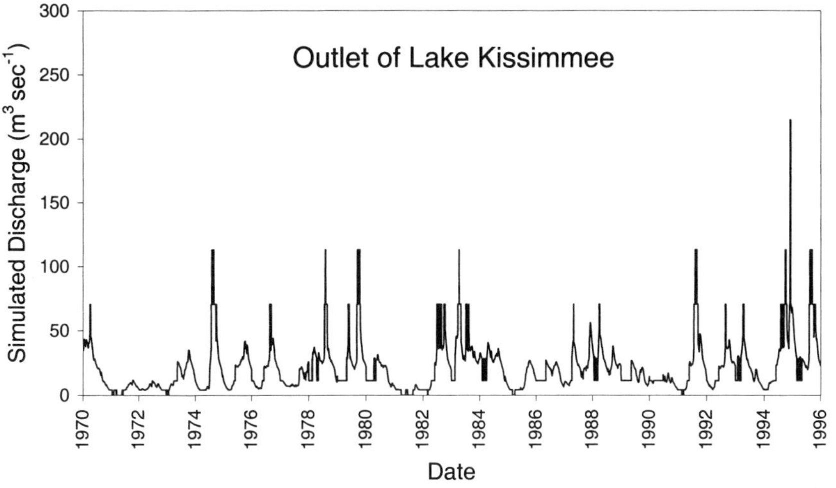

Figure 6-9. *Simulated discharge hydrograph (restored discharge pulse) at the outlet of Lake Kissimmee, based on the modified flood control regulation schedule and operation criteria for the river's headwater lakes. Simulations were based on a model that uses rainfall, temperature, and solar radiation data to generate daily lake stages and discharges (see Toth et al., 1997, for a description of model simulations.)*

with hydraulic characteristics of the historic river (Toth 1996). Flow will be redirected through the meandering river channel, which has a much lower discharge capacity than the flood control canal, and over the floodplain during moderate to high discharge periods. As the drained floodplain is reinundated, reestablished wetland vegetation communities will increase resistance to flow and thereby moderate stage recession rates and associated rates of drainage through the reconstructed system. Because the lower end of the river will remain channelized, the water control structure downstream of the reconstructed river will be used to reestablish downstream control on rates of drainage by managing water stages at the structure to replicate the historic relationship between river discharge and the water surface slope.

Restoration of historic rates of drainage will be accommodated by the elimination of the secondary canal systems that have been constructed on the floodplain since its channelization. Where feasible, drainage networks in tributary watersheds in the lower basin will also be eliminated to help restore lateral inflow regimes, which presently deliver upland runoff to the river in more rapid pulses than they did historically.

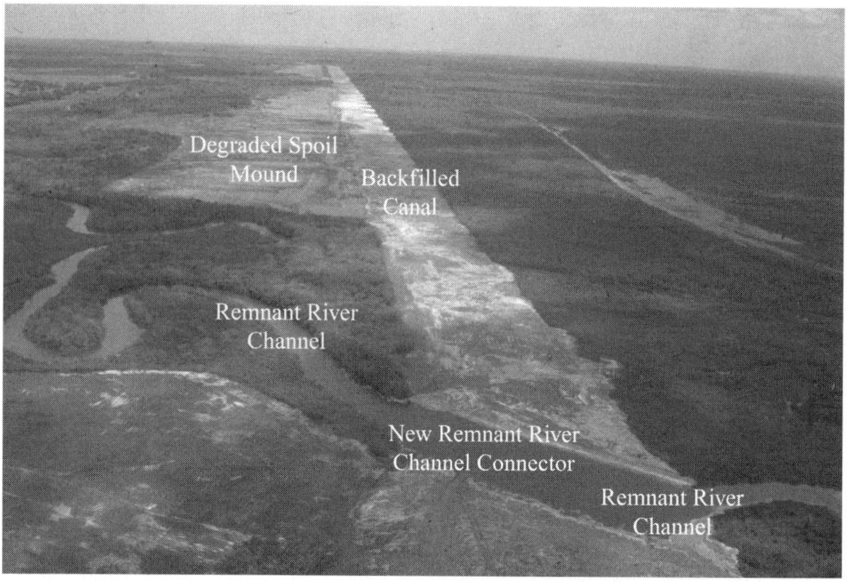

Figure 6-10. *Recently (June 1999–February 2001) backfilled flood control canal, degraded spoil mound, and reconnected remnant Kissimmee River channel. The new remnant river channel connector spans a portion of the flood control canal that was backfilled to a depth of 3 to 4 m, which is consistent with the bottom elevation of the reconnected remnant channels. (Photograph by Paul Whalen, South Florida Water Management District; used by permission.)*

RESTORATION EXPECTATIONS

The reconstruction of the river and reestablishment of historical hydrologic characteristics are expected to provide the habitat template and driving force for the restoration of the ecological integrity of the river-floodplain ecosystem. Achievement of the implicit scope of this restoration goal will be tracked by success criteria (Toth and Anderson, 1998) that include expectations for both structural and functional ecological attributes of the ecosystem. Reestablishment of these features will be driven by restoration of flood pulse characteristics, including an initial increase in stage, prolonged floodplain inundation frequencies, and slow stage recession rates (Table 6-2). Restoration expectations for the Kissimmee River are derived from a combination of data from the historical system, ecologically similar but undisturbed reference sites, and conceptual models of river-floodplain ecosystem pattern and process.

Figure 6-11. *New (recarved) river sections upstream and downstream of a remnant channel of the historic Kissimmee River. (Photograph by Paul Whalen, South Florida Water Management District; used by permission.)*

Following dechannelization, the rising limb of the discharge and stage hydrograph will begin to reestablish the hydrologic and associated biological connectivity between the river channel and floodplain that was severed by channelization (Toth, 1993). Although the initial phase of the flood pulse will provide the stimulus and conveyance pathway for recolonization of floodplain habitats by fish and invertebrate taxa that are presently confined to the flood control canal or remnant river channels, the usefulness of floodplain habitats for fish and wildlife will be dependent on spatial and temporal patterns of hydroperiods and water depths. These hydrologic conditions will be primary determinants of the structure of floodplain biological communities (Toth et al., 1995), which will be proximately tied to and mediated by the reestablishment of floodplain wetlands and the associated food web, including biological interactions such as predator-prey relationships (Lowe, 1986; Heck and Crowder,1991; Jordan et al., 1997, 1998).

TABLE 6-2. Expectations for ecological responses to restoration of the flood pulse within the restored Kissimmee River ecosystem

Restoration Expectations	Relation to Restored Flood Pulse
Floodplain Vegetation	
Reestablishment of broadleaf marsh communities, primarily composed of *Sagittaria lancifolia* (Linnaeus) and *Pontederia cordata* (Linnaeus), over 67 percent of the restored floodplain	Prolonged, deep hydroperiod
Reestablishment of wetland shrub communities dominated by buttonbush (*Cephalanthus occidentalis* Linnaeus) and willow (*Salix caroliniana* Michaux)	Prolonged, deep hydroperiod
Reestablishment of wet prairie communities, primarily composed of a diverse complement of obligate and facultative wetland grasses, sedges, rushes, and forbs, over 22 to 35 percent of the restored floodplain	Short, variable hydroperiod Consistent seasonal drying
Aquatic Invertebrates	
Aquatic invertebrate colonization of newly inundated floodplain through immigration and/or oviposition by aerial adults, drift, or active crawling across the river-channel–floodplain boundary	Initial flood pulse
Reestablishment of an aquatic invertebrate community with species richness > 125 and increased annual production within restored floodplain marshes	Prolonged, variable hydroperiod
Export of invertebrate biomass from the floodplain to the river channel	Slow stage recession
Amphibians and Reptiles	
Reestablishment of a herpetofaunal community with species richness ≥ 27 within restored floodplain marshes	Prolonged, variable hydroperiod
Predictable seasonal succession of larval amphibians within restored floodplain marshes	Prolonged, variable hydroperiod
Fishes	
Immediate movement of fish from river channel to floodplain habitats	Initial flood pulse
Use of restored floodplain habitats by large-bodied fish species, including at least four game fish species, which will constitute approximately 30 percent of the floodplain fish assemblage	Prolonged, deep hydroperiod
Reestablishment of resident forage fish communities, dominated by Cyprinodontidae and Poeciliidae and constituting 40 to 60 percent of total fish abundance within restored floodplain habitats	Prolonged hydroperiod

TABLE 6-2. (Continued)

Avian Community	
Seasonal increase in waterfowl populations, primarily blue-winged teal (*Anas discors* Linnaeus) and resident mottled ducks (*Anas fulvigula* Ridgway), within restored wet prairie habitats	Short hydroperiod
Greater than 100 percent increase in annual densities of wading birds on the restored floodplain	Prolonged, variable hydroperiod
Reestablishment of two to five historic breeding colonies, consisting of at least four species with > 50 nesting pairs	Prolonged, variable hydroperiod Slow stage recession rates
Increased (> 40 percent) abundance of passerine species within restored broadleaf marsh habitats	Prolonged, deep hydroperiod

Source: Toth et al., unpublished.

Following the initial reestablishment of the flood pulse, prolonged hydroperiods with historical spatial and temporal depth patterns will be the primary driver for the achievement of restoration expectations related to the structural and functional integrity of the floodplain. Restoration of floodplain habitat structure will begin with the reestablishment of broadleaf marsh, wetland shrub, and wet prairie communities as the dominant wetland community types (areal coverage of 10,000 to 11,000 ha), with spatial distributions that replicate those of the historic floodplain (Pierce et al., 1982; Toth et al., 1995).

Broadleaf marsh and wetland shrub communities are expected to cover 65 to 75 percent of the restored floodplain in areas where prolonged, deep hydroperiods will select for low plant species richness dominated by community indicator species. Within reestablished broadleaf marshes, *Sagittaria lancifolia* and/or *Pontederia cordata* will account for > 50 percent of live vegetation cover. These and associated secondary indicator species, including *Panicum hemitomon, Leersia hexandra, Sacciolepis striata, Alternanthera philoxeroides, Nuphar lutea, Polygonum punctatum, Bacopa caroliniana, Hydrocotyle umbellata, Cephalanthus occidentalis,* and *Ludwigia peruviana,* will likely account for > 90 percent of the vegetation cover.

Wetland shrub communities are expected to be dominated by buttonbush *(Cephalanthus occidentalis)* or willow *(Salix caroliniana).* In buttonbush communities, *Cephalanthus* will account for > 25 percent of the live ground or canopy cover, and *Salix* will form and dominate a relatively closed canopy (total canopy cover > 80 percent) within willow shrub habitats. These and other obligate and facultative wetland species (Reed, 1988), including *S. lancifolia, P. cordata, P. hemitomon,* and *A. philoxeroides* and shade-tolerant ferns (e.g., *Blechnum serrulatum, Osmunda* spp., *Thelypterus* spp., and *Woodwardia* spp.), are expected to account for > 75 percent of the species composition within these habitats.

The spatial distribution of reestablished wet prairie communities is expected to be restricted to higher elevations, particularly along the periphery of the floodplain, where shorter and more variable hydroperiods with consistent seasonal drying will provide for the highest plant species diversity on the restored floodplain. Wet prairie communities composed of *Panicum hemitomon, Leersia hexandra, Rhynchospora inundata, Cyperus* spp., *Fimbristylis* spp., and a complex of other obligate and facultative wetland grasses, sedges, and forbs are likely to cover 22 to 35 percent of the floodplain within the restored system.

Colonization and succession of an aquatic invertebrate community are expected to occur within one to three years following reestablishment of historic vegetation coverage, associated periphyton communities, and the detrital food base within floodplain habitats. Expectations for the restoration of aquatic invertebrate community structure within floodplain habitats are based on invertebrate community characteristics within natural flatwoods marshes of central Florida (Evans et al., 1999), the Everglades (Rader and Richardson, 1992; Rader, 1994, 1999; Jordan et al., 1997), and remnant river channels and floodplain habitats of the channelized Kissimmee River (Anderson et al., 1998).

Initial colonization of the newly inundated floodplain by aquatic invertebrates is expected to occur through immigration and/or oviposition by aerial adults, or from invertebrate drift and active crawling across the river-channel–floodplain boundary (Kwak, 1988; Smock, 1994). Taxa likely to drift from the river channel to the floodplain on the initial flood pulse include amphipods *(Hyalella azteca),* mayflies *(Caenis diminuta),* chironomids, aquatic mites, and microcrustaceans. These taxa account for the greatest proportion of invertebrates within littoral river channel habitats (e.g., *Nuphar lutea, Polygonum densiflorum,* and snags) and should be rapidly transported to the floodplain with rising stages. Taxa likely to crawl or actively swim to the floodplain following the initial pulse include grass shrimp *(Palaemonetes paludosus),* crayfish *(Procambarus fallax),* odonates, hemipterans, and gastropods.

The macroinvertebrate community of restored marshes is expected to consist of more than 125 species and exhibit increased annual production within three to five years after the drained floodplain has been reinundated. Aquatic invertebrate community production within restored floodplain habitats will be facilitated by habitat stability provided by extended hydroperiods, and due to high standing stock biomass of keystone species (e.g., crayfish, grass shrimp, amphipods, and odonates) and rapid turnover rates of dominant taxa (primarily chironomids). Restoration of the floodplain invertebrate community will reestablish a critical link between primary producers and secondary consumers that is necessary to sustain the productivity of higher trophic levels.

Although fish will also use the newly inundated floodplain during the initial flood pulse (Welcomme, 1979; Furse et al., 1996), densities, community structure, and stability of floodplain fish assemblages will be determined largely by the amplitude and duration of the pulse. Based on data from the historical Kissimmee (Florida Game and Fresh Water Fish Commission, 1957) and reference ecosystems (Guillory, 1979; Dineen, 1968, Kushlan, 1974; Trexler, 1995; Jordan et al., 1997), fish assemblages on the restored floodplain are expected to include resident small-bodied species and both young and adults of larger-bodied species that are typically found in the main river channel. Resident "forage fish" communities are expected to be dominated by Cyprinodontidae (primarily *Lucania goodei*) and Poeciliidae *(Gambusia holbrooki* and *Heterandria formosa)* and to constitute between 40 and 60 percent of total fish abundance on the floodplain. As the amplitude of the flood pulse heightens, the relative abundance of forage fish is expected to decline as the floodplain habitats are colonized by the young of large-bodied fish species (Table 6-3), including at least four game fish (Centrarchidae and Esocidae) species, whose relative abundance should constitute about 30 percent of the floodplain fish assemblage.

Based on historical data (Carr, 1940; Carr and Goin, 1955) and the geographic distribution and habitat requirements of amphibians and reptiles of the Florida peninsula (Tennant, 1997; Conant and Collins, 1991; Bartlett and Bartlett, 1999), at least 27 herpetofaunal species (Table 6-4) are expected to recolonize the restored floodplain and include a predictable seasonal succession of larval amphibians (Mount,1975; Conant and Collins, 1991) (Table 6-5).

Expectations for the avian community include a significant increase in waterfowl populations, which will be due primarily to reestablished use of

TABLE 6-3. Indicator species for evaluating the expectation for use of the restored Kissimmee River floodplain wetlands by large-bodied fishes

Amia calva Linnaeus (bowfin)
Anguilla rostrata Lesueur (American eel)
Dorosoma cepedianum Lesueur (gizzard shad)
Erimyzon sucetta Lacepede (lake chubsucker)
Esox americanus Lesueur (redfin pickerel)
Esox niger Lesueur (Chain pickerel)
Ictalurus punctatus Rafinesque (channel catfish)
Micropterus salmoides Lacepede (largemouth bass)
Lepisosteus osseus Linnaeus (longnose gar)
Lepisosteus platyrhincus De Kay (Florida gar)
Lepomis auritus Linnaeus (redbreast sunfish)
Lepomis gulosus Cuvier (warmouth)
Lepomis machrochirus Rafinesque (bluegill)
Lepomis microlophus Gunther (redear sunfish)
Lepomis punctatus Valenciennes (spotted sunfish)
Pomoxis nigromaculatus Lesueur (black crappie)

Source: Lawrence Glenn, unpublished, based on Florida Game and Fresh Water Fish Commission, 1957; Guillory 1979.

the floodplain, particularly restored wet prairie habitats (Chamberlain, 1960; Weller, 1995), by migratory blue-winged teal *(Anas discors)* and resident mottled ducks *(Anas fulvigula).* Annual densities of wading birds are expected to increase by > 100 percent, and two to five historic breeding colonies (National Audubon Society, 1936–1959) will redevelop and include at least four species (Table 6-6) with > 50 nesting pairs. Reestablishment of breeding colonies will be linked to the reappearance of suitable nest sites in restored willow, buttonbush, and cypress *(Taxodium distichum)* habitats and to increased availability of prey items (e.g., forage fish, juvenile game fish, amphibians, and reptiles) near potential colony sites (Ogden et al., 1980; Burger, 1982; Kushlan, 1986; Frederick, 1995). Based on predictions using simulated hydroperiods, isolated drying pools with concentrated prey resources may develop even more frequently on the restored floodplain than they did historically (Table 6-1) and should stimulate nest initiation by wading birds. Restoration of marsh habitat (Toth et al., 1998) is also expected to lead to increased (> 40 percent) abundance of passerine species, such as common yellowthroats *(Geothlypis trichas),* red-winged blackbirds *(Agelaius phoeniceus),* and sedge wrens *(Cistothorus platenis).*

TABLE 6-4. Indicator species for evaluating the expectation for amphibian and reptile utilization of the restored Kissimmee River floodplain marshes

Amphibians
Acris gryllus Le Conte (southern cricket frog)
Gastrophryne carolinensis Holbrook (narrow-mouthed toad)
Hyla cinerea Schneider (green tree frog)
Hyla femoralis Bosc (pine woods tree frog)
Hyla squirella Bosc (squirrel tree frog)
Pseudacris ocularis Bosc and Daudin (little grass frog)
Rana catesbeiana Shaw (bullfrog)
Rana grylio Stejneger (pig frog)
Rana sphenocephala Cope (southern leopard frog)
Eurycea quadridigitata Holbrook (dwarf salamander)
Notopthalmus viridescens piaropicola Schwartz and Duellman (peninsular newt)
Amphiuma means Garden (two-toed amphiuma)
Pseudobranchus axanthus axanthus Netting and Goin (narrow-striped dwarf siren)
Siren intermedia Barnes (lesser siren)
Siren lacertina Linnaeus (greater siren)
Reptiles
Alligator mississippiensis Daudin (American alligator)
Anolis carolinensis Voigt (green anole)
Agkistrodon piscivorous Lacepede (eastern cottonmouth)
Farancia abacura Holbrooke (eastern mud snake)
Lampropeltis getula floridana Blanchard (Florida kingsnake)
Nerodia fasciata pictiventris Cope (Florida water snake)
Nerodia floridana Goff (Florida green water snake)
Nerodia taxispilota Holbrook (brown water snake)
Regina alleni Garman (striped crayfish snake)
Seminatrix pygaea cyclas Dowling (South Florida swamp snake)
Sistrurus miliarius barbouri Gloyd (dusky pigmy rattlesnake)
Storeria dekayi victa Hay (Florida brown snake)
Thamnophis sauritus sauritus Linnaeus (eastern ribbon snake)
Chelydra serpentina osceola Stejneger (Florida snapping turtle)
Deirochelys reticularia chrysea Schwartz (Florida chicken turtle)
Kinosternon baurii Garman (striped mud turtle)
Kinosternon subrubrum steindachneri Siebenrock (Florida mud turtle)

Source: Koebel, unpublished, based on Connant and Collins, 1991; Tennant, 1997; and Bartlett and Bartlett, 1999.

In addition to prolonging foraging opportunities (e.g., drying pools) for wading birds and waterfowl, the restoration of slow stage recession rates and the associated drainage of floodwaters will reestablish a protracted export of autochthonous primary and secondary production from the floodplain to the river channel. The import of this coarse and fine particulate

TABLE 6-5. Florida breeding periods of amphibian species likely to colonize restored Kissimmee River floodplain marsh habitats

Indicator Species	Spring	Summer	Autumn	Winter
Acris gryllus dorsalis Harlan (Florida cricket frog)	X	X	X	X
Gastrophryne carolinensis Holbrook (narrow-mouthed toad)	X	X	X	
Hyla cinerea Schneider (green tree frog)	X	X	X	
Hyla gratiosa Le Conte[a] (barking tree frog)	X	X		
Hyla femoralis Bosc[a] (pine woods tree frog)	X	X	X	
Hyla squirella Bosc[a] (squirrel tree frog)	X	X	X	
Pseudacris ocularis Bosc and Daudin (little grass frog)	X	X	X	X
Rana catesbeiana Shaw (bullfrog)	X	X	X	
Rana grylio Stejneger (pig Frog)	X	X	X	X
Rana sphenocephala Cope (southern leopard frog)	X	X	X	X
Eurycea quadridigitata Holbrook (dwarf salamander)	X		X	X
Amphiuma means Garden (two-toed amphiuma)	X			
Siren intermedia Barnes (lesser siren)	X			
Siren lacertina Linnaeus (greater siren)	X			
Pseudobranchus axanthus axanthus Netting and Goin[b] (narrow-striped dwarf siren)				

[a] Likely to occur near upland edges of floodplain.

[b] Breeding habits unknown.

Source: Koebel, unpublished, based on Connant and Collins, 1991.

matter, fishes, and invertebrates will complete the reestablishment of the trophic/energy link between the river and the floodplain and provide fuel for riverine food chains.

CONCLUSIONS

The attributes of flood pulses of the historic Kissimmee River were unique among North American river-floodplain systems. The Kissimmee's flood pulse was controlled by wet season precipitation, dense floodplain vegetation, and hydrogeomorphic factors including (1) the limited outflow capacity of the upper basin, (2) poorly developed drainage networks in contributing watersheds (both upper basin and lateral watersheds), (3) low valley slope, (4) morphological features such as floodplain constrictions and major changes in the course of the river, and (5) a bottleneck to the in-

TABLE 6-6. Indicator wading bird and waterbird species likely to establish colonies in restored wetland habitat within the Kissimmee River (KR) floodplain and tributaries

Indicator Species	Historic Colony Location
Eudocimus albus Linnaeus (white ibis)	KR and tributaries
Plegadis falcinellus Linnaeus (glossy ibis)	Tributaries
Casmerodius albus Linnaeus (great egret)	KR and tributaries
Ardea herodius Linnaeus (great blue heron)	Tributaries
Egretta caerulea Linnaeus (little blue heron)	KR and tributaries
Egretta tricolor Muller (tricolored Heron)	Tributaries
Egretta thula Molina (snowy egret)	KR and tributaries
Butorides striatus Linnaeus (green heron)	[a]
Nycticorax nycticorax Linnaeus (black-crowned night heron)	Tributaries
Nycticorax violacea Linnaeus (yellow-crowned night heron)	Tributaries
Bubulcus ibis Linnaeus (cattle egret)	[a]
Mycteria americana Linnaeus (wood stork)	Tributaries
Aramus guarauna Linnaeus (limpkin)	KR and tributaries
Anhinga anhinga Linnaeus (anhinga)	KR and tributaries
Phalacrocorax auritus Lesson (double-crested cormorant)	[a]

[a] Not documented in historic data, although likely to nest in mixed-species colonies in the restored system.
Source: Stefani Melvin, unpublished, based on National Audubon Society, 1936–1959.

flow to Lake Okeechobee. The ecological tenets of the flood pulse concept and the utility of application of the natural flood pulse restoration approach (Junk et al., 1989; Bayley, 1991) will be verified by a restoration evaluation program that will track the progress of the project. Success will be determined by achievement of explicit expectations representing a composite of structural and functional components of the ecosystem that are linked to the reestablishment of the river's flood pulse.

ACKNOWLEDGMENTS

Restoration expectations were developed by Bruce Dugger (waterfowl) and staff of the Kissimmee Division of the South Florida Water Management District, including J. Lawrence Glenn (fish), Stefani Melvin (avian species), Laura Carnal (vegetation), L. A. Toth (vegetation), and J. W. Koebel Jr. (aquatic invertebrates, herpetofauna). Renee Clark and Bonnie Rose assisted in the analysis of hydrogeomorphic data and drafted figures. David Anderson, Sandra Cooper, J. Lawrence Glenn, Paul Hudson, Steve Lin, Stefani Melvin, Beth Middleton, Luis R. Soler-

Lopez, Al Steinman, Todd Tisdale and others provided helpful comments on drafts of this chapter.

REFERENCES

Anderson, D. H., J. W. Koebel Jr., and L. M. Rojas. 1998. *Baseline Assessment of Aquatic Invertebrate Community Structure in the Kissimmee River, Florida.* Final deliverable (Contract C-6625) to the South Florida Water Management District, West Palm Beach, FL. Unpublished report.

Bartlett, R. D., and P. P. Bartlett. 1999. *A Field Guide to Florida Reptiles and Amphibians.* Gulf Publishing, Houston, TX.

Bayley, P. B. 1991. The flood pulse advantage and restoration of river-floodplain systems. *Regulated Rivers: Research and Management* **6:**75–86.

Bond, P., L. Smith, and W. F. Tanner. 1981. Structural patterns in south Florida. *Gulf Coast Association of Geological Societies Transactions* **31:** 239–242.

Burger, J. 1982. An overview of proximate factors affecting reproductive success in colonial birds: Concluding remarks and summary of panel discussion. *Colonial Waterbirds* **5:**58–65.

Carr, A. F., Jr. 1940. *A Contribution to the Herpetology of Florida.* Vol. 3. Biological Science Series. University of Florida Press, Gainesville.

Carr, A. F., Jr., and C. J. Goin. 1955. *Guide to the Reptiles, Amphibians, and Fresh-water Fishes of Florida.* University of Florida Press, Gainesville.

Chamberlain, E. B., Jr. 1960. *Florida Waterfowl Populations, Habitats, and Management.* Technical Bulletin 7. Florida Game and Fresh Water Fish Commission, Tallahassee.

Conant, R., and J. T. Collins. 1991. *A Field Guide to Reptiles and Amphibians—Eastern and Central North America.* Houghton Mifflin, New York.

Dahm, C. N., K. W. Cummins, H. M. Valett, and R. L. Coleman. 1995. An ecosystem view of the restoration of the Kissimmee River. *Restoration Ecology* **3:** 225–238.

Dineen, J. W. 1968. Determination of the effects of fluctuating water levels on the fish population of Conservation Area III, Final Report, Dingel-Johnson Project F-16-R. Florida Game and Fresh Water Fish Commission. Tallahassee, FL.

Evans, D. L., W. J. Streever, and T. L. Crisman. 1999. Natural flatwoods marshes and created freshwater marshes of Florida: Factors influencing aquatic invertebrate distribution and comparisons between natural and created marsh communities. In *Invertebrates in Freshwater Wetlands of North America: Ecology and Management,* ed. by D. P. Batzer, R. B. Rader, and S. A. Wissinger, pp. 81–104. John Wiley & Sons, New York.

Florida Game and Fresh Water Fish Commission. 1957. *Recommended Program for Kissimmee River Basin.* Florida Game and Fresh Water Fish Commission, Tallahassee.

Frederick, P. C. 1995. *Wading Bird Nesting Success Studies in the Water Conservation Areas of the Everglades, 1992–1995.* Final Report. South Florida Water Management District, West Palm Beach.

Furse, J. B., L. J. Davis, and L. A. Bull. 1996. Habitat use and movement of largemouth bass associated with changes in dissolved oxygen and hydrology in the Kissimmee River, Florida. *Proceedings of the American Conference of the Southeast Association of Fish and Wildlife Agencies* **50:**12–25.

Guillory, V. 1979. Utilization of an inundated floodplain by Mississippi River fishes. *Florida Scientist* **42:**222–228.

Heck, K. L., Jr., and L. B. Crowder. 1991. Habitat structure and predator-prey interactions in vegetated aquatic systems. In *Habitat Structure: The Physical Arrangement of Objects in Space,* ed. by S. S. Bell, E. D. McCoy, and H. R. Mushinsky. pp. 281–299. Chapman and Hall, New York.

Huber, W. C., J. P. Heaney, P. B. Bedient, and J. P. Bowden. 1976. *Environmental Resource Management Studies of the Kissimmee River Basin.* Final Report. University of Florida Department of Environmental Engineering Science. Report ENV-05-76-2. Gainesville, FL.

Jordan, F., K. J. Babbit, and C. C. McIvor. 1998. Habitat use by freshwater marsh fishes in the Blue Cypress Marsh Conservation Area, Florida. *Ecology of Freshwater Fish* **7:** 159–166.

Jordan, F., S. Coyne, and J. C. Trexler. 1997. Sampling fishes in vegetated habitats: Effects of habitat structure on sampling characteristics of the 1-m^2 throw trap. *Transactions of the American Fisheries Society* **126:**1012–1020.

Junk, W. J., P. B. Bayley, and R. E. Sparks. 1989. The flood pulse concept in river-floodplain systems. In *Proceedings of the International Large River Symposium* (LARS), ed. by D. P. Dodge, pp. 110–127. Special Publication of Fisheries and Aquatic Sciences 106. Department of Fisheries and Oceans, Ottawa.

Koebel, J. W., Jr. 1995. An historical perspective on the Kissimmee River restoration Project. *Restoration Ecology* **3:** 149–159.

Kushlan, J. A. 1974. Effects of a natural fish kill on the water quality, plankton, and fish population of a pond in the Big Cypress Swamp, Florida. *Transactions of the American Fisheries Society* **103:**235–243.

Kushlan, J. A. 1986. Responses of wading birds to seasonally fluctuating water levels: Strategies and their limits. *Colonial Waterbirds* **9:**155–162.

Kwak, T. J. 1988. Lateral movement and use of floodplain habitats by fishes of the Kankakee River, Illinois. *The American Midland Naturalist* **120:**241–249.

Lowe, E. F. 1986. The relationship between hydrology and vegetational pattern within the floodplain marsh of a subtropical Florida lake. *Florida Scientist* **49:** 213–233.

Mount, R. H. 1975. *The Reptiles and Amphibians of Alabama.* Agricultural Experiment Station, Auburn University, Auburn, AL.

National Audubon Society. 1936–1959. *Audubon Warden Field Reports.* Everglades National Park, South Florida Research Center, Homestead, FL.

Ogden, J. C., H. W. Kale II, and S. A. Nesbitt. 1980. The influence of annual variation in rainfall and water levels on nesting by Florida populations of wading birds. *Transactions of the Linnaean Society of New York* **9:**115–126.

Pierce, G. J., A. B. Amerson, and L. R. Becker Jr. 1982. *Pre-1960 Floodplain Vegetation of the Lower Kissimmee River Valley, Florida.* Biological Services Report 82-3. Environmental Consultants, Dallas, TX.

Rader, R. B. 1994. Macroinvertebrates in the northern Everglades: Species composition and trophic structure. *Florida Scientist* **57:**22–33.

Rader, R. B. 1999. The Florida Everglades: Natural variability, invertebrate diversity, and foodweb stability. In *Invertebrates in Freshwater Wetlands of North America: Ecology and Management,* ed. by D. P. Batzer, R. B. Rader, and S. A. Wissinger, pp. 25–54. John Wiley & Sons, New York.

Rader, R. B., and C. J. Richardson. 1992. The effects of nutrient enrichment on algae and macroinvertebrates in the Everglades: A review. *Wetlands* **12:**121–135.

Reed, P. B., Jr. 1988. *National List of Plant Species that Occur in Wetlands: 1988 Florida.* NERC-88/18.09. U.S. Fish and Wildlife Service. St. Petersburg, FL.

Smock, L. A. 1994. Movement of invertebrates between stream channels and forested floodplains. *Journal of the North American Benthological Society* **13:**524–531.

Tennant, A. 1997. *A Field Guide to Snakes of Florida.* Gulf Publishing, Houston, TX.

Toth, L. A. 1993. The ecological basis for the Kissimmee River restoration plan. *Florida Scientist* **56:**25–51.

Toth, L. A. 1995. Principles and guidelines for restoration of river/floodplain ecosystems-Kissimmee River, Forida. In *Rehabilitating Damaged Ecosystems,* 2d ed., ed. by J. Cairns Jr., pp. 49–73. Lewis Publishers/CRC Press, Boca Raton, FL.

Toth, L. A., and D. H. Anderson. 1998. Developing expectations for ecosystem restoration. In *Transactions of the 63rd North American Wildlife and Natural Resources Conference,* ed. by K. G. Wadsworth, pp. 122–134. Wildlife Management Institute, Washington, DC.

Toth, L. A., D. A. Arrington, and G. Begue. 1997. Headwater restoration and reestablishment of natural flow regimes: Kissimmee River of Florida. In *Watershed Restoration: Principles and Practices,* ed. by J. E. Williams, C. A. Wood, and M. P. Dombeck, pp. 425–442. American Fisheries Society, Bethesda, MD.

Toth, L. A., D. A. Arrington, M. A. Brady, and D. A. Muszick. 1995. Conceptual evaluation of factors potentially affecting restoration of habitat structure within the channelized Kissimmee River ecosystem. *Restoration Ecology* **3:**160–180.

Toth, L. A., S. L. Melvin, D. A. Arrington, and J. Chamberlain. 1998. Hydrologic manipulations of the channelized Kissimmee River. *BioScience* **48:** 757–764.

Trexler, J. 1995. Restoration of the Kissimmee River: A conceptual model of past and present fish communities and its consequences for evaluating restoration success. *Restoration Ecology* **3:**195–210.

Warne, A. G., L. A. Toth, and W. A. White. 2000. Drainage-basin-scale geomorphic analysis to determine reference conditions for ecologic restoration—Kissimmee River, Florida. *Geological Society of America Bulletin* **112:** 884–899.

Welcomme, R. L. 1979. The *Fisheries Ecology of Floodplain Rivers.* Longman Group Limited, London.

Weller, M. W. 1995. Use of two waterbird guilds as evaluation tools for the Kissimmee River restoration. *Restoration Ecology* **3:**211–224.

White, W. A. 1970. *The Geomorphology of the Florida Peninsula.* Bureau of Geology Bulletin 51. Florida Department of Natural Resources, Tallahassee.

7

Flood Pulsing in the Regeneration and Maintenance of Species in Riverine Forested Wetlands of the Southeastern United States

Beth A. Middleton

National Wetlands Research Center, USGS, Lafayette, Louisiana

Across the southeastern United States, flood pulses from rivers and streams have been cut off from the floodplains of forested wetlands by a wide variety of regulation projects, and these have had severe effects on floodplain biota whose life history strategies are adapted to seasonal

Flood Pulsing in Wetlands: Restoring the Natural Hydrological Balance, Edited by Beth A. Middleton.
ISBN 0 471-41807-2

Figure 7-1. Buttonland Swamp, a cypress-tupelo swamp near Perks, Illinois. (Photograph by Beth Middleton.)

wet/dry dynamics. The early life history stages of plants are especially sensitive to water level fluctuation and movement patterns, and thus restoration success is dependent on the reestablishment of flood-pulsed conditions on the floodplain. This chapter discusses restoration approaches for riverine forested wetlands of the southeastern United States that incorporate reengineering measures to recreate flood-pulsed conditions.

HYDROLOGIC REENGINEERING OF FORESTED WETLANDS

In forested wetlands on floodplains of the southeastern United States (Figure 7-1), the flood pulse emanating from regulated rivers and streams is constrained by levees (Humburg et al., 1996; Arkansas State Highway and Transportation Department, 1997), dams (Hesse, 1995), channelization with downcutting (Shields and Hoover, 1991; Sengupta, 1995; Brookes et al., 1996; Shear et al., 1996), and interbasin water transfer (the reengineering of water flow from one distinct river or river catchment to another)

(Pearlstine et al., 1985; Demissie et al., 1990; Muir et al., 1995; Wichmann, 1996). Activity along the Missouri River illustrates a common scenario of river regulation and associated floodplain alteration in forested wetlands. After extensive reengineering for navigation and flood control, the nature of water discharge onto the floodplain was much changed from its original state. The Missouri River had once been a wide river with many sandbars (Humburg et al., 1996), and high and low discharges were synchronized with winter and summer climatic patterns. After its regulation by dams for flood control, discharges became uniform, so that summer water levels were much higher than they had been historically. Impoundment and channelization have virtually eliminated the flood pulse from the floodplain and severely impacted the spawning and feeding opportunities of fish outside the main channel (Hesse, 1995). These alterations in the continuity between the channel and the floodplain have profoundly affected our ability to maintain and restore forested wetlands along rivers in the southeastern part of the United States. There are few examples of unregulated rivers other than in the tropics, and their study is essential in creating the proper hydrology in restoration projects (Benke et al., 2000).

Specific types of water regulation affect the biota of the floodplain in different ways. One of the most all-pervasive changes to floodplains in the southeastern United States is related to stream channelization and associated channel incision. Downcutting of altered waterways causes water movement from channels to become restricted, and thus floodplains are disassociated from channels because of the absence of flood pulsing (Shields and Hoover, 1991; Sengupta, 1995; Brookes et al., 1996; Middleton, 1999). This lack of flood pulsing causes a drying of the floodplain, which can alter the successional trajectories of riverine forests. Unfortunately, this problem is nearly ubiquitous in the Southeast, so that few forests may resemble their original state even if they have not been directly altered (Shear et al., 1996).

Discharge onto floodplain forests below hydroelectric dams can be abnormally high during the summer growing season (Petts, 1984; Williams and Wolman, 1984) and can disrupt the regeneration of forest species (Schneider et al., 1989; Sharitz et al., 1990). Because discharges from a hydroelectric dam were erratic, trees on a floodplain below a dam in Alton, Illinois, completely failed to regenerate (Kurt Schulz, personal communication). Upstream of dams and within levees, trees die in floodplain forests because of permanent inundation (Eggler and Moore, 1961; Harms et al., 1980; Klimas et al., 1981; Francis, 1983; Conner and Day, 1988; Klimas, 1988; Osment-DeLoach and Moore, 1996). Leveed floodplains become abnormally dry because they are cut off from the flood pulse of water from the channel except in catastrophic flooding events when the embankment is breached (Humburg et al., 1996). After levees are constructed, forests are removed and floodplains are typically converted to agriculture or urbanized (Allen, 1997; Middleton, 1999).

Interbasin water transfer may cause the floodplain to dry downstream of the diversion (Davies et al., 1992; Dynesius and Nilsson, 1994) or to be flooded, at inappropriate times of year, downstream of points where irrigation return water reenters the channel (Middleton, 1999). Although floodplain trees are relatively tolerant of drying (Klimas et al., 1981), tree production can be reduced by severe floodplain drying and declining groundwater levels. Exotic species sometimes invade dried floodplains; after water drainage activities in the early 1900s, the exotic Brazilian pepper (*Schinus-terebinthifolius*) invaded a floodplain in south Florida (Weller, 1995).

Not surprisingly, most of the remaining riverine forested wetlands in the southeastern United States have an altered hydrologic regime because of river regulation (Dynesius and Nilsson, 1994). Unfortunately, the flood pulse, which is critical in these forests because it supports floodplain functions such as productivity, carbon cycling, and/or diversity (Figure 1-1) (Klimas, 1988; Junk et al., 1989; Sparks et al., 1990; Mitsch et al., 1991; Smock and Gilinsky, 1992; Junk and Piedade, 1993; Bayley, 1995; Junk, 1997; Nielson and Chick, 1997; Middleton, 1999) is lacking, so that overall function of these otherwise unaltered floodplain forests has been impaired (Battaglia et al., 1995; Allen, 1997). The vegetation of remnant forested wetlands of the southeastern United States is often vastly altered from its original composition, as based on Government Land Office records, and this is likely due to landscape level changes caused by the reengineering of waterways to eliminate flood pulsing. The present composition of naturally revegetated forests of the American Bottoms along the Mississippi River near St. Louis, Missouri, shows little correspondence to the original tree species (Table 7-1) (Giedeman, 1999). *Ulmus americana,* once the dominant tree there, is now the 10th-ranking tree, presumably in the aftermath of Dutch elm disease. Various species of oaks and other species with heavy-seeded fruits were once common in the region but are now relatively rare. *Salix* spp., now the most common woody

TABLE 7-1. Compositional changes in bottomland hardwood forests during early settlement vs. modern times

Ranks of species are given in order of importance for the American Bottoms, Illinois, 1849–1994. Basal area (1849) is based on data from Government Land Office records and mean cover (1984) is based on estimates using a spherical crown densiometer.

	1849		1994	
	Basal Area	Rank	Mean Cover	Rank
Ulmus americana (elm)	16.2	1	0.3	10
Fraxinus pennsylvanica (ash)	14.7	2	0.9	7
Populus deltoides (cottonwood)	13.7	3	6.9	3
Quercus alba (oak)	12.9	4	-	-
Quercus velutina (oak)	9.5	5	-	-
Platanus occidentalis (sycamore)	6.9	6	0.8	8
Carya illinoinensis (pecan)	4.9	7	-	-
Carya cordiformis (hickory)	4.4	8	-	-
Acer negundo (box elder)	4.1	9	3.7	4
Celtis occidentalis (hackberry)	3.5	10	0.2	11.5
Salix spp. (willow)	2.5	11	9.6	1
Quercus rubra (oak)	1.8	12	0.3	10
Prunus serotina (black cherry)	1.0	13.5	0.2	11.5
Juglans nigra (walnut)	1.0	13.5	0.3	10
Acer saccharinum (maple)	0.7	15.5	7.3	2
Acer rubrum (maple)	0.7	15.5	1.1	6
Gleditsia triacanthos (locust)	0.5	17.5	> 0.1	14.5
Quercus macrocarpa (oak)	0.5	17.5	0.1	13
Quercus palustris (oak)	0.4	19	> 0.1	14.5
Cornus florida (dogwood)	< 0.1	20.5	2.2	5
Cercis canadensis (redbud)	< 0.1	20.5	-	-

Source: Giedeman, 1999 (copyright © Charles Giedeman; used by permission).

species, ranked 11th in 1849 at the time of the Government Land Office survey (Giedeman, 1999).

Aside from changes in hydrology due to water regulation, another reason for altered composition in southeastern forested wetlands may have to do with the relative inability of heavy-seeded species to reinvade after logging or farming (Shear et al., 1996; Allen, 1997). Reforestation projects typically plant a few oak species, but on land that originally held forests of *Celtis laevigata* and *Liquidambar styraciflua*. Because of these planting practices, oak species have greatly increased in the Kentucky Lakes region and elsewhere (Table 7-2) (Ouchley et al., 2000).

Fragmentation also damages the remaining forested wetlands of the southeastern United States (Rudis, 1995) and presents obstacles to the regeneration of both fauna and flora. Fragmentation via impoundment

TABLE 7-2. Compositional changes in the forest at Bayou Cocodrie National Wildlife Refuge, Louisiana, 1830–1994

	Average No. of Trees Acre^{-1}		Average Basal Area Individual^{-1} (in^2)		Average Basal Area Acre^{-1} (in^2 acre^{-1})	
	1830	1994	1830	1994	1830	1994
Ash	7.8	16.1	196.8	53.0	1526.4	854.7
Cypress	1.1	-	1356.8	-	1429.0	-
Elm	4.6	13.1	226.6	171.4	1039.6	2251.0
Sugarberry	34.6	17.3	89.2	37.7	3085.0	651.9
Hawthorn	8.1	1.5	23.7	21.4	192.6	31.9
Locust	4.2	2.4	53.4	212.0	226.5	506.2
Maple	1.8	6.0	78.9	79.6	139.0	475.2
Oaks	9.5	40.0	249.3	240.7	2378.4	9628.8
Pecan	-	5.7	-	90.8	-	515.0
Persimmon	3.9	0.6	80.6	29.0	312.5	17.3
Sweetgum	11.7	17.3	737.7	118.0	8597.8	2043.3
Total (all spp.)	95.7	120.6	-	-	19101.5	16986.9

Source: Ouchley, Keith, et al. "Historic and Present-Day Forest Conditions: Implications for Bottomland Hardwood Forest Restoration." *Ecological Restoration,* Vol. 18, No. 1, © 2000. Reprinted by permission of the University of Wisconsin Press.

clearly affects the regeneration of species dispersed by hydrochory, the dispersal of seeds and propagules in water, as discussed in the next section of this chapter.

Wetland faunal species may be very sensitive to fragmentation, human created patches within a formerly connected matrix. Species such as the Ivory-Billed Woodpecker and the Carolina Parakeet were lost from these forests many decades ago (Harris, 1989). Tracts of forested wetland must be large enough to maintain regenerative processes in songbirds. Cowbird parasitism of Wood Thrushes was higher at the edge of Heron Pond, Illinois, than nearer the core of the area (200 to 700 m vs. 500 to 2000 m from agriculture) suggesting that declines in songbird nesting may be related to the effects of small forest size due to fragmentation (Robinson and Hoover, 1995). Many tracts of suitable forested wetland for songbird nesting in the southeastern United States do not have a core area that exceeds the minimum size. As compared with the central Missouri Ozarks and the Hoosier National Forest of Indiana, however, Illinois forests have much higher levels of nest parasitism by Cowbirds (10, 20, and 70 to 80 percent nest parasitism, respectively) (Robinson et al., 1995). Information regarding the effects of fragmentation is incomplete for fauna other than birds. Species richness of amphibians was related less to the width of the riparian strip between the core and the upland than to the availability of upland and fishless pools in their habitat (Burbrink et al., 1998). There are many gaps in our knowledge, but fragmentation undoubtedly contributes to the decline of certain species in southeastern forested wetlands.

Forested Wetland Loss in the Southeastern United States

In 1937 there were 11.8 million ha of forested wetland in the Lower Mississippi Alluvial River Plain from southern Illinois to just north of the Gulf of Mexico (MacDonáld et al., 1979). By 1991 the remaining forests covered only 1.9 million ha (Llewellyn et al., 1996). More than 81 percent of the forested wetland in the United States has been converted to other uses (Clewell and Lea, 1990), and 96 percent of this conversion has been for agriculture (Allen, 1997). The initial destruction began with water control efforts after the 1927–28 floods along the Mississippi River (Newling, 1990), and such efforts intensified after 1972 as the demand for soybean meal increased with the failure of the Peruvian upwelling and low anchovy harvest (Barber et al., 1978). Late in the 1970s, as the price of soybeans dropped, some of these low-lying fields were abandoned, but the forest was already gone (Newling, 1990). Recent wetland loss has slowed; since the mid-1980s, only 1408 ha (2.8 percent) of the original wetlands have been lost along the Flint River in Georgia, as reported by detailed Geographic Information Systems (GIS) studies (Houhoulis and Michener, 2000). Representing a turnaround in this trend toward wetland loss, 89,000 ha of land were scheduled to be reforested between 1999 and 2003 (King and Keeland, 1999).

REGENERATION PROBLEMS FOR PLANT SPECIES ON FLOODPLAINS WITH ALTERED HYDROLOGY

Plant regeneration on reengineered floodplains is limited because the early life history of many plant species—from seed dispersal to the seedling recruitment stage—is dependent on the flood pulse. Paradoxically, for water-dispersed species in forested wetlands, although the seeds rely on a high pulse of water to carry them to safe sites on the floodplain, germination and seedling recruitment will not occur without an extended drawdown during the growing season (Figure 1-1) (Middleton, 1999). The environmental sieve (Harper, 1977) regulates regeneration because the environ-

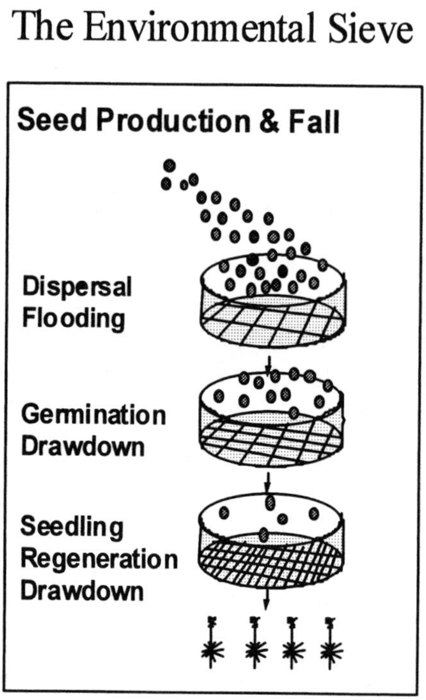

Figure 7-2. The environmental sieve concept in the early life history stages of species in bottomland forests. Each stage is dependent on either flood-pulsed or drawdown conditions of the floodplain.

ment on the floodplain becomes increasingly restrictive for each successive phase in the life history of plant species—seed production, seed germination, seedling recruitment (Figure 7-2) (Middleton, 2000). The ultimate contribution of seeds to regeneration depends on many factors, including the amounts of seeds produced, method of seed transport, timing of seed release, and dispersal distance (van der Valk, 1992).

Seed production in trees generally does not take place for many years after the individuals become established. Mast seeding is common in many tree species of forested wetlands (King and Allen, 1996). Many oak species do not produce seeds until they are 25 to 30 years old, with peak production at 50 to 75 years (Fredrickson and Batema, 1992) although *Quercus nuttallii* begins producing seeds as early as 5 years of age (Table 7-3). *Nyssa aquatica* also begins to produce seeds at about 30 years of age (Haynes et al., 1993), and in many species, the age at seed bearing is even greater (Table 7-3). Viable seed production of species may be lower in the northern than in the southern portions of their ranges, so that for a southeastern swamp species such as *Taxodium distichum* (Mattoon, 1915) natural regeneration at the northern edge of its range is potentially more difficult.

Seed Dispersal

Dispersal is critical to the ability of plant species to colonize (van der Pijl, 1982) whether they are in intact forested wetlands or in restoration sites. The delivery of live seeds to safe sites in forested wetlands can be accomplished by water (hydrochory), air (anemochory), or animals such as birds, fish, or insects (zoochory) (Barrat-Segretain, 1996; Middleton, 1999). The dominant species of baldcypress-tupelo swamps are mainly dispersed by water (Table 7-3) (Schneider and Sharitz, 1988; Middleton, 1995a, b, 1999, 2000; Wilker, 1999). Hydrochory dynamics for species in cypress swamps vary seasonally (Schneider and Sharitz, 1988; Middleton, 2000), as well as from year to year (Figure 7-3 *A* and *B* and Table 7-4) (Middleton, 2000).

Although the oak species of forested wetlands are generally considered to be animal-dispersed (Middleton, 1999), water can move them by pushing them along the soil surface in flooded wetlands (King and Allen, 1996). Viable seeds of at least one oak species, *Quercus lyrata* (Table 7-4), float on the surface of the water for a few days after seed fall (Middleton, 1995a, 2000). The seeds of many woody species of less frequently flooded forested wetlands are wind-dispersed, including *Fraxinus pensylvanica, Salix nigra, Ulmus americana,* and *Ulmus rubra* (Newling, 1990). Herbs and vines of baldcypress-tupelo swamps (e.g., *Bidens* spp., *Brunnichia ovata, Campsis radicans, Carex* spp.) are variously dispersed, with water as only one mode of transport (Schneider and Sharitz, 1988; Middleton, 1995a, 1999, 2000; Wilker, 1999). Vegetative propagules are also transported on the water (Cellot et al., 1998), and these contribute to the regeneration process (van der Valk, 1987; Newling, 1990).

The distance traveled by aquatically dispersed species and, ultimately, their success in establishment are critically linked to the magnitude and timing of the flood pulse. In flowing water, seeds may move quite far from their source, inasmuch as dispersal distance is mainly related to flotation time and the speed of the water (Nilsson et al., 1991; Danvind and Nilsson,

TABLE 7-3. Dispersal vectors of species in southeastern bottomland hardwood forests

Species	Propagule Type	Age at Seed-bearing	Dispersal Mechanism: Wind	Water	Bird	Animal	References
Acer barbatum	Seeds		+		+	+	Haynes and Moore, 1988
Acer negundo	Seeds		+		+	+	Haynes and Moore, 1988
Acer rubrum	Seeds	35	+	+	+	+	Schneider and Sharitz, 1988; Haynes and Moore, 1988
Acer saccharinum[a]	Seeds		+	+	+	+	Weitzman and Hutnik 1965; Middleton, 2000
Agrostis perennans	Seeds			+			Middleton, 2000
Alnus serrulata	Seeds			+			Schneider and Sharitz, 1988
Asimina triloba	Seeds				+	+	Haynes et al., 1988
Berchemia scandens	Seeds			+			Schneider and Sharitz, 1988
Betula nigra	Seeds	40	+	+			Haynes and Moore, 1988
Bidens discoidea	Seeds			+			Middleton, 2000
Bidens frondosa	Seeds			+			Middleton, 2000
Brunnichia cirrhosa	Seeds			+			Middleton, 2000
Calopogon pallidus	Seeds					+	Brewer, 1998
Campsis radicans	Seeds			+			Schneider and Sharitz, 1988; Middleton, 2000
Carex sp.	Achenes			+	+		Vivian-Smith and Stiles 1994; Middleton, 2000
Carpinus caroliniana	Seeds			+	+		Schneider and Sharitz, 1988; Haynes and Moore, 1988
Carya aquatica	Seeds	20		+	+	+	Schneider and Sharitz, 1988; Haynes and Moore, 1988
Carya illinoienensis	Seeds	20			+	+	Haynes and Moore, 1988
Carya laciniosa	Seeds	40			+	+	Haynes and Moore, 1988
Carya ovata	Seeds	40				+	Haynes and Moore, 1988
Celtis laevigata	Seeds			+	+	+	Haynes and Moore, 1988
Celtis occidentalis	Seeds	15		+	+	+	Haynes and Moore, 1988
Cephalanthus occidentalis	Seeds, stem fragments		+	+			McKnight et al.,1981; Schneider and Sharitz, 1988 Middleton, 2000
Cornus drummondii	Seeds				+	+	Haynes and Moore, 1988
Cornus florida	Seeds				+	+	Haynes and Moore, 1988

+ = present, blank = not documented.

Crataegus spp.	Seeds			+			Middleton, 2000
Diospyros virginiana	Seeds	10			+	+	Haynes and Moore, 1988
Eriocaulon compressum	Seeds					+	Brewer, 1998
Fagus grandifolia	Seeds	40				+	Haynes and Moore, 1988
Forestiera acuminata	Seeds			+	+		Haynes and Moore, 1988
Fraxinus americana	Seeds		+	+			Haynes and Moore, 1988
Fraxinus caroliniana	Seeds		+	+			Haynes and Moore, 1988; Schneider and Sharitz, 1988
Fraxinus pennsylvanica[a]	Seeds		+	+			Putnam et al., 1960; Middleton, 2000; Schneider and Sharitz, 1988
Fraxinus profunda	Seeds		+	+			Haynes and Moore, 1988
Gleditsia aquatica	Seeds	10	+	+	+	+	Putnam et al., 1960; Middleton, 2000; Haynes et al., 1988
Gleditsia triacanthos	Seeds	10	+	+	+	+	Haynes and Moore, 1988
Hypoxis micrantha	Seeds					+	Brewer, 1998
Ilex opaca	Seeds				+	+	Haynes and Moore, 1988
Ilex spp.	Seeds			+			Schneider and Sharitz, 1988
Juglans nigra	Seeds	8				+	Haynes and Moore, 1988
Lachnanthes caroliniana	Seeds					+	Brewer, 1998
Lachnocaulon anceps	Seeds					+	Brewer, 1998
Liquidambar styraciflua	Seeds	20			+		Schneider and Sharitz, 1988
Liriodendron tulipifera	Seeds	15	+				Haynes and Moore, 1988
Lophiola aurea	Seeds					+	Brewer, 1998
Magnolia grandiflora	Seeds	10			+	+	Haynes and Moore, 1988
Magnolia virginiana	Seeds				+	+	Haynes and Moore, 1988
Morus rubra	Seeds				+	+	Haynes and Moore, 1988
Nyssa aquatica	Seeds	30		+		+	Schneider and Sharitz, 1988; Middleton, 2000 Haynes and Moore, 1988
Nyssa ogeche	Seeds			+	+	+	Haynes and Moore, 1988
Nyssa sylvatica	Seeds	30		+		+	Haynes and Moore, 1988
Nyssa sylvatica var. *biflora*	Seeds			+	+		Haynes and Moore, 1988; Schneider and Sharitz, 1988
Persea borbonia	Seeds				+		Haynes and Moore, 1988
Persea borbonia var. *pubescens*	Seeds			+			Haynes and Moore, 1988

TABLE 7-3. (continued)

Species	Propagule Type	Age at Seed-bearing	Dispersal Mechanism				References
			Wind	Water	Bird	Animal	
Pinus taeda	Seeds			+			Schneider and Sharitz, 1988
Planera aquatica	Seeds			+			Middleton, 2000
Platanus occidentalis	Seeds	25	+	+	+		Schneider and Sharitz, 1988; Haynes and Moore, 1988
Pogonia ophioglossoides	Seeds					+	Brewer, 1998
Populus deltoides	Seeds	10	+	+			Haynes and Moore, 1988
Populus heterophylla	Seeds	10	+	+			Haynes et al., 1988
Prunus serotina	Seeds	10			+	+	Haynes and Moore, 1988
Quercus alba	Seeds	50				+	Haynes and Moore, 1988
Quercus bicolor	Seeds	35		+		+	Haynes and Moore, 1988
Quercus laurifolia	Seeds	15		+	+	+	Haynes and Moore, 1988
Quercus lyrata	Seeds	25		+	+	+	Morris, 1965; Middleton, 2000; Haynes and Moore, 1988
Quercus macrocarpa	Seeds	35		+		+	Haynes and Moore, 1988
Quercus michauxii	Seeds	25				+	Haynes and Moore, 1988
Quercus nigra	Seeds	20		+	+	+	Haynes and Moore, 1988
Quercus nuttallii	Seeds	5		+		+	Haynes and Moore, 1988
Quercus pagoda	Seeds	25			+	+	Haynes and Moore, 1988
Quercus palustris	Seeds	15		+	+	+	Haynes and Moore, 1988
Quercus phellos	Seeds	20		+		+	Haynes and Moore, 1988

Quercus shumardii	Seeds	25		+		+	Haynes and Moore, 1988
Quercus stellata var. *paludosa*	Seeds	25	+	+	+		Haynes and Moore, 1988
Quercus virginiana	Seeds				+	+	Haynes and Moore, 1988
Quercus spp.	Seeds			+			Schneider and Sharitz, 1988
Rumex orbiculatus	Seeds			+			Middleton, 2000
Salix exigua	Seeds	10	+	+			Haynes and Moore, 1988
Salix nigra	Seeds	10	+	+			Haynes and Moore, 1988
Sassafras albidum	Seeds	10			+		Haynes and Moore, 1988
Smilax spp.	Seeds			+			Schneider and Sharitz, 1988
Taxodium distichum	Seeds			+			Schneider and Sharitz, 1988; Middleton, 2000
Taxodium distichum var. nutans	Seeds			+			Haynes and Moore, 1988
Tofieldia racemosa	Seeds					+	Brewer, 1998
Ulmus alata	Seeds		+	+			Haynes and Moore, 1988; Schneider and Sharitz, 1988
Ulmus americana	Seeds	15	+	+			Guilkey, 1965; Middleton, 2000; Haynes and Moore, 1988
Ulmus crassifolia	Seeds		+	+			Haynes and Moore, 1988
Ulmus rubra	Seeds	15	+	+			Haynes and Moore, 1988
Xyris ambigua	Seeds					+	Brewer, 1998
Xyris baldwiniana	Seeds					+	Brewer, 1998
Zigadenus glaberrimus	Seeds					+	Brewer, 1998

[a] Includes closely related species.

Source: Middleton 2000 (based on Barrat-Segretain, 1996, and Middleton, 1999; copyright © Barrat-Segretain, 1996).

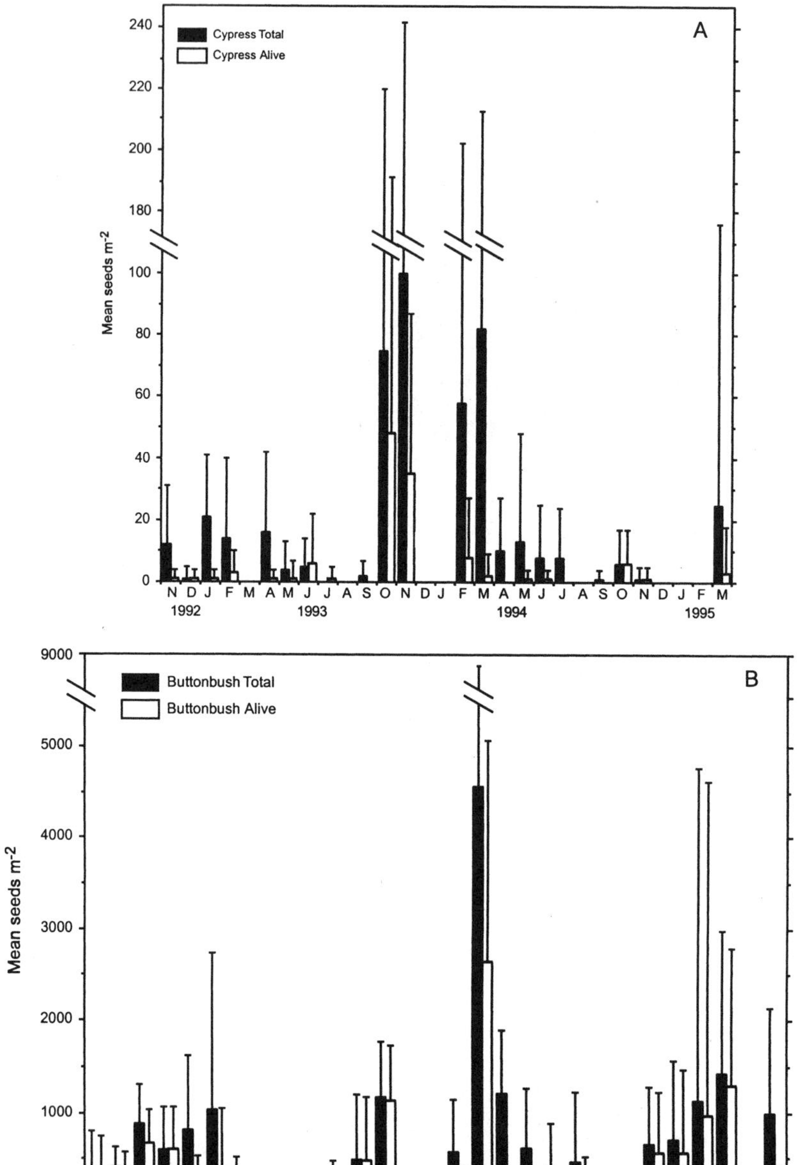

Figure 7-3. *Mean seeds m^{-2} yr^{-1} of the total (viable and dead: solid bar) and alive (open bar) water-dispersed seeds of* Taxodium distichum *(A: cypress) and* Cephalanthus occidentalis *(B: buttonbush) from November 1991 through July 1994, Buttonland Swamp, Perks, Illinois. (From Middleton, 2000, copyright © Kluwer; Figure 5 reprinted with kind permission from Kluwer Academic Publishers.)*

TABLE 7-4. Annual water dispersal of seeds and baldcypress swamps

Mean (x ± SE) number of water-dispersed seeds (m^{-2} yr^{-1}) and total seeds (alive and dead) collected in aquatic and aerial traps[a], June 3, 1993, through May 30, 1995, in Buttonland Swamp, Illinois

	1993–94	1994–95	Total
	Aquatic		
Acer saccharinum	2±7	2±14	2
Agrostis sp.	0±0	1±7	1
Andropogon virginicus	0±2	0±0	0
Bidens discoidea	6±25	19±75	12
Bidens frondosa L.	3±12	4±20	4
Brunnichia cirrhosa	325±1979	13±56	169
Campsis radicans	0±2	1±5	1
Carex sp.	10±26	57±355	34
Cephalanthus occidentalis	8356±13742	5207±10362	6782
Crataegus sp.	0±0	34±229	17
Fraxinus americana	3±11	1±9	2
Gleditsia aquatica	3±13	6±38	4
Hordeum sp.	1±4	0±0	1
Nyssa aquatica	15±33	1±7	8
Planera aquatica	7±17	0±3	4
Quercus lyrata	3±10	4±16	3
Rosa palustris	40±170	0±2	20
Rumex crispus	4±22	1±7	3
Taxodium distichum	98±153	46±149	72
Ulmus americana	2±10	65±177	34
Other	1±3	24±130	12
	Aerial		
Acer saccharinum	1±7	0±3	1
Agrostis sp.	6±32	4±28	5
Bidens discoidea	1±3	3±17	2
Bidens frondosa	2±7	0±0	1
Brunnichia cirrhosa	1±7	2±13	2
Campsis radicans	1±4	0±0	1
Carex sp.	1±6	21±128	11
Cephalanthus occidentalis	1212±3520	69±205	640
Crataegus sp.	1±3	4±26	3
Elymus sp.	1±4	0±3	1
Fraxinus americana	9±29	4±28	7
Gleditsia aquatica	6±29	0±0	3
Hordeum sp.	1±8	4±22	2
Nyssa aquatica	0±0	1±6	1
Quercus lyrata	11±39	3±14	7
Rosa palustris	215±642	4±22	109
Rumex crispus	0±0	1±8	1
Taxodium distichum	110±338	3±26	57
Ulmus americana	212±187	98±108	155
Other	44±213	16±93	30

[a] Other water-dispersed seeds collected in the survey with < 1.0 seeds m^{-2} yr^{-1} were *Carex* (three species), *Carya aquatica, Carya laciniosa, Carya* sp., *Echinochloa crusgalli, Eleocharis* sp., *Fraxinus profunda, Juglans cinerea, Jussisaea* sp., *Leersia oryzoides, Malva* sp., *Panicum* sp., *Platanus occidentalis, Quercus bicolor, Quercus palustris, Rhynchospora corniculata, Rosa palustris, Salix* sp., and *Scutellaria lateriflora*.

Source: Middleton, 2000. Copyright © Kluwer; Table 1 reprinted with kind permission from Kluwer Academic Publishers.

1997). Although the distance seeds travel is situation-dependent, it is likely that they can move for great distances along unregulated rivers. The seeds of certain species of baldcypress-tupelo swamps, such as *Cephalanthus occidentalis, Nyssa aquatica,* and *Taxodium distichum,* float for periods longer than one year (Wilker and Middleton, unpublished data). Other species float for only short periods of time. In an experimental study, the majority of *Planera aquatica* seeds (64 percent) floated for less than 1 week, although some seeds floated for up to 24 weeks (Rayner, 1976), but these patterns are likely to be dependent on season. In Buttonland Swamp along the Cache River in Illinois at the northern tip of the Gulf Coastal Plain, *Planera aquatica* was dispersed from May through September (Wilker, 1999).

Many species that are considered wind-dispersed also disperse readily by water. These species include *Acer rubrum, Fraxinus caroliniana, Liquidambar stryaciflua,* and *Ulmus alata* (Schneider and Sharitz, 1988), as well as *Acer saccharinum, Fraxinus pennsylvanica,* and *Ulmus americana* (Tables 7-3 and 7-4) (Middleton, 2000). Wind-dispersed species in Buttonland Swamp, Illinois, disappeared from the surface of the water sheet in the month following peak dispersal, suggesting that these seeds may not float for very long (Wilker, 1999). Unimpeded dispersal of seeds and propagules is important in restoration, because they can disperse to naturally revegetate damaged areas (Allan and Flecker, 1993; Middleton, 1999, 2000).

When dams restrict the flow of water, the flow of seeds to and from forested wetlands on floodplains is impacted downstream. Along the Roanoke River in Virginia, regeneration is affected by dispersal limitations because of river regulation downstream. Along six rivers, the impacts of dams on regeneration were still apparent 65 km downstream of the dams, although the impacts decreased farther downstream (Schneider et al., 1989).

Impoundments restrict the dispersal of seeds, because dams fragment corridors and slow the flow of water above the dams (Jansson et al., 2000a, b). Because opportunities for establishment are restricted in this setting, each reservoir develops characteristic flora (Figure 1-2) (Jansson et al., 2000a). In European studies, the shorelines of impoundments were found to have more wind-dispersed species than those of freely flowing rivers (Nilsson et al., 1997; Jansson et al., 2000b). Shorelines are colonized by species and individuals from the same impoundment (Jansson et al.,

2000a), so that the populations in impoundments may also suffer from restricted gene flow.

Dispersal pathways in impounded cypress swamps are altered in ways that are likely to affect regeneration. Across Buttonland Swamp, the spatial distribution of aquatically dispersed seeds from species such as *Taxodium distichum* was primarily spherical or exponential (Figures 7-4 and 7-5 and Table 7-5) (Wilker, 1999; Middleton, 2000). Both of these patterns of distribution suggest that in impounded swamps, seeds do not disperse very far from the trees where they were produced. Spherical and exponential models are similar in that they describe patterns that cease to be autocorrelated over a specific distance, but differ in that for the exponential, the model and the sill never actually converge. The distances over which these spherical distribution patterns operated (A_0) were relatively small, further suggesting that seeds do not move very far from the source (14 to 225 m; Table 7-5) (Middleton, 2000). In flowing rivers, seed dispersal is more likely have a linear spatial pattern—that is, to become less related to each other moving away from the sample point (Middleton, 2000).

Across a swamp/old field interface in Buttonland Swamp, the dispersal of the seeds of water-dispersed species such as *Taxodium distichum* and *Cephalanthus occidentalis* ended abruptly at the shoreline of the swamp

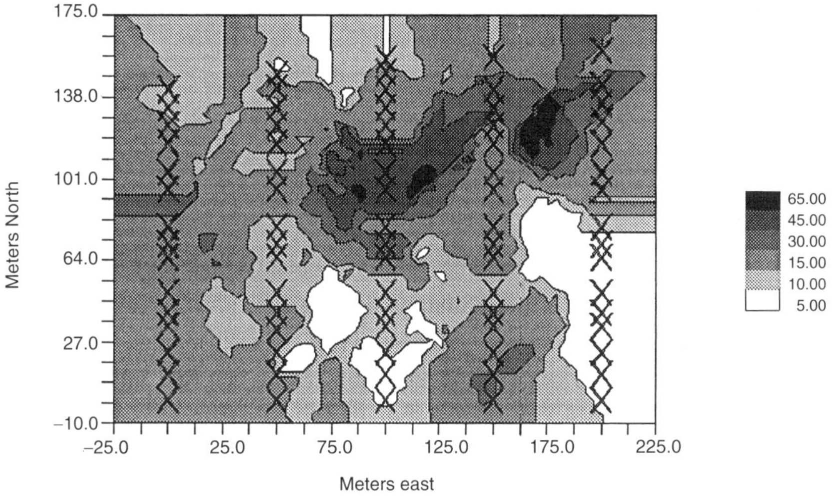

Figure 7-4. *Distribution of water-dispersed seeds m^{-2} for* Taxodium distichum, *Buttonland Swamp, December 1993, as generated via a semivariogram model using geostatistics. (Reprinted from Wilker, 1999, copyright © by John Wilker, with permission.)*

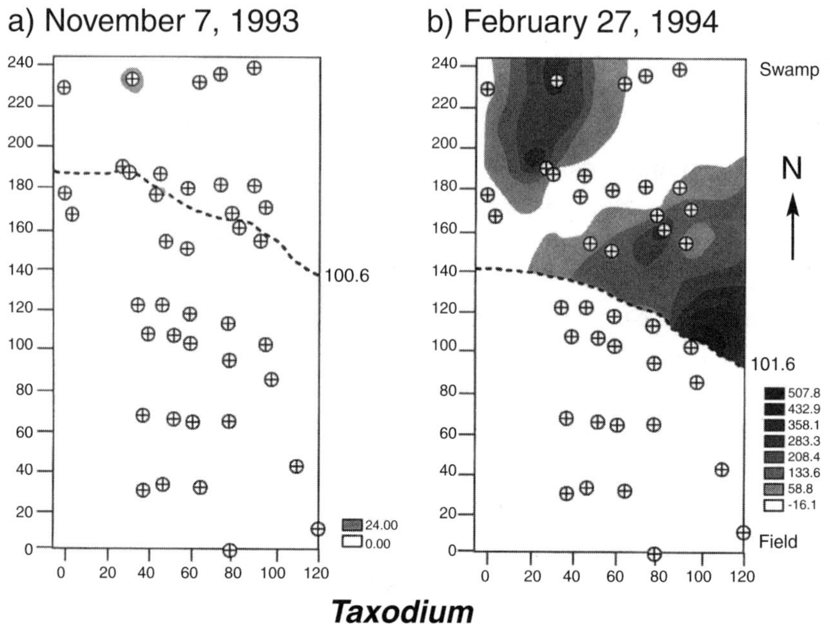

Figure 7-5. Kriged maps of water-dispersed seeds m^{-2} *of* Taxodium distichum *trapped along the boundary of a baldcypress swamp and abandoned farm field. Maps are based on semivariance models using an active lag of 100 m and a step interval of 10 m on November 7, 1993, and February 27, 1994. On August 28, 1993, there were too few seeds to fit a model. The line traversing the map indicates the position of the peak water level for a given month and is visually interpolated along a contour based on the elevations of the seed traps. Peak water levels, indicated by the dashed line, occurred on October 1, 1993, and January 29, 1994, respectively. (From Middleton, 2000, copyright © Kluwer; Figure 3 reprinted with kind permission from Kluwer Academic Publishers.)*

(Figures 7-5 and 7-6, respectively). This also was true for certain water-dispersed species of bryophytes, including *Ephemerum crassinervium, Riccia hirta, Riccia huebeneriana,* and *Ricciocarpos natans* (Conrad, 1997), and likely also for aquatically dispersed insects and their larva (personal observation). Limits to the dispersal of aquatic species by water impoundment, particularly because of altered flood pulsing patterns, impact regeneration in forested wetlands in the southeastern United States.

Seed Banks and Seed Longevity

An unimpeded dispersal pathway is critical to the regeneration and restoration of natural communities, because it is essential in the resupply

TABLE 7-5. Spatial characteristics of seed dispersal in baldcypress swamps

Semivariance models (isotropic) for water- and gravity-dispersed seeds (mean number seeds m^{-2} yr^{-1}) using square root transformed data with an active lag of 140 m and a step interval of 10 m, Buttonland Swamp, Illinois, June 3, 1993 to May 30, 1995. The nugget (C_0) is the proportion of the sample variation that occurs at a scale smaller than measured in the study; the sill (C_0 + C) is the maximum variance described by the model; $C_0/(C_0 + C)$ is the proportion of the sample variance explained by the spatial structure of the model at the scale defined for the study; the range (A_0) represents the distance over which the spatial dependence described by the model is apparent , and r^2 is the coefficient of determination that describes the statistical fit of the model.

Species	Best Fit Model	C_o	C_o+C	C/C_o+C)	A_o	r^2
			Gravity-Dispersed			
Acer saccharinum	Spherical	0.0	0.2	1.0	30	0.34
Agrostis sp.	Spherical	0.0	6.8	1.0	22	0.47
Bidens frondosa	Spherical	0.0	1.1	0.9	107	0.49
Carex sp.	Exponential	0.0	0.3	1.0	174	0.59
Cephalanthus occidentalis	Spherical	0.0	730.9	1.0	287	0.73
Fraxinus americana	Exponential	0.0	5.1	1.0	47	0.41
Quercus lyrata	Exponential	0.0	3.7	1.0	45	0.35
Rumex crispus	Spherical	0.0	1.5	1.0	18	0.10
Taxodium distichum	Spherical	0.0	82.0	0.9	291	0.83
Ulmus americana	Exponential	0.0	33.9	0.6	776	0.33
			Water-Dispersed			
Acer saccharinum	Spherical	0.0	1.1	1.0	14	0.01
Bidens discoidea	Spherical	0.0	5.8	1.0	16	0.09
Bidens frondosa	Spherical	0.0	0.5	1.0	14	0.02
Brunnichia cirrhosa	Linear	0.1	188.7	1.0	26	0.12
Carex sp.	Spherical	0.0	5.2	1.0	65	0.68
Cephalanthus occidentalis	Spherical	0.0	4801.0	1.0	225	0.87
Elymus sp.	Spherical	0.0	0.2	1.0	17	0.18
Fraxinus americana	Spherical	0.0	1.0	1.0	13	0.02
Gleditsia aquatica	Spherical	0.0	1.2	1.0	22	0.15
Planera aquatica	Exponential	0.0	2.9	1.0	32	0.38
Taxodium distichum	Spherical	0.0	105.2	1.0	195	0.89
Unknown grass	Spherical	0.0	17.1	1.0	26	0.15

Source: Middleton, 2000. Copyright © Kluwer; Table 2 reprinted with kind permission from Kluwer Academic Publishers.

of seeds to depleted seed banks (Bakker et al., 1996). Thus, a high pulse of water may be needed to facilitate the movement of seeds to the seed banks of riverine forested wetlands. Studies of restoration failure often cite an inadequate seed supply due to limitations of dispersal and/or the seed bank (Allen and Kennedy, 1989; van der Valk and Pederson, 1989; Allen, 1990; Newling, 1990; Middleton, 1995a, c, 2000). This problem may be at least

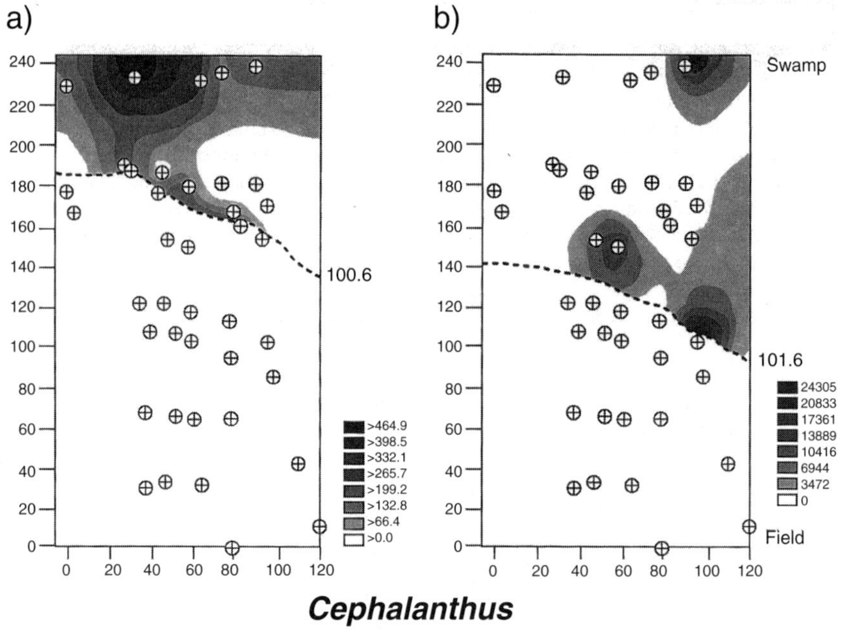

Figure 7-6. *Kriged maps of water-dispersed seeds m*$^{-2}$ *of* Cephalanthus occidentalis *trapped along the boundary of a baldcypress swamp and abandoned farm field. Maps are based on semivariance models using an active lag of 100 m and a step interval of 10 m on November 7, 1993, and February 27, 1994. (From Middleton, 2000, copyright © Kluwer; Figure 4 reprinted with kind permission from Kluwer Academic Publishers.)*

partially related to the extensive reengineering of rivers and streams that restricts the flow of water onto the floodplain.

Seeds of many forested wetland species in the southeastern United States live for relatively short periods of time, underscoring the need for frequent episodes of flood pulsing to replenish the supply of seeds. For example, *Taxodium distichum* may enter the seed bank only briefly. After 100 days, few of this species were alive, and after 359 days, only 4 percent remained viable in Buttonland Swamp. Many seeds were lost to germination and subsequent mortality as a result of summer reflooding (Middleton, 2000). Seeds of other species are viable for only a short period. After fall dispersal, seeds of various species became less viable in the seed bank during the course of the year in a baldcypress-tupelo swamp along the Savannah River, Georgia (Schneider and Sharitz, 1986), and *Taxodium dis-*

tichum had low seed viability, which may have limited regeneration (Sharitz and Lee, 1985a). In Buttonland Swamp most aquatically dispersing seeds of swamp species (e.g., *Taxodium distichum* and *Cephalanthus occidentalis*) were dead by June or July of the following growing season (Figure 7-3).

Seeds that remain submerged may stay viable for a longer period of time than those that are not submerged. Although seeds of *Taxodium distichum* germinated after 7 weeks of submergence in University Lake near the campus of Louisiana State University, they did not germinate after 14 months (Applequist, 1959). In the same study, a large number of seeds of *Nyssa aquatica* were viable after 7 and 14 months, but not after 21 months of submergence. In another study, most seeds of *Taxodium distichum* were dead after 30 months of storage in water (Demaree, 1932). After 15 weeks of floating in water, the majority of seeds of *Planera aquatica* (82.9 percent) germinated when transferred to drawndown conditions (Rayner, 1976). Submerged seeds of *Quercus pagoda* had a lower viability than this after 34 days, but seeds of *Quercus nuttallii* were unaffected (Briscoe, 1961).

The seeds of herbaceous species and liverworts in cypress swamp seed or spore banks may or may not be longer-lived than those of woody species, but many of these species do not rely only on water for dispersal (Table 7-3) (Middleton, 2000). Aquatic liverworts do rely primarily on flood pulsing to renew their spore bank, at least in baldcypress-tupelo swamps. After one year in soil in plots adjacent to Buttonland Swamp, liverwort spore viability dropped by 33 to 50 percent. Old field species of bryophytes relied on local dispersal rather than hydrochory (Conrad, 1997). Knowledge of the specific life history dynamics of herbaceous and liverwort species and their relationship to the water regime is critical to their successful restoration in forested wetlands.

Studies of the seed banks of farmed land that was originally cypress swamp also suggest that seeds do not live very long in the soil (Table 7-6). In an old field adjacent to Buttonland Swamp, the viable seed bank of water-dispersed species ends abruptly near the high-water mark (e.g., *Cephalanthus occidentalis*) (Figure 7-6). After one to three years of farming, most seeds of woody species are dead and no longer stored in the seed bank (Newling, 1990; Middleton, 1995a, 1996, 1999). In contrast, at least some species of prairie wetlands survive many decades of farming (Dahl and Nomsen, 1987; Wienhold and van der Valk, 1989; Warne, 1992), even though these wetlands are no longer interconnected and thus water dispersal is not effective (Galatowitsch and van der Valk, 1996).

Paradoxically, although a high pulse of water facilitates dispersal (Middleton, 1999, 2000), germination of the emergent woody and bryophyte species of forested wetland is dependent on drawdown (DeBell and Auld, 1971; Conner et al., 1981; Conti and Gunther, 1984; Ewel, 1990; Huenneke and Sharitz, 1990; Shankman and Drake, 1990; Baskin et al., 1993; Middleton, 1996, 1999, 2000; Conrad, 1997). Even with their longevity, the seeds of many species of forested wetlands will not germinate until dormancy can be overcome by cold stratification and/or exposure to specific temperature or light regimes (Baskin and Baskin, 1998) along with drawdown conditions.

The normal hydrology of forested wetlands in the Mississippi River Alluvial Valley includes high water levels during the winter and spring followed by a progressive drawdown during the summer growing season (Wax, 1977). These conditions allow the germination and recruitment of species at least in some years. On the other hand, recruitment is almost nonexistent in permanently impounded forested wetlands, so that because of the lack of seed germination and recruitment, old trees are not replaced (Kozel et al., 1988; Loftus, 1994; Middleton, 1999, 2000). Over 50 to 100 years, the permanently impounded Buttonland Swamp will become a shallow open pond, as predicted by a model (SWAMP) with the addition of an algorithm for regeneration parameters (Xiao et al., in press).

Whether a seed successfully passes through the environmental sieve to the seedling stage is obviously dependent on the seeds being moved to a suitable environment where its germination requirements can be met. An intact flood pulse across the floodplain of forested wetlands can facilitate the success of a seed in moving to its next life history stage. On the other hand, water regulation activities such as impoundment severely limit the number of seeds that reach seed banks with environments in which germination can be successful, such as the shoreline of an impoundment.

Seedling and Sapling Recruitment and Adult Flood Tolerance

The spatial zone in which seedlings can regenerate, survive, and grow in a swamp depends to a large extent on the nature of the flood pulse. If seedlings of tree species become flooded too soon after germination, they die (Figures 7-7 and 7-8 and Table 7-7), but saplings become more toler-

TABLE 7-6. Seed bank composition and annual seed dispersal along the edge of a swamp and old field site.

Number of water-dispersed seeds[a] in the seed bank (mean number of seeds m^{-2} ± S.E.) versus dispersal (mean of two combined year totals in mean number of seeds m^{-2}) of dominant species (species with number of seeds > 1 seed m^{-2} $year^{-1}$) at six elevations from June 3, 1993 through May 30, 1995, Crawford Tract, Perks, Illinois. An '*' indicates that all of the seeds sifted from the seed bank at the end of the study were dead; the lack of an '*' indicates that all of the seeds were alive. Totals for yearly mean seed dispersal reflect the column total; totals for the seed bank include species that are not aquatically dispersed. Species scores are from a Canonical Correspondence Analysis of mean species density in the seed bank with elevation.

	Yearly Mean Seed Dispersal	Species Score	Seed Bank Elevation (m msl)					
Species			97.84	99.59	99.22	101.32	101.57	102.00
Acer saccharinum	2	0.5	3±7	0	0	0	3±7	0
Agrostis perennans	1	-	0	0	0	0	0	0
Andropogon virginicus	<1	-	0	0	0	0	0	0
Bidens discoidea	12	-	0	0	0	0	0	0
Bidens frondosa	4	0.1	0	0	3±7	3±7*	0	0
Brunnichia cirrhosa	169	-	3±7*	0	0	0	0	0
Campsis radicans	1	-	0	0	0	0	0	0
Carex sp.	34	-	0	0	0	0	0	0
Cephalanthus occidentalis	6782	0.8	13±29*	10±9	6±9	6±9	0	0
Crateagus sp.	17	-	0	0	0	0	0	0
Elymus sp.	<1	-	0	0	0	0	0	0
Fraxinus americana	2	-	0	0	0	0	0	0
Gleditsia aquatica	4	-	0	0	0	0	0	0
Hordeum sp.	1	-	0	0	3±7*	0	3±7*	0
Nyssa aquatica	8	-	0	3±7*	0	0	0	0
Planera aquatica	4	-	0	0	0	0	0	0
Quercus lyrata	3	-	0	0	0	0	0	0
Rumex crispus	3	-1.2	0	0	0	0	0	0
Taxodium distichum	72	-	19±26*	19±18*	48±41*	83±31*	3±7*	6±14*
Ulmus americana	.34	-	0	0	0	0	0	0

TABLE 7-6. (continued)

	Yearly Mean Seed Dispersal	Species Score	Seed Bank Elevation (m msl)					
Species			97.84	99.59	99.22	101.32	101.57	102.00
Other	12		-	-	-	-	-	-
Total	7165		944	1821	3229	1619	1827	1200
Sorenson's Coefficient			14%	14%	17%	10%	14%	5%

[a] Species found in the seed bank but not listed in this table that are primarily wind-dispersed (w), animal-dispersed (a) or with unknown dispersal syndrome (u) were: *Agrostis perennans* (w), *Ammannia auriculata* (u), *Ammannia coccinea* (u), *Andropogon virginicus* (a), *Aster* sp. (u), *Boeheria cylindrica* (a), *Callitriche terrestris* (u), *Conyza canadensis* (u), *Cyperus* sp. (u), *Cyperus erythrorhizos* (u), *Digitaria ischaemum* (u), *Digitaria sanguinalis* (u), *Echinochloa crusgalli* (u), *Eclipta prostrata* (u), *Eleocharis obtusa* (u), *Eragrostis poaeoides* (u), *Erigeron philadelphicus* (u), *Festuca* sp. (u), *Gratiola neglecta* (u), *Gratiola virginiana* (u), *Heteranthera dubia* (u), *Hypericum mutilum* (u), *Ipomoea* sp. (u), *Juncus tenuis* (u), *Krigia oppositifolia* (u), *Leptochloa panicoides* (u), *Ludwigia palustris* (u), *Ludwigia polycarpa* (u), *Melilotus alba* (u), *Mollugo verticillata* (u), *Oenothera fructicosa* (u), *Oxalis stricta* (a), *Panicum* sp. (u), *Panicum dichotomiflorum* (u), *Paspalum fluitans* (u), *Penthorum sedoides* (u), *Phyla nodiflora* (u), *Platanus occidentalis* (w), *Poa* sp. (u), *Polygonum pennsylvanicum* (a), *Ranunculus abortivus* (u), *Rorippa sessiliflora* (u), *Rumex verticillata* (u), *Sagittaria* sp. (u), *Sida spinosa* (u), *Stellaria* sp. (u), *Taraxacum officinale* (w), *Trifolium hybridum* (u), *Verbena* sp. (u), *Veronica peregrina* (u) and 22 unknown spp. *Leersia oryzoides* was found in the seed bank and is both bird and water dispersed, but was not captured in aquatic seed traps in this study and thus is not listed. Dispersal information from Ridley, 1930 and Middleton, 1999.

Source: Middleton, 2000.

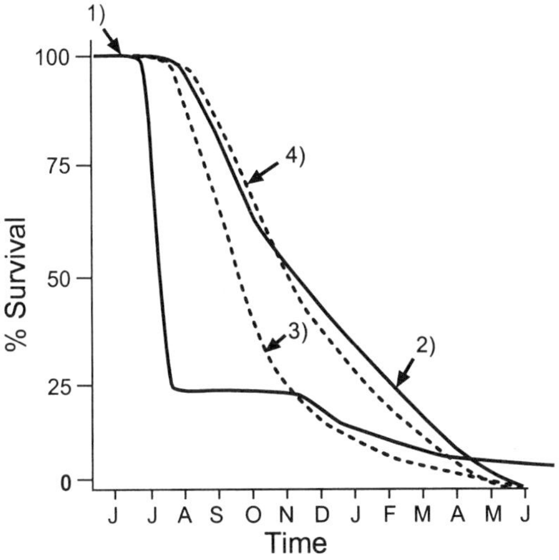

Figure 7-7. Percent (%) survival of four cohorts of newly germinated seedlings of Taxodium distichum in a swamp-field boundary of the Crawford Tract and on a point bar in Buttonland Swamp, Illinois. Cohorts were marked on (1) June 7, 1994 (72 seedlings), (2) July 31, 1994 (15 seedlings), (3) August 2, 1992 (106 seedlings), and (4) July 31, 1994 (10 seedlings). (From Middleton, 2000; copyright © Kluwer; Figure 7 reprinted with kind permission from Kluwer Academic Publishers.)

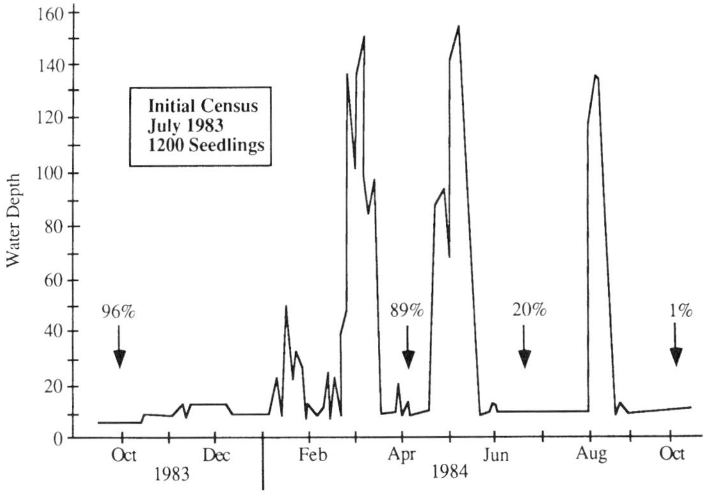

Figure 7-8. Relationship between percent seedling survival (shown by numbers and arrows) of Taxodium distichum and Nyssa aquatica and flood events in the Savannah River floodplain, South Carolina. (Reprinted from Sharitz et al., 1990, copyright © by Lewis Publishers, by permission.)

TABLE 7-7. Effect of water depth on seedling survivorship of woody species in bottomland hardwood forests of the southeastern United States.

Studies were conducted in a greenhouse unless otherwise indicated.

Species	Survival					
	Soil Saturated	Flooded Below Tip	Flooded Above Tip	Experiment Length (Day)	Reference	Comments
Acer negundo	-	Yes	No	32	Hosner, 1958, 1960	Died if submerged more than 32 days.
Acer rubrum	-	Yes	-	21	DeShield et al., 1994	
	-	-	No	30	Hosner, 1960	Died if submerged more than 20 days.
	Yes	-	-	15–60	Hosner and Boyce, 1962	
Acer saccharinum	-	-	Yes	30	Hosner, 1958, 1960	Seedlings died if submerged for 4 days.
	Yes	-	-	15–60	Hosner and Boyce, 1962	
Betula nigra	-	Yes	-	21	DeShield et al., 1994	
Celtis laevigata	Yes	-	-	15–60	Hosner and Boyce, 1962	6.7% mortality after 60 days. All roots died except primary roots.
Celtis occidentalis	-	-	No	30	Hosner, 1960	
	Yes	-	-	15–60	Hosner and Boyce, 1962	20% mortality after 60 days. Some mortality of secondary roots.
Cephalanthus occidentalis	-	-	Yes	30	Hosner, 1960	
Fraxinus caroliniana	-	Yes	-	Putnam et al., 1960		Seedling established in moist soil from April to July after water receded
Fraxinus pennsylvanica	Yes	-	No	15–60	Hosner, 1958; Hosner and Boyce 1962	Seedling died if completely submerged for 32 days.
	Yes	Yes	-	285	Good and Patrick, 1987	
	-	Yes	-	30	Baker, 1977	91% survival one week after flooding.
Fraxinus profunda	Yes	-	-	15–60	Hosner and Boyce, 1962	
Liquidambar styraciflua	-	-	No	2–32	Hosner, 1958	
	Yes	-	-	15–60	Hosner and Boyce, 1962	Secondary roots died.

Liriodendron tulipifera	-	No	-	21	DeShield et al., 1994	Died after more than 7 days of flooding.
	-	-	Yes	30	Baker, 1977	68% survival one week after flooding.
Nyssa aquatica	Yes	-	-	84	Dickson et al., 1965	Lost height in treatments with tips held under the water.
	Yes	Yes	Yes	120–180	Kennedy, 1970	
	-	-	Yes	30	Baker, 1977	95% survival one week after flooding.
	Yes	-	-	15–60	Hosner and Boyce, 1962	
Nyssa sylvatica var. biflora	Yes	Yes	-	210	Hook et al., 1970	Flooded treatment was 20 cm, seedlings one year old.
Pinus taeda	-	Yes	-	21	DeShield et al., 1994	Some reduction in survival if flooded late in growing season.
	Yes	-	-	60	Tolliver et al., 1997	40% mortality at 10 g l^{-1};100% above 20 g l^{-1}.
	-	-	No	180	Williston, 1962	Seedlings submerged in a pool died after 6 to 7 months.
Planera aquatica	Yes	No	Yes	28	Rayner, 1976	10% mortality after 28 days of complete submergence. 85% of the seedlings died if grown in indirect sunlight.
Platanus occidentalis	-	-	No	30	Hosner, 1960	
	-	No		30	Baker, 1977	35% survival one week after flooding.
	Yes	-	-	15–60	Hosner and Boyce, 1962	26.7% died after 60 days in saturated soil.
Populus deltoides	-	-	No	2–32	Hosner, 1958	Seedlings died if flooded for longer than 16 days.
	-	No		30	Baker, 1977; Scott, et al., 1993, Segelquist et al., 1993	Cuttings used. 24% survival one week after flooding. Recruitment less successful where groundwater lower than root level.
Quercus falcata	-	-	No	16	Hosner, 1960	var. *pagodifolia* used in study.
	No	-	-	15–60	Hosner and Boyce, 1962	86.7% mortality after 60 days in saturated soil.
Quercus nigra	Yes	Yes	-	285	Good and Patrick, 1987	
	-	Yes	-	21	DeShield et al., 1994	
Quercus nuttallii	-	No	-	21	DeShield et al., 1994	Died after more than 7 days of flooding.
Quercus palustris	-	-	No	30	Hosner, 1960	
	Yes	-	-	15–60	Hosner and Boyce, 1962	Some secondary shoots died.
Quercus phellos	-	Yes	-	21	DeShield et al., 1994	Some reduction in survival after 21 days of flooding, particularly late in the growing season.
	Yes	-	-	15–60	Hosner and Boyce, 1962	Some mortality of secondary roots.

TABLE 7-7. (continued)

Species	Survival				Reference	Comments
	Soil Saturated	Flooded Below Tip	Flooded Above Tip	Experiment Length (Day)		
Quercus shumardii	-	-	No	30	Hosner, 1960	
	No	-	-	15–60	Hosner and Boyce, 1962	33.3% mortality after 60 days in saturated soil. Some mortality of secondary roots.
Taxodium distichum	-	Yes	-	21	DeShield et al., 1994	Died if submerged during growing season. Required a 2-year drawdown for establishment to occur. Recruitment less successful in shade Seedlings successful at the high-water mark (Eggler, 1961).
	Yes	No	No	14	Demaree, 1932; Shelford, 1954; Eggler and Moore 1961; Shankman 1991	
Taxodium distichum var. *nutans*					*McKnight et al., 1981*	*Similar to var.* distichum.
	Yes	*Yes*	*No*	*~30*	*Middleton and Fessel, upublished*	*Seedlings had limited ability to survive if submerged during the senescent season.*
	Yes	*No*	-	*44*	*Mattoon, 1916*	*Two-week-old seedlings died within 2 weeks in an inch of water*
	Yes	*Yes*	*No*	*~10*	*Conner and Flynn, 1989*	*Field study: in deepest water treatment, water level June–September not above 10 cm.*
Ulmus alata	*Yes*	-	-	*0–32*	*McDermott, 1954*	
Ulmus americana	*Yes*	-	*No*	*30*	*Hosner, 1960; Guilkey, 1965*	
	Yes	-	-	*15–60*	*Hosner and Boyce 1962*	*6.7% mortality after 60 days.*
Ulmus crassifolia	*Yes*	-	-	-	*McKnight et al., 1981*	

Source: Adapted from Middleton, 1999, as loosely based on Whitlow and Harris, 1979, and others.

ant of flooding (Table 7-8). For example, seedlings of *Taxodium distichum* die if flooded over the tips of the plants during the growing season (Middleton, 2000) for periods as short as three days (Figures 7-7 and 7-8) (Sharitz et al., 1990). After Lake Chicot, Louisiana, was impounded, seedlings that were fully submerged died, whereas those that were taller than the water surface survived (Penfound, 1949). In Brushy Lake, Arkansas, no regeneration was noted after the construction of levees that inundated a baldcypress-tupelo swamp (Osment-DeLoach and Moore, 1996). Conversely, at Pompano Beach, Florida, no seedlings were observed until water was pumped into a dewatered cypress swamp (Weller, 1995), suggesting that sites must be moist for germination to occur.

In an artificially inundated swamp such as Buttonland Swamp, the flood-pulsed zone is very narrow, so seedling and sapling recruitment is successful only at elevations represented by the highest winter flood levels above summer drawdown levels (Figure 7-9). This creates a spatial distribution pattern of regeneration that is linear in *Taxodium distichum,* due to the constraints imposed by flooding on seed dispersal and seedling recruitment because of the confined nature of the flood pulse (Middleton, 2000). The seeds do not live very long, so dispersal during the winter is necessary to deliver seeds to the seed bank at the annual high-water mark of the swamp. Seeds germinated only during the following growing season in the zone of drawdown at the edge of the swamp. The lowest elevations reflooded during the summer growing season, so that only seedlings at the highest elevations of winter flooding survived, and thus grew in a linear pattern in an old field adjacent to Buttonland Swamp. The spatial and temporal variability of seed rain (input of seeds to the seed bank from above ground sources), dispersal, and longevity, as well as seedling recruitment, were also determinants of the distribution of recruited seedlings in forests of *Betula allegheniensis* in northern forests (Houle, 1998). These early life history dynamics are likely to be important determinants of forest spatial distribution and dynamics, although scarcely studied in the majority of bottomland forest types.

Seedlings of *Taxodium distichum* can withstand overtopping by water while senescent in winter, as well as flooding below the tips of the plants during the growing season (Bennett et al., 2001). Although *T. distichum* is considered to be one of the most water-tolerant species of forested wetlands, its seedlings grow best in moist, well-aerated conditions (Mattoon, 1916). Another species, *Nyssa aquatica,* withstood complete inundation

TABLE 7-8. Sapling survivorship and flood depth in bottomland forests of the southeastern United States.

Information given includes the flooding depth above root crown (cm) and inundation period (days)

Species	Flood Depth (cm) Above Root Crown	Inundation Period (days)	Survive	Reference	Comment
Acer negundo	240–480	105	Yes	Noble and Murphy, 1975	
Acer saccharinum	> 61–122	9	Yes	Harris et al., 1975	Trees cut to 61–122 cm; complete inundation.
Fraxinus pennsylvanica	90	210	Yes	Broadfoot, 1967	Flooded trees grew 17% more than control.
	> 61–122	83	Yes	Harris et al., 1975	Trees cut to 61–122 cm; complete inundation.
Gleditsia triacanthos	240–480	105	No	Noble and Murphy, 1975	
	> 61–122	53	Yes	Harris et al., 1975	Trees cut to 61–122 cm; complete inundation.
Ilex decidua	240–480	105	Yes	Noble and Murphy, 1975	
Liquidambar styraciflua	90	210	Yes	Broadfoot, 1967	Flooded trees grew 82% more than control.
	240–480	105	No	Noble and Murphy, 1975	
Platanus occidentalis	> 61–122	44–56	No	Harris et al., 1975	Trees cut to 61–122 cm; complete inundation.
Populus x *canadensis*	> 61–122	55–102	No	Harris et al., 1975	Trees cut to 61–122 cm; complete inundation.
Salix sp.	> 61–122	114–155	Yes	Harris et al., 1975	Trees cut to 61–122 cm; complete inundation.
Salix terminis	> 61–122	125	Yes	Harris et al., 1975	Trees cut to 61–122 cm; complete inundation.
Taxodium distichum	10	~730	Yes	Young et al., 1993	Periodically flooded saplings developed false growth rings; continuously flooded had higher radial growth.
	> 61–122	171	No	Harris et al., 1975	Trees cut to 61–122 cm; complete inundation.

Source: Adapted from Middleton, 1999, as based on Whitlow and Harris, 1979, and others

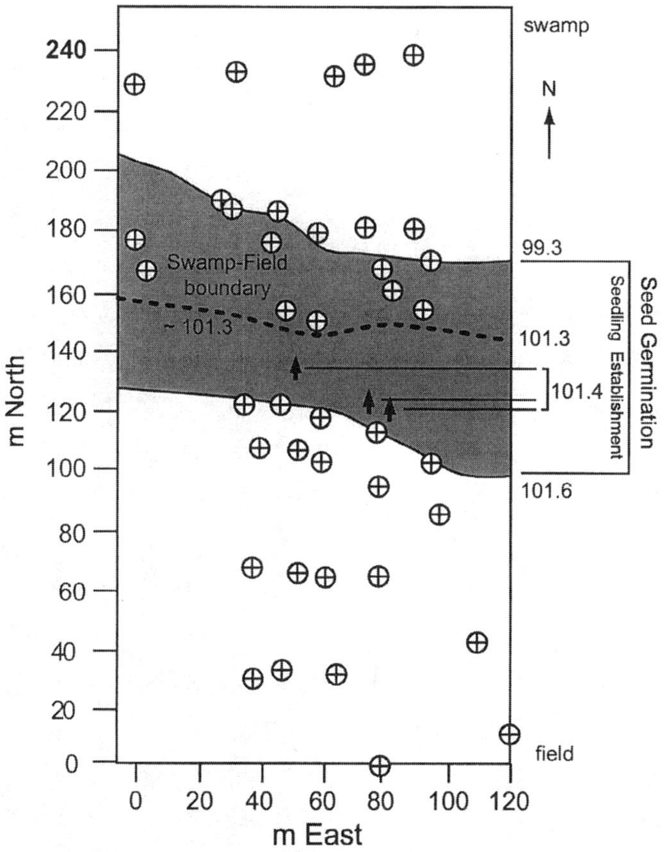

Figure 7-9. *Seedling establishment for* Taxodium distichum *in a swamp/field boundary at the highest level of winter flooding in the impounded Buttonland Swamp, Illinois. The germination zone is broad (elevation 99.4–101.6 m NGVD; National Geodetic Vertical Datum), but the seedling establishment zone is narrow (elevation 101.4 m NGVD) because of recurring summer flooding in the impoundment. (From Middleton, 2000, copyright © Kluwer; Figure 6 reprinted with kind permission from Kluwer Academic Publishers.)*

during the senescent period from late February to early July (87 percent survival), but when the flooding was continued until early August, survival dropped to 32 percent (Kennedy, 1970). Most other seedlings of woody wetland species are less tolerant of flooding (Table 7-7). In saturated conditions all but the primary roots of seedlings of *Celtis laevigata* died after 60 days (Hosner and Boyce, 1962). In forested wetlands these differences in the flood tolerances of seedlings contribute to the sorting of adults by species along elevational gradients (Bedinger, 1978).

Nevertheless, other species of baldcypress, such as *Planera aquatica,* in swamps withstood complete submergence for a period of time if in direct sunlight (Rayner, 1976). In addition to the flood tolerance that is related to elevation, seedlings also sort along gradients of shade tolerance. For example, *Taxodium distichum* is intermediate in shade tolerance, and cherrybark oak and sweetgum are intolerant of shade (Hall and Harcombe, 1998). It is likely the interactions of flood and shade tolerances influence the spatial distribution of seedling recruitment in complex ways.

Substrates also influence seedling recruitment. Seedlings were distributed across a wide variety of microsites in a mixed cypress swamp forest with natural hydrology along the Savannah River (Figure 7-10), but the highest growth rates were found on consolidated muck substrates (Sharitz et al., 1990). In general, seedlings that germinate earlier in the growing season have higher survivorship, but these can be killed by late-season flooding (Streng et al., 1989; Jones et al., 1994). Seedlings of *Taxodium distichum* planted in flooded or intermittently flooded sites in September or March had higher survival rates (>70 percent) than those planted in unflooded sites in March (40 percent) (Conner and Flynn, 1989). Sudden flooding, as is experienced along the Mississippi River during storms, can kill seedlings in forests and plantations (Kennedy and Krinard, 1974).

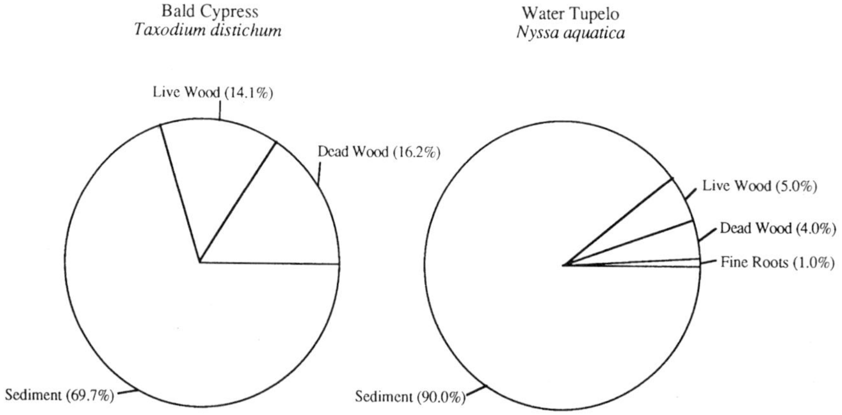

Figure 7-10. *Distribution of* Taxodium distichum *and* Nyssa aquatica *on microsites in a natural mixed deciduous swamp forest, Savannah River, South Carolina. Microsites include soil (loose or consolidated muck or muck over wood and near a tree, stump, knee, log, or branch), live wood (tree, knee, shrub), dead wood (stump, log, branch, twigs), and fine roots. (Reprinted from Sharitz et al., 1990, copyright © by Lewis Publishers, by permission.)*

Saplings are more tolerant of flooding than seedlings. Although there are many gaps in research, the available data suggest that most but not all species of saplings can survive periods of flooding above the root crown for at least short periods of time (Table 7-8). However, even though never completely submerged, many saplings more than 3 m tall died on the delta of Pigeon Creek, Illinois, as a result of the 1993–94 Mississippi flood (Loftus, 1994). Saplings less tolerant of flooding (e.g., *Acer negundo*) survived a flood of 240 to 480 cm above the root crown for 105 days, while even less tolerant ones *(Gleditsia triacanthos)* did not (Noble and Murphy, 1975). Saplings of the most flood-tolerant species, such as *Taxodium distichum,* grow best if they are shallowly flooded for most of the year (Conner and Flynn, 1989). For some species, such as *Planera aquatica,* prolonged flooding of two-to-three-year-old saplings, especially during the winter season, does not affect survival. However, because it is relatively shade intolerant, an open canopy is essential for this species and saplings are found only in relatively open areas in the swamp (Rayner, 1976).

Flood-tolerance mechanisms develop with age (Patrick et al., 1981), so that the adult period is the most flood tolerant of the life history stages (Table 7-9) (Whitlow and Harris, 1979; McKnight ct al., 1981; Wharton et al., 1982; Hook, 1984; Theriot, 1993). The most flood tolerant of the woody species is the shrub *Cephalanthus occidentalis,* and it increases in abundance with water depth up to 130 cm (Faber-Langendoen and Maycock, 1989). However, *Cephalanthus occidentalis* died after it was completely covered with water in the impoundment of Lake Chicot, Louisiana, in 1942 (Penfound, 1949).

In 63-year-old baldcypress-tupelo forests in the Atchafalaya River basin, Louisiana, *Nyssa aquatica* grew faster than *Taxodium distichum* in permanently flooded stands, and *Taxodium distichum* grew faster in seasonally flooded stands (Dicke and Toliver, 1990). Prolonged flooding in all species, however, results in decreased shoot growth, decreased respiration, leaf chlorosis, downward leaf curling, leaf abscission, absence of seed production, branch dieback, and increased susceptibility to pathogens or herbivores (Broadfoot and Williston, 1973). Tolerance to flooding is much lower in other species. *Quercus pagoda* died after one year of continuous flooding in less than 12 inches of water because of depressed root respiration (Broadfoot and Williston, 1973).

Impounded forested wetlands with permanent water regimes in the

TABLE 7-9. North American water tolerance ratings for adult trees

Species Rating	Common Name	Water Tolerance
Cephalanthus occidentalis	Buttonbush	Most tolerant
Forestiera acuminata	Swamp-privet	Most tolerant
Fraxinus caroliniana	Carolina ash	Most tolerant
Fraxinus profunda	Pumpkin ash	Most tolerant
Nyssa aquatica	Water tupelo	Most tolerant
Nyssa sylvatica var. biflora	Swamp tupelo	Most tolerant
Planera aquatica	Water elm	Most tolerant
Salix nigra	Black willow	Most tolerant
Taxodium distichum	Baldcypress	Most tolerant
Taxodium distichum var. *nutans*	Pondcypress	Most tolerant
Carya aqautica	Water hickory	Highly tolerant
Gleditsia aquatica	Waterlocust	Highly tolerant
Quercus lyrata	Overcup oak	Highly tolerant
Acer negundo	Boxelder	Moderately tolerant
Acer rubrum	Red maple	Moderately tolerant
Acer saccharinum	Silver maple	Moderately tolerant
Betula nigra	River birch	Moderately tolerant
Chamaecyparis thyoides	Atlantic white cedar	Moderately tolerant
Crataegus spp.	Hawthorn	Moderately tolerant
Diospyros virginiana	Persimmon	Moderately tolerant
Fraxinus pennsylvanica	Green ash	Moderately tolerant
Gleditsia triacanthos	Honeylocust	Moderately tolerant
Gordonia lasianthus	Loblolly-bay	Moderately tolerant
Ilex decidua	Possumhaw	Moderately tolerant
Liquidambar styraciflua	Sweetgum	Moderately tolerant
Magnolia virginiana	Sweetbay	Moderately tolerant
Persea borbonia	Redbay	Moderately tolerant
Pinus elliottii	Slash pine	Moderately tolerant
Pinus serotina	Pond pine	Moderately tolerant
Pinus taeda	Loblolly pine	Moderately tolerant
Platanus occidentalis	American sycamore	Moderately tolerant
Populus deltoides	Eastern cottonwood	Moderately tolerant
Quercus nuttalli	Nuttall oak	Moderately tolerant
Quercus palustris	Pin oak	Moderately tolerant
Quercus phellos	Willow oak	Moderately tolerant
Ulmus americana	American elm	Moderately tolerant
Ulmus crassifolia	Cedar elm	Moderately tolerant
Carpinus caroliniana	American hornbeam	Weakly tolerant
Carya illinoensis	Pecan	Weakly tolerant
Carya laciniosa	Shellbark hickory	Weakly tolerant
Celtis laevigata	Sugarberry	Weakly tolerant
Celtis occidentalis	Hackberry	Weakly tolerant
Ilex opaca	American holly	Weakly tolerant
Juglans nigra	Black walnut	Weakly tolerant
Magnolia grandiflora	Southern magnolia	Weakly tolerant
Morus rubra	Red mulberry	Weakly tolerant
Nyssa sylvatica	Blackgum	Weakly tolerant
Pinus glabra	Spruce pine	Weakly tolerant
Quercus falcata var. *pagodifolia*	Cherrybark oak	Weakly tolerant
Quercus laurifolia	Laurel oak	Weakly tolerant

TABLE 7-9. (continued)

Species Rating	Common Name	Water Tolerance
Quercus michauxii	Swamp chestnut oak	Weakly tolerant
Quercus shumardii	Shumard oak	Weakly tolerant
Quercus virginiana	Live oak	Weakly tolerant
Ulmus alata	Winged elm	Weakly tolerant
Asimina triloba	Pawpaw	Least tolerant
Cornus florida	Flowering dogwood	Least tolerant
Fagus grandifolia	American beech	Least tolerant
Juniperus virginiana	Eastern red cedar	Least tolerant
Liriodendron tulipifera	Yellow poplar	Least tolerant
Ostrya virginiana	Eastern hophornbeam	Least tolerant
Pinus echinata	Shortleaf pine	Least tolerant
Prunus serotina	Black cherry	Least tolerant
Quercus alba	White oak	Least tolerant
Sassafras albidum	Sassafras	Least tolerant
Ulmus rubra	Slippery elm	Least tolerant

Source: Adapted from Table 1, "Summary: Waterlogging-Tolerance Ratings," in "Waterlogging Tolerance of Lowland Tree Species of the South," by Donal D. Hook, *Southern Journal of Applied Forestry,* volume 8, number 2, pages 138–144. Published by the Society of American Foresters, 5400 Grosvenor Lane, Bethesda, MD 20814-2198. Not for further reproduction.

southeastern United States have a high level of tree mortality over time (Penfound, 1949; Eggler and Moore, 1961; Harms et al., 1980; Conner and Day, 1988, 1992; Heitmeyer et al., 1989; Osment-DeLoach and Moore, 1996). The length of time species can be flooded above the root crown and not suffer mortality varies widely between species, but most trees die if permanently flooded above the root collar for one to six years (Table 7-10) (Yeager, 1949). *Acer negundo* survived for 1460 days with flooding above the root crown, but *Gleditsia aquatica* died earlier (Yeager, 1949). All species eventually died in a mixed baldcypress swamp behind the Rodman Dam along the Oklawaha River, Florida, in water depths exceeding 1.3 m; however, many trees were crushed before the impoundment was built. As the depth of water increased in the impoundment, the relative amount of senescent root tissue increased as the amount of starch in the living root tissues decreased (Harms et al., 1980). Some species can recover if drawdown is reestablished; *Taxodium distichum* individuals that have been impounded for one (but not five) years recover rapidly (Duever and McCollom, 1986). In Lake Chicot four years after the construction of a dam, although most of the *Taxodium distichum* and *Planera aquatica* were still alive in 0.6 to 3.0 m of water (97 and 83 percent, respectively), only 34 percent of the *Nyssa aquatica* were still alive (Penfound, 1949).

TABLE 7-10. Adult tree and shrub survivorship in bottomland forests of the southeastern United States.

Information given includes survivorship and inundation length (days) with depth flooded above root crown (cm). The flooding depth above the root crown is not known unless given in parentheses.

	Survival					
Species	Saturated Soil	Flooded to Root Crown	Flooded Above Root Crown	Flood Inundation Length (day)	Reference	Comments
Acer negundo	Yes	Yes	Yes	1460	Yeager, 1949	
	?	?	?	189	Bell and Johnson, 1974	
	-	-	Yes	73	Harris et al., 1975	99% survival after 73 days.
Acer rubrum	-	-	Yes	?	Hall and Smith, 1955	Survived if flooded less than 41% of growing season.
Acer saccharinum	Yes	Yes/no	No	730	Yeager, 1949	64% survival after 730 days.
Betula nigra	Yes	Yes	No	730	Yeager, 1949	
Carya cordiformis	-	-	No	148	Brunk et al., 1975	
Carya illinoiensis	Yes	Yes	No	1460	Yeager, 1949	11% survival after 730 days.
	-	-	No	73	Harris et al., 1975	55% survival after 73 days of flooding.
Carya ovata	-	-	No	114	Bell and Johnson, 1974	38% survival after 114 days.
Carya tomentosa	-	-	No	149	Bell and Johnson, 1974	35% survival after 109 days.
Catalpa speciosa	-	-	No	67	Harris et al., 1975	60% survival after 67 days.
	-	-	No	?	Hall and Smith, 1955	Survived if flooded for less than 18% of growing season.
Celtis occidentalis	Yes	No	No	730	Yeager, 1949	In wet but not flooded, 66% survival after 730 days.
	-	-	No	189	Bell and Johnson, 1974	
Cephalanthus occidentalis	-	-	No	1460	Yeager, 1949	40% survival.
Cercis canadensis	Yes	No	No	240	Yeager, 1949	
Cornus florida	-	-	No	?	Hall and Smith, 1955	Survived if flooded less than 2% of growing season.

Diosporos viriginiana	Yes	No	No	1460	Yeager, 1949	20% survival if wet but not flooded after 1460 days.
Fagus grandifolia	-	-	No	?	Hall and Smith, 1955	Survived if flooded less than 4% of growing season.
Forestiera acuminata	Yes	Yes	No	730	Yeager, 1949	30% survival after 730 days.
Fraxinus americana	Yes	Yes	No (51 cm)	2555	Yeager, 1949	7% survival.
Gleditsia aquatica	Yes	No	No	1460	Yeager, 1949	8% survival in flooded after 730 days.
Gleditisia triacanthos	-	-	Yes	189	Bell and Johnson, 1974	
	-	-	No	730	Broadfoot and Williston, 1973	None survived after 2 yrs inundation.
Gymnocladus dioicus	Yes	Yes	Yes	1460	Yeager, 1949	
Ilex decidua	Yes	Yes	No	1460	Yeager, 1949	
Juglans nigra	-	-	No	149	Bell and Johnson, 1974	
	-	-	No	180	Kennedy and Krinard, 1974	
Juniperus virginiana	-	-	No	365	Hall and Smith, 1955	All died in one year.
	-	Yes/no	-	73	Harris, 1975	47% survival after 73 days of flooding.
Larix laricina	-	Yes	-	2190	Denyer and Riley, 1964	Survived 6 years of flooding.
Liquidambar styraciflua	-	-	No (< 30 cm)	1460	Broadfoot and Williston, 1973	
Liriodendron tulipifera	-	-	No	60	Kennedy and Krinard, 1974	11-year-old plantation.
Morus rubra	Yes	No	No	240	Yeager, 1949	
Nyssa aquatica	-	Yes	Yes (50–86)	365	Keeland and Sharitz, 1995	Maximum growth in deep, periodically flooded riverine sites.
Nyssa sylvatica var. *biflora*	-	Yes	Yes (50–86)	365	Keeland and Sharitz, 1995	Maximum growth in deep, periodically flooded riverine sites.
Pinus echinata	-	-	No (46 cm)	105	Williston, 1962	
Pinus resinosa	-	-	Yes	48	Ahlgren and Hansen, 1957	90% survival after 48 days of flooding.

TABLE 7-10. (continued)

	Survival					
Species	Saturated Soil	Flooded to Root Crown	Flooded Above Root Crown	Flood Inundation Length (day)	Reference	Comments
Pinus taeda	-	Yes	No	180	Williston, 1962	79% of trees survived 30 days of submergence followed by 90 days of root submergence.
	-	-	Yes	60	Kennedy and Krinard, 1974	
Platanus	Yes	No	No	240	Yeager, 1949	
occidentalis	-	-	Yes	169	Bell and Johnson, 1974	
Populus deltoides	Yes	Yes	No	730	Yeager, 1949	
	-	-	Yes	210	Broadfoot, 1967	
	-	Yes	?	189	Bell and Johnson, 1974	
Prunus serotina	-	-	No	?	Green, 1947	Died if flooded for more than 2% of growing season.
Quercus alba	-	-	No	149	Bell and Johnson, 1974	
Quercus bicolor	-	-	Yes	149	Bell and Johnson, 1974	
	-	-	Yes	60	Kennedy and Krinard, 1974	
Quercus falcata	-	-	No	1095	Green, 1947	Died if flooded for more than 3 years.
Quercus imbricaria	-	-	No	149	Bell and Johnson, 1974	
	-	-	No	129	Brunk et al., 1975	Died if flooded for more 129 days.
Quercus	Yes	-	No	730	Yeager, 1949	
macrocarpa	-	-	No (< 30 cm)	1460	Broadfoot and Williston,	

					1973	
	-	-	Yes	189	Bell and Johnson, 1974	
Quercus muehlenbergii	-	-	No	67	Harris et al, 1975	10% survival after 67 days of flooding.
Quercus nigra	-	-	No (< 30 cm)	1460	Broadfoot and Williston, 1973	
	-	-	Yes	60	Kennedy and Krinard, 1974	
Quercus nuttalli	-	-	Yes	60	Kennedy and Krinard, 1974	
Quercus palustris	Yes	No	No (< 76.2 cm)	730	Yeager, 1949	
	-	-	Yes	109	Bell and Johnson, 1974	
Quercus rubra	-	-	Yes	50	Bell and Johnson, 1974	
Quercus stellata	-	-	No	-	Harris et al., 1975	45% survival after 73 days.
Quercus velutina	-	-	No	109	Bell and Johnson, 1974	
Robinia pseudoacacia	-	-	No	73	Harris et al., 1975	5% survival after 73 days of flooding.
Salix nigra	-	-	Yes	1460	Yeager, 1949	56% survival.
	-	-	No	1095	Green, 1947	All died after 3 years of flooding.
Sassafras albidum	-	-	No	149	Bell and Johnson, 1974	
Taxodium distichum	-	Yes	Yes (50–86)	365	Keeland and Sharitz, 1995	Maximum growth with shallow permanent flooding (mean 39 cm).
Ulmus americana	Yes	No	No	730	Yeager, 1949	49% survival in wet soil after 1460 days.
	-	-	Yes	189	Bell and Johnson, 1974	91% survival after 189 days.

Source: Adapted from Middleton, 1999, as based on Whitlow and Harris, 1979, and others.

Eventually, all of the hardwood trees in the lake died (Eggler and Moore, 1961).

In cypress swamps overall, production levels are lower in impounded situations than in natural swamps with flood pulsing (Table 7-11) (Broadfoot and Williston, 1973; Conner et al., 1981; Mitsch et al., 1991; Conner and Day, 1992; Megonigal et al., 1997; Conner and Buford, 1998; Watt and Golladay, 1999; Middleton and McKee, unpublished). The most important factor in production may have to do with level of root aeration, rather than nutrient input, as the subsidy-stress hypothesis would suggest (Megonigal et al., 1997).

Only a few studies have assessed the ability of herbaceous species of forested wetlands to survive flooding; however, most of the species studied tolerate saturated soil conditions (Table 7-12). Typically, understory species are found where the soil is not flooded (Brinson, 1990). Ferns and herbaceous species grow on emergent objects, such as stumps, logs, or the bases of trees in cypress strands (Lugo et al., 1984), in cypress swamps (Anderson and White, 1970), and at the edges of swamps near springs (Phillips, 1996). Bryophytes in baldcypress swamps are typically found either on emergents or on the swamp surface during drawdown (Conrad, 1997). The fact that herbaceous species cling to high places and are unusual in flooded wetlands suggests that they are limited by flooding (Brinson, 1990).

RESTORATION APPROACHES

Successful restoration is a reflection of the extent to which we understand the environmental setting necessary for the regeneration of species. Thus far, restoration in forested wetland has met with mixed success, but apparently is improving, particularly if flood pulsing is incorporated into the restoration setting. More than 75,000 ha of land, mostly abandoned agricultural fields, have been restored to forested wetland in the south central United States (Allen, 1997). Most of the sites are on high-risk farmland or in national forests, wildlife refuges, management areas, flood-control projects, or public lands where the restoration of forested wetland serves the public interest (Haynes and Moore, 1988). Large-scale forested wetland restoration projects include sites in the Yazoo National Wildlife Refuge (west central Mississippi), Tensas River National Wildlife Refuge (northeastern

Louisiana), Ouachita Wildlife Management Area (central Louisiana), and the Delta Experimental Forest (Stoneville, Mississippi) (Sharitz, 1992).

The key factor in restoration is reestablishing natural hydrology (Sharitz, 1992) to regain floodplain function (Petts et al., 1992; Middleton, 1999). It is more important to establish the physical environment (water and soil) than the biological components of the community (Shear and Malcom, 1997), which can reestablish naturally if the conditions are appropriate (Trepagnier et al., 1995; Shear et al., 1996; Allen, 1997; Middleton, 1999, 2000). Without attention to hydrology, the long-term prospects of restoration are not good. In forested wetlands in southwestern Kentucky, planted by the Tennessee Valley Authority 50 years ago, stands changed from hydric to mesic types because of the stream downcutting associated with channelization. This phenomenon also occurred in naturally revegetated forests in the vicinity (Shear et al., 1996) and has been documented in many other places in the Southeast (Shields and Hoover, 1991; Sengupta, 1995). At least for tidal wetlands, it is recognized that the physical setting, including both tidal pulsing and organic matter levels, may be very difficult to recreate (Zedler, 1996b; Haltiner et al., 1997; Zedler and Callaway, 1999).

Hydrologic Reconstruction via Reengineering

Simple dams or weirs are often constructed across streams or rivers to increase water levels in forested wetlands as one of the commonest but least effective forms of restoration. The problem with simple dams is that they maintain water levels behind them at unnaturally high levels throughout the year and do not allow drawdown at some time during the growing season to increase regeneration and production levels (Middleton, 1999). In Buttonland Swamp, a water control structure was added to the Cache River channel to increase the summer water levels in a mature baldcypress swamp. The project was an attempt to ameliorate the effects of a 1916 upstream interbasin water transfer (Post Creek Cutoff) that redirected the Cache into the Ohio River (Muir et al., 1995). Because the dam created a permanent impoundment, regeneration was restricted to the margins of the swamp at the highest elevations of winter flooding (Figure 7-9) (Middleton, 2000) and production levels decreased in the permanently impounded portions of the swamp from 1992 to 1998 (also see Table 7-11) (Middleton and McKee, in press). After a dam was constructed for a similar

TABLE 7-11. Leaf litter production in cypress swamps of the southeastern United States

Includes location and flooding characteristics of study sites for leaf litter production (litterfall) excluding wood production (Middleton and McKee, unpublished).

Latitude	Location	Hydrologic Status	Forest Type	Litterfall g m^{-2} yr^{-1}	Reference
37.8	Henderson Slough, KY	Natural[a]	*Taxodium distichum, Fraxinus pensylvanica*	136	Mitsch et al., 1991
37.8	Henderson Slough, KY	Natural[b]	*T. distichum*	253	Mitsch et al., 1991
37.4	Cypress Creek floodplain, KY	Impounded[c]	*T. distichum*	63	Mitsch et al., 1991
37.2	Heron Pond, IL	Natural	*T. distichum, N. aquatica*	348	Dorge et al., 1984
37.2	Buttonland Swamp, IL	Impounded[d]	*T. distichum, N. aquatica*	81	Middleton and McKee, n.d.
37.2	Buttonland Swamp, IL	Impounded[e]	*T. distichum, N. aquatica*	134	Middleton and McKee, n.d.
37.2	Buttonland Swamp, IL	Impounded[f]	*T. distichum, N. aquatica*	162	Middleton and McKee, n.d.
37.2	Buttonland Swamp, IL	Impounded[g]	*T. distichum, N. aquatica*	261	Middleton and McKee, n.d.
36.1	Great Dismal Swamp, NC/VA	Natural[h]	*T. distichum*	568	Gomez and Day, 1982
36.1	Great Dismal Swamp, NC/VA	Natural	*T. distichum*	342[i]	Day, 1979
35.5	Tar River floodplain, Pitt County, NC	Natural	*N. aquatica, T. distichum*	524	Brinson et al., 1980
33.3	Myers Branch, Steel Creek, SC (site 1)	Natural	*T. distichum*	679[j]	Megonigal et al., 1997
33.3	Myers Branch, Steel Creek, SC, (site 2)	Natural	*T. distichum*	905[j]	Megonigal et al., 1997
33.3	Myers Branch, (stream) Steel Creek, SC	Natural	*T. distichum*	544	Muzika et al., 1987

33.3	Myers Branch, (river) Steel Creek, SC	Natural	*T. distichum*	466	Muzika et al., 1987
33.3	Upper Three Creeks Run, (1) Savannah River, GA	Natural	*N. aquatica, T. distichum*	695[j]	Megonigal et al., 1997
33.3	Upper Three Creeks Run, (2) Savannah River, GA	Natural	*T. distichum, N. sylvatica*	772[j]	Megonigal et al., 1997
33.3	Savannah River, GA (site 1)	Impounded[k]	*N. aquatica, T. distichum*	477[j]	Megonigal et al., 1997
33.3	Savannah River, GA (site 2)	Impounded[k]	*N. aquatica, T. distichum*	398[j]	Megonigal et al., 1997
32.0	Ogeechee River, GA (east)	Natural[l]	*T. distichum Liquidamber styraciflua*	808[m]	Cuffney, 1988
32.0	Ogeechee River, GA (west)	Natural[n]	*T. distichum, Liquidamber styraciflua*	637	Cuffney, 1988
31.2	Ichauway Ecological Reserve, GA	Natural	*T. distichum, T. distichum* var. *ascendens, N. biflora*	410-582[i]	Watt and Golladay, 1999
30.6	Apalachicola River, FL	Natural	*Nyssa* sp., *T. distichum*	464	Elder and Cairns, 1982
30.3	Okefenokee Swamp, GA	Natural	*T. distichum* var. *nutans*	317[o]	Schlesinger, 1978
30.3	Pearl River, LA (site 1)	Natural	*T. distichum, N. sylvatica*	713[j]	Megonigal et al., 1997
30.3	Pearl River, LA (site 2)	Natural	*T. distichum, N. sylvatica*	620[j]	Megonigal et al., 1997
29.9	Lac des Allemands, LA	Natural	*T. distichum, N. aquatica*	584	Day et al., 1977
29.9	Verret Basin, LA (site 1)	Impounded[p]	*T. distichum, Diospyros virginiana*	350[j]	Megonigal et al., 1997
29.9	Verret Basin, LA (site 2)	Impounded[p]	*T. distichum, Diospyros virginiana*	317[j]	Megonigal et al., 1997
29.9	Verret Basin, LA	Natural[q]	*T. distichum, Fraxinus pennsylvanica*	350–534[j]	Conner et al., 1993
29.9	Bayou Chevreuil, LA	Impounded[r]	*T. distichum, N. aquatica*	293	Conner et al., 1981, Conner and Day, 1992

TABLE 7-11. (continued)

Latitude	Location	Hydrologic Status	Forest Type	Litterfall g m^{-2} yr^{-1}	Reference
29.9	Bayou Chevreuil, LA	Natural	*T. distichum, N. aquatica*	405	Conner et al., 1981, Conner and Day, 1992
29.9	Bayou Chevreuil, LA	Natural	*T. distichum, N. aquatica*	620	Conner and Day, 1976
29.8	Barataria Basin, LA (site 1)	Natural	*N. aquatica., T. distichum*	487[j]	Megonigal et al., 1997
29.8	Barataria Basin, LA (site 2)	Natural	*N. aquatica, T. distichum*	489[j]	Megonigal et al., 1997
29.7	Prairie Creek, near Gainesville, FL	Natural[s]	*T. distichum, Fraxinus caroliniana*	500	Brown, 1981
29.7	Austin Cary Forest, near Gainesville, FL	Natural	*T. distichum*[t]	380	Deghi et al., 1980
29.7	Hatchett Creek, *Austin Cary Forest, near Gainesville, FL*	Natural	*T. distichum*[t]	*578*[i]	*Richardson et al., 1983*
27.0	*Corkscrew Swamp, FL*	*Natural*	T. distichum[t]	*501*	*Duever, 1984*
27.0	*Corkscrew Swamp, FL*	*Natural*	T. distichum[t]	*591*	*Duever, 1984*
26.4	*Big Cypress, FL (site 2)*	*Natural*	T. distichum[t]	*315*	*Burns, 1984*
26.4	*Big Cypress, FL (site 2)*	*Natural*	T. distichum[t]	*345*	*Burns, 1984*
26.2	*Collier County, FL*	*Natural*	T. distichum *var.* nutans	*177*	*Brown et al., 1984*

	Natural		Impounded	
	All Sites	*T. distichum* Dominant	All Sites	*T. distichum* Dominant
Maximum	905	905	477	350
Minimum	136	136	63	63
Range	769	769	414	287
Mean	512	528	254	208
S.E.	31	35	44	39
Median	524	556	278	212

[a] Slowly flowing, semipermanently flooded.

[b] Slowly flowing, permanently flooded.

[c] Impounded within a levee.

[d] Mean of Transect 3 with a mean minimum water depth of 49 cm.

[e] Mean of Transect 4 with a mean minimum water depth of 43 cm.

[f] Mean of Transect 1 with a mean minimum water depth of 5 cm; not included in the latitudinal comparison because of the low mean minimum water depths.

[g] Mean of Transect 2 with a mean minimum water depth of <1 cm; not included in the latitudinal comparison because of the low mean minimum water depths.

[h] Continuously flooded for 6 months.

[i] Not included in the latitudinal comparison because the amount of woody material included in the leaf litter was not given.

[j] Based on means of two years of data.

[k] Sites permanently flooded to 60 cm because of subsidence.

[l] Flooded for 60–90% of the year.

[m] Based on 95% ash-free dry weight.

[n] Flooded for 90% of the year.

[o] Not included in the latitudinal comparison because some of the sites may have been permanently impounded; the Suwanee River was regulated to 34.8 m above sea level. Litter estimates based on a mean of 17 sites.

[p] Permanently flooded because of subsidence.

[q] Not included in the latitudinal comparison because the site was subsiding and many dead trees were noted.

[r] Although this site was originally impounded, it was not included in the latitudinal comparison because cuts were made in the levee to allow exchange of water between the bayou and the swamp. Subsequently, water levels fell to 10 cm depth during parts of the growing season.

[s] Streamflow in Prairie Creek controlled by a dam.

[t] Probably a mixture of *T. distichum* and *T. distichum* var. *nutans*.

TABLE 7-12. Herbaceous species flood tolerance in bottomland hardwood forests of the southeastern United States.

Information includes survivorship and water depth for studies were conducted in greenhouse conditions unless otherwise indicated.

Species	Survival				Reference	Comments
	Soil Saturated	Flooded Below Tip	Flooded Above Tip	Experiment Length (days)		
Aster ontarionis	Yes	-	-	70	Smith and Moss, 1998	94% survival.
Aster pilosus	Yes/no	-	-	70	Smith and Moss, 1998	61% survival.
Boltonia decurrens	Yes	-	-	56	Stoecker et al., 1995	
	Yes	-	-	70	Smith and Moss, 1998	
Conyza canadensis	Yes/no	-	-	56	Stoecker et al., 1995	About 50% survival after 56 days of surface saturation.
	No	-	-	70	Smith and Moss, 1998	28% survivorship.
Eleocharis interstincta	-	Yes	-	730	Newman et al., 1996	Water levels drawn down to 5 cm in dry season, otherwise held at 60 cm.
Ludwigia peploides	Yes	Yes	-	120	Mathis and Middleton, 1999	Plants died within 1 week when clipped at ground level in freely drained conditions.
Nelumbo nucifera	-	Yes	No	147	Nohara and Kimura, 1997	Experiment conducted in concrete ponds; maximum biomass at 0.5 m; some plants grown at 3–5 m water depths died.
Panicum hemitomon	-	Yes	-	58	Lessman et al., 1997	Water levels raised slowly in the experiment to maximum of 39 cm.
Phragmites australis	Yes	Yes	Yes	455	Hellings and Gallagher, 1992	Cutting and salinity reduced biomass.
	Yes	Yes	-	120	Mathis, 1999	Plants died within 1 week when clipped at ground level in 15 cm of water.
Scirpus olneyi	Yes	Yes	-	37	Broome et. al., 1995	Salinity above 10 ppt reduced growth.
Typha domingensis	Yes	Yes	-	68	Glenn et al., 1995	Compared with freshwater treatment; dry mass reduction 50% at 3.5 ppt., 90% at 6 ppt. salinity; at 9 ppt., growth negligible, though plants survived.
	-	Yes	-	730	Newman et al., 1996	Water levels drawn down to 5 cm in dry season, otherwise held at 60 cm; higher biomass at higher water and nutrient levels.
Typha latifolia	Yes	Yes	-	120	Mathis and Middleton, 1999	Plants died within 1 week when clipped at ground level in 15 cm of water.

Source: Adapted from Middleton, 1999, as based on Whitlow and Harris, 1979.

purpose from Dead Lake along the Chipola River, Florida, sport fishing increased briefly, but then declined ten years later (Young and Crew, 1982). After this dam was removed, water quality improved and sport fishing increased (Hill et al., 1994).

In sites that have been destroyed after being impounded by dams for hydroelectric or other public works projects, dam removal is sometimes considered as an option. This practice has been successful in allowing restoration in several instances in the North—for example, the Woolen Mills Dam removal near Milwaukee, Wisconsin (National Research Council, 1992; Middleton, 1999). However, there can be problems associated with dam removal, including the release of toxic chemicals (Tofflemire, 1986) and the movement of sediment plumes downstream (Simons and Simons, 1991). As part of the construction of the Cross Florida Barge Canal, the Rodman Dam and Reservoir in north central Florida was created (Harms et al., 1980). This dam has been considered for removal for years (Shuman, 1995), and in July, 2000, Governor Jeb Bush signed an order to remove the Kirkpatrick Dam and restore the reservoir area (Nelson, 2000). No dam lives forever, as the reservoir fills in with sediment, but it is still very difficult to restore areas that have been impounded by dams (Cairns and Palmer, 1993).

In cases where water has been diverted from forested wetland, water pumping has been used to restore a more natural hydrology. The swamp in Pompano Beach, Florida, was cut off from its water source in the early 1900s. Water pumping into the swamp followed a schedule mimicking wet/dry seasons and wetland function, at least as judged by the presence of key plants and animals that were reestablished in three years (Weller, 1995). Systems that require a high level of human maintenance are not self-sustainable (Eggers, 1992; Zedler, 1996a; Middleton, 1999). However, as shown at the Pompano Beach swamp, pumping may rehabilitate disturbed wetlands surrounded by human population centers where a high degree of water level control is necessary (Weller, 1995).

The best restorations in riparian wetlands have recreated naturally fluctuating water levels synchronized with seasonal wet/dry cycles—that is, the flood pulse. With their efforts at dechannelization and remeandering, the Kissimmee River Project, Florida (Kondolf, 1996), and the Lodi Marsh Project, Wisconsin, are exemplary attempts to restore flood-pulsed conditions to waterways in North America (Middleton, 1999), but the results do not directly pertain to forested wetlands. The idea of dechannelizing rivers is rather new, but it holds great promise for the restoration and revitalization of southeastern forested wetlands.

Flood pulsing has been restored to some forested wetlands through the removal of levees in scattered projects in the Southeast; many of these projects are just getting under way. In the Brushy Lake Mitigation Site, Arkansas, naturally fluctuating water levels have been reestablished in a forested oxbow through the removal of a levee (Figure 7-11) (Osment-DeLoach and Moore, 1996). The levee was constructed in 1963 to create agriculture fields in a palustrine wetland (U.S. Army Corps of Engineers, 1996; Arkansas State Highway and Transportation Department, 1997). After cultivation was discontinued in 1995 (Arkansas State Highway and Transportation Department, 1997), *Taxodium distichum* and *Nyssa aquatica* were reestablished by planting. Two natural lakes reappeared (U.S. Army Corps of Engineers, 1996). The monitoring of the area will include studies of functional attributes (e.g., production studies) of these mitigated wetlands by P. J. Young and B. Keeland (Osment-DeLoach and Moore, 1996). Another approach related to levee breaking is to set back the levee to widen the functional floodplain (Welcomme, 1994) and thus reduce the

Figure 7-11. *Breached levee, Brushy Lake Mitigation Site, Arkansas Department of Transportation, Arkansas. Henry Langston pictured. (Photograph by Phillip Moore; used by permission.)*

mean depth of the water in the forest between the levee and the river channel. This idea also holds promise to improve the hydrological conditions for remnant forested wetlands that have been impounded between levees.

To Plant or Not to Plant

Restoration of forested wetland in the southeastern United States has often focused on replanting oak species, resulting in low-diversity communities that resemble plantations more than natural communities (Allen, 1997). In future efforts at restoration of forested wetlands, attention should be given to the seeding of shrubs and herbaceous plants (King and Keeland, 1999) as well as a wider variety of tree species (Allen, 1997).

Planted species can interfere with the establishment of naturally invading species, because the plantings tend to be taller. In a study in west central Mississippi, planted trees were 58 percent taller than naturally invading trees, reducing the chances of a diverse stand ever developing (Table 7-13) (Allen, 1990, 1997). Because restoration attempts to restore original conditions, an additional problem is that most of these deforested wetlands replanted to oak originally supported forest of a different composition, such as *Celtis laevigata* and *Liquidambar styraciflua* (Ouchley et al., 2000).

Deforested wetlands can regenerate naturally if the hydrologic constraints of the species are not exceeded (Sharitz and Lee, 1985b; Middleton, 1999). In baldcypress-tupelo swamps, regeneration can occur if there is an upstream seed source, a large number of seeds produced in the previous year, a drawdown in two consecutive growing seasons, and water levels that do not exceed the height of the seedlings and saplings (Keeland and Conner, 1999). Extensive natural regeneration of species such as *Cephalanthus occidentalis, Nyssa aquatica, Salix nigra,* and *Taxodium distichum* occurred in Lake Chicot after an 18-month drawdown starting in 1986 to repair the water control structure (Table 7-14). This drawdown spanned two growing seasons and was sufficient to allow the seedlings to become tall enough to survive reflooding (Keeland and Conner, 1999). Similarly, a regeneration pulse of *Taxodium distichum* and *Nyssa aquatica* occurred in the absence of summer flooding between 1985 and 1989 (including a winter flood in early 1987) along Pen Branch Creek, South Carolina. Previously, there was near complete mortality in the mixed baldcypress-tupelo forest in a thermal discharge area of the Savannah River

TABLE 7-13. Tree invasion in old fields in west central Mississippi

Mean number and height of planted and invader trees are given for the 10 stands evaluated by Allen (1990).[a]

	Stand Number									
Species	1	2	3	4	5	6	7	8	9	10
	Mean Number ha^{-1}									
Planted Oaks	741	657	613	781	494	497	2265	395	252	215
Invader Species										
Liquidambar styraciflua	14,329	49	3	7	7	141	32	12	3	5
Fraxinus pennsylvanica	247	12	64	17	17	99	15	3	235	0
Ulmus spp.	141	37	5	126	54	5	0	67	10	49
Platanus occidentalis	413	0	3	20	0	0	0	0	0	0
Diospyros virginiana	7	42	158	15	12	25	20	17	20	59
Acer negundo	12	3	0	37	0	0	0	7	185	72
Celtis laevigata	106	7	3	5	32	0	3	0	12	0
Gleditsia spp.	17	5	0	7	27	7	42	0	0	0
Quercus nigra/Quercus phellos	37	0	0	0	0	3	0	3	0	0
Acer rubrum	0	0	7	5	0	0	0	3	3	5
Carya aquatica	0	0	0	0	0	5	10	0	0	0
Quercus lyrata	0	0	0	0	0	0	12	0	0	0
Other trees	7	5	0	10	0	302	25	0	0	5
Shrubs	3	0	0	27	3	72	111	3	7	5
Total Invaders	15,319	160	243	276	152	659	270	115	475	200
	Average Height (m)									
Planted Trees	5.84	2.95	2.81	4.76	2.64	1.39	2.10	1.24	1.02	0.54
Standard Error	0.11	0.05	0.06	0.07	0.07	0.02	0.02	0.04	0.07	0.05
Invaders	4.03	2.25	1.76	2.60	1.55	1.47	1.41	1.18	1.97	0.51
Standard Error	0.02	0.12	0.08	0.13	0.09	0.03	0.04	0.06	0.07	0.04

[a] Stands 1-5 were established with seedlings, stands 6-10 by direct seeding.

Source: Allen, 1990 (copyright © *Restoration Ecology,* by permission).

TABLE 7-14. Densities, mean heights, and mean diameters of saplings at Lake Chicot, Louisiana

Standard deviations are given in parentheses.

	1986		1992		1996		
	Number m^{-2}	Height (cm)	Number m^{-2}	Height (cm)	Number m^{-2}	Height (cm)	Diameter (cm)
Taxodium distichum	1.39 (0.82)[a]	75 (18)	2.11 (1.79)	315 (56)	1.96 (1.43)	476 (131)	4.3 (2.4)
Salix nigra	0.12 (0.13)	82 (29)	0.01 (0.01)	241 (70)	0.01 (0.01)	395 (147)	2.5 (1.3)
Cephalanthus occidentalis	0.21 (0.13)	69 (29)	0.06 (0.07)	373 (64)	0.02 (0.03)	279 (39)	2.3 (1.1)
Nyssa aquatica	0.03 (0.03)	56 (12)	0.01 (0.01)	404 (19)	0.01 (0.01)	714 (84)	8.4 (1.9)
Rhus copallina	0.01 (0.01)	52 (21)	0	-	0	-	-
Sapium sebiferum	0.01 (0.01)	112 (16)	0	-	0	-	-
Liquidambar styraciflua	0.02 (0.03)	32 (2)	0	-	0	-	-

[a] Seedlings m^{-2} including areas later sprayed with herbicides because of the thick growth of seedlings that prevented access to parts of the lake. In the fringe zone, where all plots were located for the 1992 and 1996 measurements, there were 2.03 seedlings m^{-2} in 1986. Values given for number, height, and diameter are the mean (± standard deviation).

Source: Keeland and Conner, 1999 (copyright © *Wetlands,* by permission).

Site (De Steven and Sharitz, 1997). Because cypress tend to regenerate in cohorts, it is likely that this species has historically regenerated during extended periods of drawdown (Mattoon, 1915; Putnam et al., 1960), which may recur every 30 to 50 years (Rathborne, 1951).

Baldcypress communities often have not reestablished after their removal. After logging in the Manchac Swamp, Louisiana, *Taxodium distichum* regeneration did not occur, probably because of high levels of herbivory by nutria, rather than because of saltwater intrusion as is generally assumed (Myers et al., 1995). However, forest community redevelopment patterns after the destruction of habitats by logging or agriculture may also be due to changes in hydrology. In the Kentucky Lake Region of Tennessee and Kentucky, less tolerant species with wind-dispersed seeds have invaded stands of *Taxodium distichum* and *Chamaecyparis thyoides* planted in the 1940s (Shear et al., 1996).

Species that disperse via hydrochory can regenerate naturally on old field sites (Newling, 1990) if there is water moving through nearby intact wetlands. This type of natural restoration was observed in an abandoned agricultural field with flood pulsing adjacent to Buttonland Swamp (Middleton, 1999, 2000). Wind-dispersed species may naturally invade restoration sites and develop into diverse stands of forested wetland (Allen, 1990) if these are less than 60 m from mature stands (Figure 7-12) (Allen, 1997). Wind is effective at dispersing species short distances, but the maximum distance varies for species such as *Acer saccharinum, Fraxinus pennsylvanica,* and *Platanus occidentalis* (Mathis 2001). However, if dispersal is not successful in reintroducing species into abandoned farm fields after one to three years of farming, very little regeneration may occur either from the seed bank or stump resprouting, so deliberate reintroduction may be necessary (Newling, 1990).

The succession of old agricultural fields on lowlands eventually progresses toward forested wetlands, even without replanting. During the 1st year after abandonment, annuals and seedlings of trees and shrubs invade the old field. From the 2nd to the 10th year, perennial herbs are dominant, and by the 10th year, trees and shrubs begin to be prominent (Hopkins and Wilson, 1974). On lower-elevation sites, annuals generally persist for longer than 1 year (Battaglia, 1991; Battaglia et al., 1995). After 25 years, the site supports a forest with large shrubs and small trees (Bonck and Penfound, 1945; Bazzaz, 1968). Old field succession is fundamentally differ-

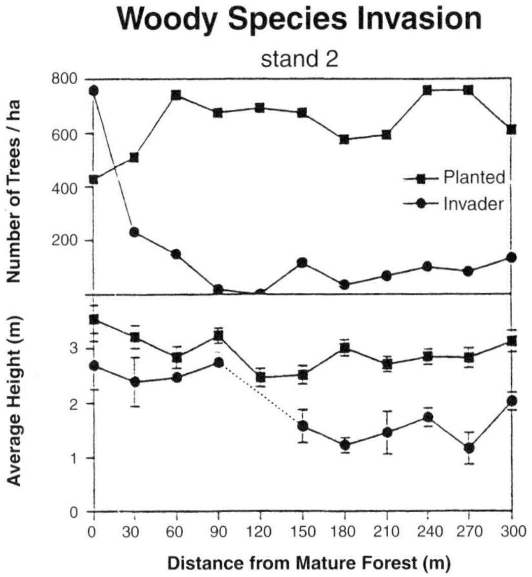

Figure 7-12. *Average number of trees ha*$^{-1}$ *and heights (m) of planted and invader trees in relation to distance from mature forest. (Reprinted from Allen 1997, copyright © by Society for Ecological Restoration, by permission.)*

ent from the revegetation of newly developed point bars that are formed by deposition in the bends of rivers (Hosner and Minckler, 1963; Shankman, 1993; Allen, 1997), where, among other differences, the herbaceous component of old field succession is bypassed because of prolonged periods of flooding.

Making Plantings More Natural

The best way to promote the development of forests with high diversity may be to broadcast seeds of many species on sites of appropriate elevations rather than to replant seedlings and saplings of a limited number of species (Allen, 1997; Streever and Perkins, 2000). Local seed sources should be used to ensure the genetic integrity of the restored forest, because of differences in gene expression from genotypes of various geographic regions (Allen and Kennedy, 1989). For example, the flowering time of the introduced genotype may preclude cross-pollination (Somers and Grant, 1981), competitive abilities may differ (Gustafson, 2000), or the genotype may be poorly adapted to the climate and soil conditions of the region (Eggers, 1992). Furthermore, studies of midwestern prairies have demonstrated that genotypic differences in species, as detected in Random Amplified Polymorphic DNA (RAPD) banding patterns, are related to choices of planting stock for restoration sites (Gustafson et al., 1999). If local genotypes are not available for planting, sources from close-by latitudes may be more critical to locate than those matched by longitude (McKenney et al., 1999; Streever and Perkins, 2000). It is not advisable to use genotypes that grow faster or are resistant to disease (Streever and Perkins, 2000), because these genotypes may have a competitive advantage.

Seed should be collected early in the season, keeping in mind that the first seeds that fall are usually not viable (except for *Quercus lyrata;* Middleton, personal observation). Acorns should be floated in a container of water; those that sink are not viable and should be discarded (Strader et al., 1994; Dulohery et al., 1995). Collected seeds should be kept in cold storage at 1.7 to 4.5° C (Strader et al., 1994).

Revegetation after the spreading of seed may be slow because of dense vegetation cover (Dunn and Sharitz, 1987) or granivory by rodents. Thus, it may be advisable to disk a field before planting (Allen and Kennedy, 1989). However, there is a trade-off, because on lowland sites disking can temporarily reduce the natural establishment of diverse seedlings (Allen and Kennedy, 1989; Allen et al., 1998). Sites with excessive vegetation may be prepared by mowing, burning, bush hogging, and scalping prior to disking (Haynes et al., 1993).

Failure in restoration often occurs because species are not matched to the site environment (Haynes et al., 1993), particularly to elevation, which determines the water regime. In a study to track the success of planted seeds and seedlings of *Taxodium distichum* and *T. distichum* var. *nutans* in logged and burned stands in Corkscrew Swamp (Florida), less than 1 percent of the seedlings survived for more than three years (Gunderson, 1984). Seeds and seedlings in forested wetlands fail to establish because of inadequate water environment (Gunderson, 1984; King and Keeland, 1999), competitive exclusion (Gunderson, 1984; Dunn and Sharitz, 1987), and herbivory by mice, beaver, deer, and rabbits (King and Keeland, 1999). Another study near the Four Mile Branch, South Carolina, determined that the elevation of planting was a much more important determinant of seedling survival than controlling nontarget vegetation (McLeod et al., 2000).

Soil preparation is useful before planting. Heavy clays should be double-disked in the summer, and spring planting sites should be plowed the previous fall and plow pans broken up (Strader et al., 1994). Fertilization should be avoided. Although one study found that planted baldcypress grew much faster after fertilization (Myers et al., 1995), it is not desirable to give planted trees too great an advantage if the objective is to create diverse forests (Allen, 1997). Other studies have found that fertilizer does not have an effect on planted seedlings (McLeod et al., 1994) or that it is an unnecessary expense (Haynes et al., 1993; Strader et al., 1994).

An alternative to broadcast seeding is planting, but gaps should be left between planted seedlings or saplings to encourage the establishment of other species (e.g., *Liquidambar styraciflua* and *Gleditsia* spp). Fewer trees should be planted ha^{-1} than are commonly planted in restoration projects. Although the planting of target species has been successful in creating stands of trees, it produces monocultures for the ensuing 40 to 50 years. Managers typically hope that new species will come in as old stands of planted mixed oak begin to break up, but modifications can be made in these stands to encourage other species to enter more quickly (Allen, 1997).

More diverse forested wetland can be restored through natural restoration after the reestablishment of flood-pulsed conditions. Even so, certain species are not likely to come back on their own, because of the high level of fragmentation in forests of the southeastern United States. Planting a wider variety of trees may help reestablish forests in this region, including species that are rarely planted such as *Arundinaria gigantea, Carya aquatica, Carya cordiformis, Carya ovata, Diosporos virginiana, Platanus occidentalis, Quercus lyrata, Quercus michauxii, Quercus lauriflora, Quercus obtusa,* and *Ulmus* spp. (other than *Ulmus americana*) (Allen, 1997). Virtually all of the herbs, shrubs, and vines of these forests have been ignored in the restoration of forested wetlands in the Southeast. The restoration of cypress swamps has been somewhat more successful than restoration of drier types of forested wetlands, perhaps because more attention has been paid to hydrology in these forest types. However, the monospecific stands of oaks that have been restored are a far cry from the diverse stands of forested wetlands that once existed in the southeastern United States. Until recently, little attention has been paid to the importance of flood pulsing in the restoration of riverine wetlands, but fortunately there is evidence that great emphasis will be given to this critical process in the future.

REFERENCES

Ahlgren, C. E., and H. L. Hansen. 1957. Some effects of temporary flooding on coniferous trees. *Journal of Forestry* **55:**647–659.

Allen, J. A. 1990. Establishment of bottomland oak plantations on the Yazoo National Wildlife Refuge Complex. *Southern Journal of Applied Ecology* **14:** 206–210.

Allen, J. A. 1997. Reforestation of bottomland hardwoods and the issue of woody species diversity. *Restoration Ecology* **5:**125–134.

Allen, J. A., and H. E. Kennedy Jr. 1989. *Bottomland Hardwood Reforestation in the Lower Mississippi Valley.* U.S. Fish and Wildlife Service, Slidell, LA, and U.S. Forest Service Southern Experiment Station, United States Department of Agriculture, Stoneville, MS.

Allen, J. A., J. McCoy, and B. D. Keeland. 1998. Natural establishment of woody species on abandoned agricultural fields in the Lower Mississippi Valley: First and second year results. In *Proceedings of the Ninth Biennial Southern Silvicultural Research Conference,* ed. by T. A. Waldrup, pp. 263–268. U.S. Department of Agriculture, Forest Service, Southern Research Station, Asheville, NC.

Allan, J. D., and A. S. Flecker. 1993. Biodiversity conservation in running waters. *BioScience* **43:**32–43.

Anderson, R. C., and J. White. 1970. A cypress swamp outlier in southern Illinois. *Transactions of the Illinois Academy of Science* **63:**6–13.

Applequist, M. B. 1959. *Longevity of Submerged Tupelogum and Baldcypress Seed.* Report Note #27. Louisiana State University, Baton Rouge, LA.

Arkansas State Highway and Transportation Department. 1997. *Report on the Brushy Lake Mitigation Bank Site (BLMBS).* Arkansas State Highway and Transportation Department, Little Rock, AK.

Baker, J. B. 1977. Tolerance of planted hardwoods to spring flooding, *Southern Journal of Applied Forestry* **1:** 23–25.

Bakker, J. P., P. Poschlod, R. J. Strykstra, R. M. Bekker, and K. Thompson. 1996. Seed banks and seed dispersal: Important topics in restoration ecology. *Acta Botanica Neerlandica* **45:**461–490.

Barber, R. T., W. W. Kirby-Smith, and P. E. Parsley. 1978. Wetland alterations for agriculture. In *Wetlands Functions and Values: The State of Our Understand-*

ing; Proceedings of the National Symposium on Wetlands, ed. by P. E. Greeson, J. E. Clark, and J. R. Clark, pp. 642–651. American Water Resources Association, Washington, DC.

Barrat-Segretain, M. H. 1996. Strategies of reproduction, dispersion, and competition in river plants: A review. *Vegetatio* **123:**13–37.

Baskin, C. C., and J. M. Baskin. 1998. *Seeds: Ecology, Biogeography, and Evolution of Dormancy and Germination.* Academic Press, San Diego, CA.

Baskin, C. C., J. M. Baskin, and E. W. Chester. 1993. Seed germination ecophysiology of four summer annual mudflat species of Cyperaceae. *Aquatic Botany* **45:**41–52.

Battaglia, L. 1991. Early secondary succession in a bottomland hardwood area, Master's thesis, Northeast Louisiana University, Monroe, LA.

Battaglia, L. L., J. R. Keough, and D. W. Pritchett. 1995. Early secondary succession in a southeastern U.S. alluvial floodplain. *Journal of Vegetation Science* **6:**769–776.

Bayley, P. B. 1995. Understanding large river-floodplain ecosystems. *BioScience* **45:**153–158.

Bazzaz, F. A. 1968. Succession on abandoned fields in the Shawnee Hills, southern Illinois. *Ecology* **49:**924–936.

Bedinger, M. S. 1978. Relation between forest species and flooding. In *Wetland Functions and Values: The State of Our Understanding;* Proceedings of the National Symposium on Wetlands, ed. by P. E. Greeson, J. R. Clark, and J. E. Clark, pp. 427–435. American Water Resources Association, Washington, DC.

Bell, D. T., and F. L. Johnson. 1974. Flood caused tree mortality around Illinois reservoirs. *Transactions of the Illinois State Academy of Science* **67:**28–67.

Benke, A. C., I. Chaubey, G. M. Ward, and E. L. Dunn. 2000. Flood pulse dynamics in an unregulated river floodplain in the southeastern U.S. coastal plain. *Ecology* **81:**2730–2741.

Bennett, D., B. Middleton, S. Kraft, C. Lant, R. Sengupta, J. Beaulieau, D. Sharpe, K. Cook, R. Burr, R. Beck, and K. Flanagan. 2001. Ecosystem function and restoration in the Cache River Bioreserve: Final Report to the Southern Illinois Field Office of the Nature Conservancy, Ullin, Illinois, Southern Illinois University, Carbondale, IL.

Bonck, J., and W. T. Penfound. 1945. Plant succession on abandoned farm land in the vicinity of New Orleans, Louisiana. *American Midland Naturalist* **33:** 520–529.

Brewer, J. S. 1998. Patterns of plant species richness in a wet slash-pine *(Pinus elliottii)* savanna. *Journal of the Torrey Botanical Society* **125:**216–224.

Brinson, M. 1990. Riverine forests. In *Ecosystems of the World 15: Forested Wetlands,* ed. by A. E. Lugo, M. Brinson, and S. Brown, pp. 87–141. Elsevier, Amsterdam, The Netherlands.

Brinson, M. M., H. D. Bradshaw, R. N. Holmes, and J. B. Elkins Jr. 1980. Litterfall, stemflow, and throughfall nutrient fluxes in an alluvial swamp forest. *Ecology* **61:**827–835.

Briscoe, C. B. 1961. Germination of cherrybark and nuttall oak acorns following flooding. *Ecology* **42:**430–431.

Broadfoot, W. M. 1967. Shallow-water impoundment increases soil moisture and growth of hardwoods, Soil Scientists of America Proceedings **31:** 562–564.

Broadfoot, W. M., and H. L. Williston. 1973. Flooding effects on southern forests. *Journal of Forestry* **71:**584–587.

Brookes, A., J. Baker, and C. Redmond. 1996. Floodplain restoration and riparian zone management. In *River Channel Restoration: Guiding Principles for Sustainable Projects,* ed. by A. Brookes and F. D. Shields Jr., pp. 201–229. John Wiley & Sons, Chichester, U.K.

Broome, S. W., I. A. Mendelssohn, and K. L. McKee. 1995. Relative growth of *Spartina patens* (Ait.) Muhl. and *Scirpus olneyi* Gray occurring in a mixed stand as affected by salinity and flooding depth. *Wetlands* **15:**20–30.

Brown, S. 1981. A comparison of the structure, primary productivity, and transpiration of cypress ecosystems in Florida. *Ecological Monographs* **51:**403–427.

Brown, S. L., E. W. Flohrschutz, and H. T. Odum. 1984. Structure, productivity, and phosphorus cycling of the scrub cypress ecosystems, In *Cypress Swamps,* ed. by K. C. Ewel, and H. T. Odum, pp. 304–317, University Presses of Florida, Gainesville, FL.

Brunk, E. L., A. R. Allmon, and G. P. Dellinger. 1975. *Mortality of Trees Caused by Flooding During the Growing Season at Two Midwest Reservoirs.* Missouri Department of Conservation, Jefferson City, MO.

Burbrink, F. T., C. A. Phillips, and E. J. Heske. 1998. A riparian zone in southern Illinois as a potential dispersal corridor for reptiles and amphibians. *Biological Conservation* **86:**107–115.

Burns, L. A. 1984. Productivity and water relations in the Fakahatchee Strand of South Florida. In *Cypress Swamps,* ed. by K. C. Ewel and H. T. Odum, pp. 318–333. University Presses of Florida, Gainesville, FL.

Cairns, J. Jr., and S. E. Palmer. 1993. Senescent reservoirs and ecological restoration: An overdue reality check. *Restoration Ecology* **1:**212–219.

Cellot, B., F. Mouillot, and C. P. Henry. 1998. Flood drift and propagule bank of aquatic macrophytes in a riverine wetland. *Journal of Vegetation Science* **9:** 631–640.

Clewell, A. F., and R. Lea. 1990. Creation and restoration of forested wetland vegetation in the southeastern United States. In *Wetland Creation and Restoration: The Status of the Science,* ed. by J. A. Kusler and M. E. Kentula, pp. 195–231. Island Press, Washington, DC.

Conner, W. H., and M. A. Buford. 1998. Southern deepwater swamps. In *Southern Forested Wetlands,* ed. by M. G. Messina and W. H. Conner, pp. 261–287. CRC Press, Boca Raton, FL.

Conner, W. H., and J. W. Day Jr. 1976. Productivity and composition of a baldcypress-water tupelo site and a bottomland hardwood site in a Louisiana swamp. *American Journal of Botany* **63:**1354–1364.

Conner, W. H., and J. W. Day Jr. 1989. Response of coastal wetland forest to human and natural changes in environment with emphasis on hydrology. In *Proceedings of the Symposium on the Forested Wetlands of the Southern United States,* ed. by D. D. Hook and R. Lea, pp. 34–43. U.S.D.A. Forest Service, Southeast Experiment Station, Asheville, NC.

Conner, W. H., and J. W. Day Jr. 1992. Water level variability and litterfall productivity of forested freshwater wetlands in Louisiana. *American Midland Naturalist* **128:**237–245.

Conner, W. H., and K. Flynn. 1989. Growth and survival of baldcypress (*Taxodium distichum* (L.) Rich.) planted across a flooding gradient in a Louisiana bottomland forest. *Wetlands* **9:**207–217.

Conner, W. H., J. G. Gosselink, and R. T. Parrondo. 1981. Comparison of the vegetation of three Louisiana swamp sites with different flooding regimes. *American Journal of Botany* **68:**320–331.

Conner, W. H., J. W. Day, Jr., and W. R. Slater. 1993. Bottomland hardwood productivity: case study in a rapidly subsiding, Louisiana, USA watershed, *Wetland Ecology and Management* **2:**189–197.

Conrad, S. T. 1997. Reproductive ecology and diaspore bank of the bryophytes of a bald cypress swamp. Master's thesis, Southern Illinois University, Carbondale, IL.

Conti, R. S., and P. P. Gunther. 1984. Relations of phenology and seed germination to the distribution of dominant plants in Okefenokee Swamp. In *The Okefenokee Swamp: Its Natural History, Geology, and Geochemistry,* ed. by A. D. Cohen, D. J. Casagrande, M. J. Andrejko, and G. R. Best, pp. 144–167. Wetland Surveys, Los Alamos, NM.

Cuffney, T. F. 1988. Input, movement and exchange of organic matter within a subtropical coastal blackwater river-floodplain system. *Freshwater Biology* **19:**305–320.

Dahl, T. E., and D. E. Nomsen. 1987. Seed bank of a drained prairie wetland. In *Proceedings of the Eighth Annual Meeting of the Society of Wetland Scientists: Wetland and Riparian Ecosystems of the American West,* Society of Wetland Scientists Lawrence, KS.

Danvind, M., and C. Nilsson. 1997. Seed floating ability and distribution of alpine plants along a northern Swedish river. *Journal of Vegetation Science* **8:** 271–276.

Davies, B. R., M. Thoms, and M. Meador. 1992. An assessment of the ecological impacts of inter-basin water transfers and their threats to river basin integrity and conservation. *Aquatic Conservation: Marine and Freshwater Ecosystems* **2:**325–349.

Day, F. P. Jr. 1979. Litter accumulation in four plant communities in the Dismal Swamp, Virginia. *American Midland Naturalist* **102:**281–289.

Day, F. P. Jr., T. J. Butler, and W. H. Conner. 1977. Productivity and nutrient export studies in a cypress swamp and lake system in Louisiana, In *Estuarine Processes,* Vol. 2, ed. by M. Wiley, pp. 255–269, Academic Press, New York.

DeBell, D. S., and I. D. Auld. 1971. *Establishment of Swamp Tupelo Seedlings After Regeneration Cuts,* U.S.D.A. Forest Service, Washington, DC. USDA Forest Service Research Note SE-164: 1–7.

Deghi, G. S., K. C. Ewel, and W. J. Mitsch. 1980. Effects of sewage effluent application on litter fall and litter decomposition in cypress swamps. *Journal of Applied Ecology* **17:**397–408.

Demaree, D. 1932. Submerging experiments with *Taxodium. Ecology* **13:** 258–262.

Demissie, M., T. W. Soong, R. Allgire, L. Keefer, and P. Makowski. 1990. *Cache River Basin: Hydrology, Hydraulics, and Sediment Transport.* Vol. 1. Illinois State Water Survey, Champaign, IL.

Denyer, W. B. G., and C. G. Riley. 1964. Dieback and mortality of tamarack caused by high water, *Forestry Chronicles* **40:**334–338.

DeShield, M. A. Jr., M. R. Reddy, S. Leonard, N. H. Assar, and W. T. Brown. 1994. Inundation tolerance of riparian plant species. *Wetland Journal* **6:**20–21.

De Steven, D., and R. R. Sharitz. 1997. Differential recovery of a deepwater swamp forest across a gradient of disturbance intensity. *Wetlands* **17:**476–484.

Dicke, S. G., and J. R. Toliver. 1990. Growth and development of bald-cypress/water-tupelo stands under continuous versus seasonal flooding. *Forest Ecology and Management* **33/34:**523–530.

Dickson, R. E., J. F. Hosner, and N. W. Hosley. 1965. The effects of four water regimes upon the growth of four bottomland tree species, *Forest Science* **11:**299–305.

Dorge, C. L., W. J. Mitsch, and J. R. Wiemhoff. 1984. Cypress wetlands in southern Illinois. In *Cypress Swamps.* ed. by K. C. Ewel and H. T. Odum, pp. 393–404. University of Florida Presses, Gainesville, FL.

Duever, M. J. 1984. Environmental factors controlling plant communities of the Big Cypress Swamp. In *Environments of South Florida: Past and Present,* ed. by P. J. Gleason, pp. 127–137. Miami Geological Society, Miami, FL.

Duever, M. J., and J. M. McCollom. 1986. Cypress tree-ring analysis in relation to wetlands and hydrology. In *Proceedings of the International Symposium on Ecological Aspects of Tree-Ring Analysis,* pp. 249–260. U.S. Department of Energy, Washington, DC.

Dulohery, N. J., C. S. Bunton, C. C. Trettin, and W. H. McKee Jr. 1995. *Reforestation, Monitoring, and Research at Pen Branch: Restoring a Thermally-Impacted Wetland Forest, Charleston, South Carolina.* Report FS-6200-7. Center for Forested Wetlands Research, U.S.D.A. Forest Service, Charleston, S.C.

Dunn, C. P., and R. R. Sharitz. 1987. Revegetation of a *Taxodium-Nyssa* forested wetland following complete vegetation destruction. *Vegetatio* **72:**151–157.

Dynesius, M., and C. Nilsson. 1994. Fragmentation and flow regulation of river systems in the northern third of the world. *Science* **266:**753–762.

Eggers, S. D. 1992. *Compensatory Wetland Mitigation: Some Problems and Suggestions for Corrective Measures, February, 1992.* U.S. Army Corps of Engineers, St. Paul, MN.

Eggler, W. A., and W. C. Moore. 1961. The vegetation of Lake Chicot, Louisiana, after eighteen years of impoundment. *Southwestern Naturalist* **6:**175–183.

Elder, J. F., and D. J. Cairns. 1982. *Production and Decomposition of Forest Litter Fall on the Apalachicola River Flood Plain, Florida.* U.S. Geological Survey Water-Supply Paper 2196-B. U.S. Department of the Interior, Washington, DC.

Ewel, K. C. 1990. Swamps. In *Ecosystems of Florida,* ed. by R. L. Myers and J. J. Ewel, pp. 281–323. University of Central Florida Press, Orlando, FL.

Faber-Langendoen, D., and P. F. Maycock. 1989. Community patterns and environmental gradients of buttonbush, *Cephalanthus occidentalis,* ponds in lowland forests of southern Ontario. *The Canadian Field-Naturalist* **103:**479–485.

Francis, J. K. 1983. *Acorn Production and Tree Growth of Nuttall Oak in a Green-Tree Reservoir.* Research Note SO-289. USDA Forest Service, New Orleans, LA.

Fredrickson, L. H., and D. L. Batema. 1992. Greentree reservoir management strategies. In *Greentree Reservoir Management Handbook,* ed. by L. H. Fredrickson and D. L. Batema, pp. 41–45. University of Missouri-Columbia, Puxico, MO.

Galatowitsch, W. M., and A. G. van der Valk. 1996. Vegetation and environmental conditions in recently restored wetland in the prairie pothole region of the USA. *Vegetatio* **126:**89–99.

Giedeman, C. 1999. Restoration of bottomland hardwood forests in the American Bottoms of the Mississippi River near St. Louis, Carbondale, IL. Ph.D. diss., Southern Illinois University, Carbondale, IL.

Glenn, E., T. L. Thompson, R. Frye, J. Riley, and D. Baumgartner. 1995. Effects of salinity on growth and evapotranspiration of *Typha domingensis* Pers. *Aquatic Botany* **52:**75–91.

Gomez, M. M., and F. P. Day Jr. 1982. Litter nutrient content and production in the Great Dismal Swamp. *American Journal of Botany* **69:**1314–1321.

Good, B. J., and W. H. Patrick Jr. 1987. Gas composition and respiration of water oak (*Quercus nigra* L.) and green ash (*Fraxinus pennsylvanica* Marsh.) roots after prolonged flooding. *Plant and Soil* **97:**419–427.

Green, W. E. 1947. Effect of water impoundment on tree mortality and growth. *Journal of Forestry* **45:**118–120.

Guilkey, P. C. 1965. American elm (*Ulmus americana* L.). In *Silvics of Forest Trees of the United States,* ed. by H. A. Fowells, pp. 725–731. Agricultural Handbook 271. USDA Forest Service, Washington, DC.

Gunderson, L. H. 1984. Regeneration of cypress in logged and burned strands at Corkscrew Swamp Sanctuary, Florida. In *Cypress Swamps,* ed. by K. C. Ewel and H. T. Odum, pp. 349–357. University Presses of Florida, Gainesville, FL.

Gustafson, D. J. 2000. Genetic and competition studies of *Andropogon gerardii, Sorghastrum nutans,* and *Dalea purpurea* from remnant and restored tallgrass prairie. Ph.D. diss., Southern Illinois University, Carbondale, IL.

Gustafson, D. J., D. J. Gibson, and D. L. Nickrent. 1999. Random amplified polymorphic DNA variation amoung remnant big bluestem (*Andropogon gerardii* Vitman) populations from Arkansas' Grand Prairie. *Molecular Ecology* **8:** 1693–1701.

Hall, R. B. W., and P. A. Harcombe. 1998. Flooding alters apparent position of floodplain saplings on a light gradient. *Ecology* **79:**847–855.

Hall, T. F., and G. E. Smith. 1955. Effects of flooding on woody plants, West Sandy Dewatering Project, Kentucky Reservoir. *Journal of Forestry* **53:** 281–285.

Haltiner, J., J. B. Zedler, K. E. Boyer, G. D. Williams, and J. C. Callaway. 1997. Influence of physical processes on the design, functioning and evolution of restored tidal wetlands in California (USA). *Wetlands Ecology and Management* **4:**73–91.

Harms, T. F., H. T. Schreuder, D. D. Hook, C. L. Brown, and F. W. Shropshire. 1980. The effects of flooding on the swamp forest in Lake Ocklawaha, Florida. *Ecology* **61:**1412–1421.

Harper, J. L. 1977. *Population Biology of Plants.* Academic Press, London.

Harris, L. D. 1989. The faunal significance of fragmentation of southeastern bottomland forests. In *The Forested Wetland of the Southern United States: Proceedings of a Symposium.* ed. by D. D. Hook and R. Lea, pp. 126–134. USDA, Forest Service, Southeastern Forest Experiment Station, Asheville, NC.

Harris, R. W., A. T. Leiser, and R. E. Fissell. 1975. *Plant Tolerance to Flooding.* RWH-200-7/1/75. Department of Environmental Horticulture, University of California, Davis, CA.

Haynes, R. J., and L. Moore. 1988. Reestablishment of bottomland hardwoods within national wildlife refuges in the Southeast. In *Proceedings of a Conference: Increasing Our Wetland Resources,* ed. by J. Zelazny and J. S. Feierabane, pp. 95–103. National Wildlife Federation, Washington, DC.

Haynes, R. J., R. J. Bridges, S. W. Gard, T. M. Wilkins, and H. R. Cook Jr. 1993. Bottomland forest reestablishment efforts of the U.S. Fish and Wildlife Service: Southeast Region. In *Proceedings: National Wetlands Engineering Workshop,* pp. 322–334. U.S. Army Corps of Engineers Waterways Experiment Station, St. Louis, MO.

Heitmeyer, M. E., L. H. Fredrickson, and G. F. Krause. 1989. *Water and Habitat Dynamics of the Mingo Swamp in Southeastern Missouri.* Fish and Wildlife Research 67, U.S. Department of the Interior, Washington, DC.

Hellings, S. E., and J. L. Gallagher. 1992. The effects of salinity and flooding on *Phragmites australis. Journal of Applied Ecology* **29:**41–49.

Hesse, L. W. 1995. Water allocation for ecosystem management of the Missouri River. *Regulated Rivers: Research and Management* **11:**299–311.

Hill, M. J., E. A. Long, and S. Hardin. 1994. Effects of dam removal on Dead Lake, Chipola River, Florida. In *Proceedings of the 48th Annual Conference of Southeastern Association of Fish and Wildlife Agencies* **48:**512–523, Tallahassee, FL.

Hook, D. D. 1984. Waterlogging tolerance of lowland tree species of the South. *Southern Journal of Applied Forestry* **8:**136–148.

Hook, D. D., C. L. Brown, and P. P. Kormanik. 1970. Lenticel and water root development of swamp tupelo under various flooding conditions, *Botanical Gazette* **131:**217–224.

Hopkins, W. E., and R. E. Wilson. 1974. Early oldfield succession on bottomlands of southeastern Indiana. *Castanea* **39:**57–71.

Hosner, J. F. 1958. The effects of complete inundation upon seedlings of six bottomland tree species, *Ecology* **39:**371–374.

Hosner, J. F. 1960. Relative tolerance to complete inundation of fourteen bottomland tree species. *Forest Science* **6:**246–251.

Hosner, J. F., and S. G. Boyce. 1962. Tolerance to water saturated soil of various bottomland hardwoods. *Forest Science* **8:**180–186.

Hosner, J. F., and L. S. Minckler. 1963. Bottomland hardwood forests of southern Illinois—Regeneration and succession. *Ecology* **44:**29–41.

Houhoulis, P. F., and W. K. Michener. 2000. Detecting wetland change: A rule-based approach using NWI and SPOT-SX data. *Photogrammetric Engineering and Remote Sensing* **66:**205–211.

Houle, G. 1998. Seed dispersal and seedling recruitment of *Betula alleghaniensis:* Spatial inconsistency in time. *Ecology* **79:**807–818.

Huenneke, L. F., and R. R. Sharitz. 1990. Substrate heterogeneity and regeneration of a swamp tree, *Nyssa aquatica. American Journal of Botany* **77:** 413–419.

Humburg, D. D., D. A. Graber, S. P. Havera, L. H. Fredrickson, and D. L. Helmers. 1996. What did we learn from the Great Flood of 1993? ed. by J. T. Ratti. In *The 7th International Waterfowl Symposium,* pp. 139–148. Institute for Wetland and Waterfowl Research, Ducks Unlimited, Stonewall, Manitoba, Canada.

Jansson, R., C. Nilsson, and B. Renöfält. 2000a. Fragmentation of riparian floras in rivers with multiple dams. *Ecology* **81:**899–903.

Jansson, R., C. Nilsson, M. Dynesius, and E. Andersson. 2000b. Effects of river regulation on river-margin vegetation: A comparison of eight boreal rivers. *Ecological Applications* **10:**203–224.

Jones, R. H., R. R. Sharitz, P. M. Dixon, D. S. Segal, and R. L. Schneider. 1994. Woody plant regeneration in four floodplain forests. *Ecological Monographs* **64:**345–367.

Junk, W. J. 1997. Structure and function of the large central Amazonian River floodplains: Synthesis and discussion. In *The Central Amazonian Floodplain,* ed. by W. J. Junk, pp. 455–472. Springer-Verlag, Berlin, Germany.

Junk, W. J., and M. T. F. Piedade. 1993. Herbaceous plants of the Amazon floodplain near Manaus: Species diversity and adaptations to the flood pulse. *Amazoniana* **12:**467–484.

Junk, W. J., P. B. Bayley, and R. E. Sparks. 1989. The flood pulse concept in river-floodplain systems. In *Proceedings of the International Large River Symposium (LARS),* ed. by D. P. Dodge, pp. 110–127. Canadian Special Publication of Fisheries and Aquatic Sciences 106. Department of Fisheries and Oceans, Ottawa, Canada.

Keeland, B. D., and W. H. Conner. 1999. Natural regeneration and growth of *Taxodium distichum* (L.) Rich. in Lake Chicot, Louisiana, after 44 years of flooding. *Wetlands* **19:**149–155.

Keeland, B. D., and R. R. Sharitz. 1995. Seasonal growth patterns of *Nyssa sylvatica* var. *biflora, Nyssa aquatica,* and *Taxodium distichum* as affected by hydrologic regime. *Canadian Journal of Forest Research* **25:**1084–1096.

Kennedy, H. E. 1970. Growth of newly planted water tupelo seedlings after flooding and siltation. *Forest Science* **16:**250–256.

Kennedy, H. E., and R. M. Krinard. 1974. *1973 Mississippi River Flood's Impact on Natural Hardwood Forests and Plantations.* SO-177. U.S. Forest Service, New Orleans, LA.

King, S. L., and J. A. Allen. 1996. Plant succession and greentree reservoir management: Implications for management and restoration of bottomland hardwood wetlands. *Wetlands* **16:**503–511.

King, S. L., and B. D. Keeland. 1999. Evaluation of reforestation in the Lower Mississippi River Alluvial Valley. *Restoration Ecology* **7:**348–359.

Klimas, C. V. 1988. River regulation effects on floodplain hydrology and ecology. In *The Ecology and Management of Wetlands.* Vol. 1. *Ecology of Wetlands,* ed. by D. D. Hook, W. H. McKee Jr., H. K. Smith, J. Gregory, V. G. Burrell Jr., M. R. DeVoe, R. E. Sojka, S. Gilbert, R. Banks, L. H. Stolzy, C. Brooks, T. D., Mathews, and T. H. Shear, pp. 40–49. Timber Press, Portland, OR.

Klimas, C. V., C. O. Martin, and J. W. Teaford. 1981. *Impacts of Flooding Regime Modification on Wildlife Habitats of Bottomland Hardwood Forests in the Lower Mississippi Valley.* Technical Report EL-81-13. U.S. Army Corps of Engineers, Vicksburg, MS.

Kondolf, G. M. 1996. A cross section of stream channel restoration. *Journal of Soil and Water Conservation* **(March-April):**119–125.

Kozel, T. R., J. Dailey, J. Craig, and K. Welborn. 1988. Some stand characteristics of baldcypress *Taxodium distichum* (L.) in an oxbow lake in extreme southwestern Indiana. *Transactions of the Kentucky Academy of Science* **49:**74–79.

Lessmann, J. M., I. A. Mendelssohn, M. W. Hester, and K. L. McKee. 1997. Population variation in growth response to flooding of three marsh grasses. *Ecological Engineering* **8:**31–47.

Llewellyn, D. W., F. P. Shaffer, N. J. Craig, L. Creasman, D. Pashley, M. Swan, and C. Brown. 1996. A decision-support system for prioritizing restoration sites on the Mississippi River Alluvial Plain. *Conservation Biology* **10:** 1446–1455.

Loftus, T. T. 1994. Status and assessment of *Taxodium distichum* (L.) Rich. and *Nyssa aquatica* L. in Horseshoe Lake, Alexander County, Illinois: Phase One—Baseline Study. Master's thesis, Southern Illinois University, Carbondale, IL.

Lugo, A. E., J. K. Nessel, and T. M. Hanlon. 1984. Root distribution in a north-central Florida cypress strand. In *Cypress Swamps,* ed. by K. C. Ewel and H. T. Odum, pp. 279–285. University Presses of Florida, Gainesville, FL.

MacDonáld, P. O., W. E. Frayer, and J. K. Clauser. 1979. *Documentation, Chronology, and Future Projections of Bottomland Hardwood Habitat Loss in the Lower Misssissippi Alluvial Plain.* U.S. Fish and Wildlife Service, Vicksburg, MS.

Mathis, M. 2001. Deer herbivory and old field succession. Ph.D. diss., Southern Illinois University, Carbondale, IL.

Mathis, M., and B. A. Middleton. 1999. Simulated herbivory and vegetation dynamics in coal slurry ponds reclaimed as wetlands. *Restoration Ecology* **7:** 392–398.

Mattoon, W. R. 1915. *The Southern Cypress.* Bulletin No. 272. United States Department of Agriculture, Washington, DC.

Mattoon, W. R. 1916. Water requirements and growth of young cypress. In *Proceedings of the Society of American Foresters,* pp. 192–197. Society of American Foresters, Washington, DC.

McDermott, R. E. 1954. Effects of saturated soil on seedling growth of some bottomland hardwood species. *Ecology* **35:**36–41.

McKenney, D. W., B. G. Mackey, and D. Joyce. 1999. Seedwhere: A computer tool to support seed transfer and ecological restoration decisions. *Environmental Modelling & Software* **14:**589–595.

McKnight, J. S., D. D. Hook, O. G. Langdon, and R. L. Johnson. 1981. Flood tolerance and related characteristics of trees of the bottomland forest of the southern United States. In *Wetlands of Bottomland Hardwood Forests, Proceedings of a Workshop on Bottomland Hardwood Wetlands of the Southeastern United*

States, ed. by J. R. Clark and J. Benforado, pp. 29–69. Elsevier Scientific Publishing Company, Amsterdam, The Netherlands.

McLeod, K. W., M. R. Reed, and T. G. Ciravolo. 1994. Selection of woody species for bottomland restoration. In *Proceedings of the 21st Annual Conference on Wetlands Restoration and Creation,* pp. 106–118. Institute of Florida Studies, Tampa, FL.

McLeod, K. W., M. R. Reed, and L. D. Wike. 2000. Elevation, competition control, and species affect bottomland forest restoration. *Wetlands* **20:**162–168.

Megonigal, J. P., W. H. Conner, S. Kroeger, and R. R. Sharitz. 1997. Aboveground production in southeastern floodplain forests: A test of the subsidy-stress hypothesis. *Ecology* **78:**370–384.

Middleton, B. A. 1995a. *The Role of Flooding in Seed Dispersal: Restoration of Cypress Swamps Along the Cache River, Illinois.* U.S. Geological Survey and Water Resources Center, Champaign, IL.

Middleton, B. A. 1995b. Sampling devices for the measurement of seed rain and hydrochory in rivers. *Bulletin of the Torrey Botanical Club* **122:**152–155.

Middleton, B. A. 1995c. Seed banks and species richness potential of coal slurry ponds reclaimed as wetlands. *Restoration Ecology* **3:**311–318.

Middleton, B. A. 1996. Characteristics of plants in forested wetlands. In *Management of Forested Wetlands in the Central Hardwoods Region,* ed. by S. D. Roberts and R. A. Rathfon, pp. 31–37. Purdue University, West Lafayette, IN.

Middleton, B. A. 1999. *Wetland Restoration, Flood Pulsing and Disturbance Dynamics.* John Wiley & Sons, New York.

Middleton, B. A. 2000. Hydrochory, seed banks, and regeneration dynamics along the landscape boundaries of a forested wetland. *Plant Ecology* **146:**169–184.

Mitsch, W. J., J. R. Taylor, and K. B. Benson. 1991. Estimating primary productivity of forested wetland communities in different hydrologic landscapes. *Landscape Ecology* **5:**75–92.

Morris, R. C. 1965. Overcup oak (*Quercus lyrata* Walt.). In *Silvics of Forest Trees of the United States,* ed. by H. A. Fowells, pp. 600–602. Agricultural Handbook 271. USDA Forest Service, Washington, DC.

Muir, D. B., R. L. Hite, M. M. King, and M. R. Matson. 1995. *An Intensive Survey of Cache River Basin.* Environmental Protection Agency, Marion, IL.

Muzika, R. M., J. B. Gladden, and J. D. Haddock. 1987. Structural and functional aspects of succession in southeastern floodplain forest following a major disturbance. *The American Midland Naturalist* **117:**1–9.

Myers, R. S., G. P. Shaffer, and D. W. Llewellyn. 1995. Baldcypress (*Taxodium distichum* (L.) Rich.) restoration in southeast Louisiana: The relative effects of herbivory, flooding, competition, and macronutrients. *Wetlands* **15:**141–148.

National Research Council. 1992. *Restoration of Aquatic Ecosystems.* National Academy Press, Washington, DC.

Nelson, B. 2000. Bush: Restore Ocklawaha. *Palatka* (Florida) *Daily News,* 17 July. <http:/www.palatkadailynews.com>.

Newling, C. J. 1990. Restoration of bottomland hardwood forests in the Lower Mississippi Valley. *Restoration and Management Notes* **8:**23–28.

Newman, S., J. B. Grace, and J. W. Koebel. 1996. Effects of nutrients and hydroperiod on *Typha, Cladium,* and *Eleocharis:* Implications for Everglades restoration. *Ecological Applications* **6:**774–783.

Nielson, D. L., and A. J. Chick. 1997. Flood-mediated changes in aquatic macrophyte community structure. *Marine Freshwater Research* **48:**153–157.

Nilsson, C., M. Gardfjell, and G. Grelsson. 1991. Importance of hydrochory in structuring plant communities along rivers. *Canadian Journal of Botany* **69:**2631–2633.

Nilsson, C., R. Jansson, and U. Zinko. 1997. Long-term responses of river-margin vegetation to water-level regulation. *Science* **276:**798–800.

Noble, R. E., and P. K. Murphy. 1975. Short term effects of prolonged backwater flooding on understory vegetation. *Castanea* **40:**22–238.

Nohara, S., and M. Kimura. 1997. Growth characteristics of *Nelumbo nucifera* Gaertn. in response to water depth and flooding. *Ecological Research* **12:** 11–20.

Osment-DeLoach, J., and P. Moore. 1996. Brushy Lake mitigation bank site. In *Proceedings: The Delta: Connecting Points of View for Sustainable Natural Resources,* pp. 297-301. US Army Corps of Engineers, Memphis, TN.

Ouchley, K., R. B. Hamilton, W. C. Barrow Jr., and K. Ouchley. 2000. Historic and present-day forest conditions: Implications for bottomland hardwood forest restoration. *Ecological Restoration* **18:**21–25.

Patrick, W. H. Jr., G. Dissmeyer, D. D. Hook, V. W. Lambou, H. M. Leitman, and C. H. Wharton. 1981. Characteristics of wetlands ecosystems of southeastern bottomland hardwood forests. In *Proceedings of the Workshop on Bottomland Hardwood Forest Wetlands of Southeastern U.S.,* ed. by J. R. Clark and J. Benforado, pp. 276–300. Elsevier, Lake Lanier, GA.

Pearlstine, L., H. McKellar, and W. Kitchens. 1985. Modeling the impact of a river diversion on bottomland forest communities in the Santee River floodplain, South Carolina. *Ecological Modelling* **29:**283–302.

Penfound, W. T. 1949. Vegetation of Lake Chicot, Louisiana, in relation to wildlife resources. *Proceedings of the Louisiana Academy of Science* **12:** 47–56.

Petts, G. E. 1984. *Impounded Rivers: Perspectives for Ecological Management.* John Wiley & Sons, Chichester, U.K.

Petts, G. E., A. R. G. Large, M. T. Greenwood, and M. A. Bickerton. 1992. Floodplain assessment for restoration and conservation: Linking hydrogeomorphology and ecology. In *Lowland Floodplain Rivers: Geomorphological*

Perspectives, ed. by P. A. Carling and G. E. Petts, pp. 217–234. John Wiley & Sons, Chichester, U.K.

Phillips, D. L. 1996. Plant species distribution of seep springs in the Cache River region of southern Illinois. Master's thesis, Southern Illinois University, Carbondale, IL.

Putnam, J. A., G. M. Furnival, and J. S. McKnight. 1960. *Management and Inventory of Southern Hardwoods.* USDA Agricultural Handbook 181. USDA, Washington, DC.

Rathborne, J. C. 1951. Cypress reforestation. *Southern Lumberman* **183:**239–240.

Rayner, D. A. 1976. A monograph concerning the water elm *Planera aquatica* (Walt.) J.F. Gmelin (Ulmaceae). Ph.D. diss., Department of Biology, University of South Carolina, Aiken, SC.

Richardson, C. J., and J. W. Gibbons. 1993. Biodiversity of the southeastern United States: lowland terrestrial communities, In *Biodiversity of the Southeastern United States: Lowland Terrestrial Communities,* ed. by W. H. Martin, S. G. Boyce, and A. C. Echternacht, pp. 257–310, John Wiley and Sons, New York.

Ridley, H. N. 1930. *The Dispersal of Plants Throughout the World.* L. Reeve & Co., Ashford, Kent, U.K.

Robinson, S. K., and J. P. Hoover. 1995. *Effects of Landscape Fragmentation on Migrant Songbirds: Implications for Floodplain Management.* Project P-001-W. Illlinois Natural History Survey, Champaign, IL.

Robinson, S. K., F. M. Thompson III., T. M. Donovan, D. R. Whitehead, and J. Faaborg. 1995. Regional forest fragmentation and the nesting success of migratory birds. *Science* **267:**1987–1990.

Rudis, V. A. 1995. Regional forest fragmentation effects on bottomland hardwood community types and resource values. *Landscape Ecology* **10:**291–307.

Schlesinger, W. H. 1978. Community structure, dynamics and nutrient cycling in the Okefenokee cypress swamp-forest, *Ecological Monographs* **48:**43–65.

Schneider, R. L., and R. R. Sharitz. 1986. Seed bank dynamics in a southeastern riverine swamp. *American Journal of Botany* **73:**1022–1030.

Schneider, R. L., and R. R. Sharitz. 1988. Hydrochory and regeneration in a bald cypress-water tupelo swamp forest. *Ecology* **69:**1055–1063.

Schneider, R. L., N. E. Martin, and R. R. Sharitz. 1989. Impact of dam operations on hydrology and associated floodplain forests of southeastern rivers. In *Freshwater Wetlands and Wildlife,* ed. by R. R. Sharitz and J. W. Gibbons, pp. 1113–1122. U.S. Department of Energy, Office of Scientific and Technical Information, Oak Ridge, TN.

Sengupta, R. 1995. Fluvial sedimentology of Cypress Creek, Union, Johnson, and Pulaski Counties, southern Illinois. Master's Thesis, Southern Illinois University, Carbondale, IL.

Shankman, D. 1991. Forest regeneration on abandoned meanders of a Coastal Plain river in western Tennessee, Castanea **56:**157–167.

Shankman, D. 1993. Channel migration and vegetation patterns in the southeastern coastal plain. *Conservation Biology* **7:**176–183.

Shankman, D., and L. G. Drake. 1990. Channel migration and regeneration of bald cypress in western Tennessee. *Physical Geography* **11:**343–352.

Sharitz, R. R. 1992. Bottomland hardwood wetland restoration in the Mississippi drainage. In *Restoration of Aquatic Ecosystems,* pp. 496–505. National Academy Press, Washington, DC.

Sharitz, R. R., and L. C. Lee. 1985a. Limits on regenerative processes in southeastern riverine wetlands. In *Riparian Ecosystems and Their Management: Reconciling Conflicting Uses: The First North American Riparian Conference,* pp. 139–143. U.S.D.A. Forest Service, General Technical Report Rmizo, Washington, DC.

Sharitz, R. R., and L. C. Lee. 1985b. Recovery processes in southeastern riverine wetlands. In *Riparian Ecosystems and Their Management: Reconciling Conflicting Uses: The First North American Riparian Conference,* pp. 499–501. University of Arizona, Tuscon, AZ.

Sharitz, R. R., R. L. Schneider, and L. C. Lee. 1990. Composition and regeneration of a disturbed river floodplain forest in South Carolina. In *Ecological Processes and Cumulative Impacts: Illustrated by Bottomland Hardwood Wetland Ecosystems,* ed. by J. G. Gosselink, L. C. Lee, and T. A. Muir, pp. 195–218. Lewis Publishers, Chelsea, MI.

Shear, T., and H. R. Malcom. 1997. *Development of Guidelines for the Restoration of Forested Wetlands in North Carolina.* The Center for Transportation and the Environment, Raleigh, NC.

Shear, T. H., T. J. Lent, and S. Fraver. 1996. Comparison of restored and mature bottomland hardwood forests of southwestern Kentucky. *Restoration Ecology* **4:**111–123.

Shelford, V. E. 1954. Some lower Mississippi Valley flood plain biotic communities: their age and elevation, *Ecology* **35:**126–142.

Shields, F. D., Jr., and J. J. Hoover. 1991. Effects of channel restabilization on habitat diversity, Twentymile Creek, Mississippi. *Regulated Rivers: Research and Management* **6:**163–181.

Shuman, J. R. 1995. Environmental considerations for assessing dam removal alternatives for river restoration. *Regulated Rivers: Research and Management* **11:**249–261.

Simons, R. K. and D. B. Simons. 1991. Sediment problems associated with dam removal, Muskegon River, Michigan. In *Hydraulic Engineering: Proceedings of the 1991 National Conference,* ed. by R. M. Shane, pp. 680–691. American Society of Civil Engineers, Nashville, TN.

Smith, M., and J. S. Moss. 1998. An experimental investigation using stomatal conductance and fluorescence of the flood sensitivity of *Boltonia decurrens* and its competitors. *Journal of Applied Ecology* 35: 553–561.

Smock, L. A., and E. Gilinsky. 1992. Coastal plain blackwater rivers. In *Biodiversity of the Southeastern United States.* ed. by C. T. Hackney, S. M. Adams, and W. H. Martin, pp. 271–311. John Wiley & Sons, New York.

Somers, G. F., and D. Grant. 1981. Influence of seed source upon phenology of flowering of *Spartina alterniflora* Loisel. and the likelihood of cross pollination. *American Journal of Botany* **68:**6–9.

Sparks, R. E., P. B. Bayley, S. L. Kohler, and L. L. Osborne. 1990. Disturbance and recovery of large floodplain rivers. *Environmental Management* **14:** 699–709.

Stoecker, M. A., M. Smith, and E. D. Melton. 1995. Survival and aerenchyma development under flooded conditions of *Boltonia decurrens,* a threatened floodplains species, and *Conyza canadensis,* a widely distributed competitor. *American Midland Naturalist* **134:**117–126.

Strader, R. W., C. Stewart, J. Wessman, and B. Ray. 1994. *Bottomland Hardwood Reforestation Guidelines.* U.S. Fish and Wildlife Service, Southeast Region, Jackson, MS.

Streever, B., and E. Perkins. 2000. *Importing Plant Stock for Wetland Restoration and Creation: Maintaining Genetic Diversity and Integrity.* U.S. Army Corps of Engineers, Vicksburg, MS.

Streng, D. R., J. S. Glitzenstein, and P. A. Harcombe. 1989. Woody seedling dynamics in an East Texas floodplain forest. *Ecological Monographs* **59:** 177–204.

Theriot, R. F. 1993. *Flood Tolerance of Plant Species in Bottomland Forests of the Southeastern United States.* WRP-DE-6. U.S. Army Corps of Engineers, Vicksburg, MS.

Tofflemire, T. J. 1986. PCB transport in the Ft. Edward area. *Northeastern Environmental Science* **3:**202–208.

Tolliver, K. S., D. W. Martin, and D. R. Young. 1997. Freshwater and saltwater flooding response for woody species common to barrier island swales, *Wetlands* **17:**10–18.

Trepagnier, C. M., M. A. Kogas, and R. E. Turner. 1995. Evaluation of wetland gain and loss of abandoned agricultural impoundments in South Louisiana, 1978–1988. *Restoration Ecology* **3:**299–303.

U.S. Army Corps of Engineers. 1996. *Cooperative banking instrument, Memphis District Corps of Engineers and Arkansas State Highway and Transportation Department Brushy Lake Mitigation Bank Site (BLMBS).* U.S. Army Corps of Engineers, Memphis, TN.

van der Pijl, L. 1982. *Principles of Dispersal in Higher Plants.* Springer-Verlag, Berlin, Germany.

van der Valk, A. G. 1987. Vegetation dynamics of freshwater wetlands: A selective review of the literature. *Archiv für Hydrobiologie Beiheft Ergebnisse der Limnologie* **27:**27–39.

van der Valk, A. G. 1992. Establishment, colonization, and persistence. In *Plant Succession: Theory and Prediction,* ed. by D. C. Glenn-Lewin, R. K. Peet, and T. T. Veblin, pp. 60–102. Chapman and Hall, London, U.K.

van der Valk, A. G., and R. L. Pederson. 1989. Seed banks and the management and restoration of natural vegetation. In *Ecology of Soil Seed Banks,* ed. by M. A. Leck, V. T. Parker, and R. L. Simpson, pp. 329–346. Academic Press, San Diego, CA.

Vivian-Smith, G., and E. W. Stiles. 1994. Dispersal of salt marsh seeds on the feet and feather of waterfowl. *Wetlands* **14:**316–319.

Warne, E. L. 1992. Seed bank and vegetation dynamics in small reconstructed wetlands. Master's thesis, University of Wisconsin, Milwaukee, WI.

Watt, K. M., and S. W. Golladay. 1999. Organic matter dynamics in seasonally inundated, forested wetlands of the Gulf Coastal Plain. *Wetlands* **19:**139–148.

Wax, C. L. 1977. An analysis of the relationship between water level fluctuations and climate, coastal Louisiana. Ph.D. diss., Louisiana State University, Baton Rouge, LA.

Welcomme, R. L. 1994. *Sustaining the Ecological Integrity of Large Floodplain Rivers: Conference Synthesis,* International Conference, Sustaining the Ecological Integrity of Large Floodplain Rivers, ed. by K. Lubinski, J. Wiener, and N. Bhowmik, pp. 1–4. La Crosse, WI.

Weller, J. D. 1995. Restoration of a south Florida forested wetland. *Ecological Engineering* **5:**143–151.

Wharton, C. H., W. M. Kitchens, E. C. Pendleton, and T. W. Sipe. 1982. *The Ecology of Bottomland Hardwood Swamps of the Southeast: A Community Profile.* FWS/OBS-81/37. U.S. Fish and Wildlife Service, Washington, DC.

Whitlow, T. H., and R. W. Harris. 1979. *Flood Tolerance in Plants: A State-of the-Art Review.* Technical Report E-79-2. U.S. Army Corps of Engineers, Vicksburg, MS.

Wichmann, R. F. 1996. A natural history of the Cache River watershed. Master's thesis, Southern Illinois University, Carbondale, IL.

Wienhold, C. E., and A. G. van der Valk. 1989. The impact of duration of drainage on the seed banks of northern prairie wetlands. *Canadian Journal of Botany* **67:**1878–1884.

Wilker, J. R. 1999. Hydrochory and seed density patterns within an alluvial cypress-tupelo swamp in southern Illinois. Master's thesis, Southern Illinois University, Carbondale, IL.

Williams, G. P., and M. G. Wolman. 1984. *Downstream Effects of Dams on Alluvial Rivers.* Geological Survey Professional Paper 1286. U.S. Geological Survey, Washington, DC.

Williston, H. L. 1962. Pine planting in a water impoundment area, U.S Forest Service Experiment Station Southern Forestry Note 137.

Xiao, N., D. Bennett, B. Middleton, and K. Fessel. In press. SISM: A multiscale model for cypress swamp regeneration. *Geographical and Environmental Modeling.*

Yeager, L. E. 1949. Effect of permanent flooding in a river-bottom timber area, Illinois. *Natural History Survey Bulletin* **25:**33–65.

Young, N., and J. Crew. 1982. Fish management report, Northwest region. *Annual Report to the Florida Game and Fresh Water Fish Commission* **3:**1–15.

Young, P. J., J. P. Megonigal, R. R. Sharitz, and F. P. Day. 1993. False ring formation in baldcypress *(Taxodium distichum)* saplings under two flooding regimes. *Wetlands* **13:**293–298.

Zedler, J. B. 1996a. Ecological function and sustainability in created wetlands. In *Restoring Diversity: Strategies for Reintroduction of Endangered Plants,* ed. by D. A. Falk, C. I. Millar, and M. Olwell, pp. 331–342. Island Press, Washington, DC.

Zedler, J. B. 1996b. *Tidal Wetland Restoration: A Scientific Perspective and Southern California Focus.* Report T-038. University of California, La Jolla, CA.

Zedler, J. B., and J. C. Callaway. 1999. Tracking wetland restoration: Do mitigation sites follow desired trajectories? *Restoration Ecology* **7:**69–73.

Index